Specialty Polymers

Specialty Polymers

Second Edition

Edited by
R.W. DYSON

Senior Lecturer in Polymer Technology
School of Polymer Technology
University of North London
UK

BLACKIE ACADEMIC & PROFESSIONAL
An Imprint of Chapman & Hall
London · Weinheim · New York · Tokyo · Melbourne · Madras

Published by Blackie Academic & Professional, an imprint of Thomson Science, 2–6 Boundary Row, London SE1 8HN, UK

Thomson Science, 2–6 Boundary Row, London SE1 8HN, UK

Thomson Science, 115 Fifth Avenue, New York, NY 10003, USA

Thomson Science, Suite 750, 400 Market Street, Philadelphia, PA 19106, USA

Thomson Science, Pappelallee 3, 69469 Weinheim, Germany

First edition 1987

Reprinted 1992

Second edition 1998

Softcover reprint of the hardcover 2nd edition 1998

Thomson Science is a division of International Thomson Publishing I(T)P®

Typeset in 10/12 Times by Blackpool Typesetting Services Limited, UK

ISBN-13:978-94-010-6499-6 e-ISBN-13: 978-94-009-0025-7
DOI: 10.1007/978-94-009-0025-7

A catalogue record for this book is available from the British Library

Library of Congress Catalog Card Number: 97-73987

∞ Printed on acid-free text paper, manufactured in accordance with ANSI/NISO Z39.48-1992 (Permanence of Paper)

Contents

List of contributors ix
List of abbreviations xi
List of trade names xv

Introduction 1
R.W. DYSON

1 Polymer structures and general properties 3
R.W. DYSON

1.1 Introduction 3
1.2 Polymer structures 3
1.3 Morphology and softening behaviour 5
1.4 Amorphous polymers and softening behaviour 7
1.5 Semi-crystalline polymers and softening behaviour 8
1.6 Molecular weight 10
1.7 Copolymers 11
1.8 Chemical behaviour 13
1.9 Solubility 16
1.10 Electrical properties 17
1.11 Conclusion 18
Further reading 18

2 Polymerization 19
R.W. DYSON

2.1 Introduction 19
2.2 Addition polymerization 19
2.2.1 Monomers 19
2.2.2 Free radical polymerization 20
2.2.3 Ionic polymerization 22
2.2.4 Catalytic surface polymerizations 23
2.3 Step-growth polymerization 23
2.4 Copolymer formation 25
2.4.1 Step-growth polymerization 26
2.4.2 Random copolymers 26
2.4.3 Block copolymers 29
2.4.4 Graft copolymers 31
2.5 Chemical conversion 32
2.6 Polyblends 34
2.7 Conclusion 35

3 High-temperature and fire resistant polymers 37
M.S.M. ALGER

3.1 Introduction 37
3.2 The need for thermally resistant polymers 39
3.3 Improving low-performance polymers for high-temperature use 39
3.4 The need for fire resistant polymers 41
3.5 Polymers for low fire-hazards 44
3.6 Polymers for high-temperature resistance 45
3.6.1 Fluoropolymers 48
3.6.2 Aromatic polymers 48

3.6.3 Hydrocarbon polymers 51
3.6.4 Polyethers 52
3.6.5 Polyphenylene sulphide 53
3.6.6 Polysulphones 54
3.6.7 Polyesters 56
3.6.8 Polyamides 58
3.6.9 Polyketones 61
3.7 Heterocyclic polymers 62
3.7.1 Polyimides 62
3.7.2 Other polymers 64
Further reading 66

4 Hydrophilic polymers **67**
C.A. FINCH

4.1 Introduction 67
4.2 Natural polymers 70
4.2.1 Carbohydrates 70
4.2.2 Proteins 74
4.3 Semisynthetic polymers 75
4.4 Synthetic polymers 78
4.4.1 Hydrogel polymers 78
4.4.2 Polyacrylamide hydrophilic polymers 82
4.4.3 Polyvinyl alcohol 84
4.4.4 Polyvinyl pyrrolidone 84
4.5 Superabsorbent polymers 85
Further reading 89

5 Polymers with electrical and electronic properties **90**
M.T. GOOSEY

5.1 Introduction 90
5.2 Conducting polymers 91
5.2.1 Introduction 91
5.2.2 Conductive polymers via controlled pyrolysis 91
5.2.3 Conduction mechanisms 92
5.2.4 Polyacetylene 93
5.2.5 Polyparaphenylenes 96
5.2.6 Polypyrrole 97
5.2.7 Polyaniline 98
5.2.8 Polyphenylene sulphide and its analogues 98
5.2.9 Other important conducting polymers 99
5.2.10 Applications for conducting polymers 100
5.3 Photoconducting polymers 101
5.4 Polymers in optoelectronics 103
5.5 Polymers with piezoelectric, pyroelectric and ferroelectric properties 105
5.5.1 Introduction 105
5.5.2 Polyvinylidene fluoride 106
5.5.3 Polyvinylidene fluoride–trifluoroethylene copolymers 108
5.5.4 Other materials 108
5.5.5 Applications 108
5.6 Photoresists for semiconductor fabrication 110
5.6.1 Introduction 110
5.6.2 Negative photoresists 113
5.6.3 Positive photoresists 114
5.6.4 New developments in resist technology 115
5.7 Summary and conclusions 119
References 119

6 Ionic layers **122**
B.N. HENDY

6.1 Introduction 122
6.2 Classification 122
6.2.1 Type of bound ion 122
6.2.2 Position of bound ion 123
6.2.3 Amount of bound ion 123
6.2.4 Type of counter ion 123
6.2.5 The backbone 124
6.3 Synthesis 124
6.4 Physical properties and applications 126
6.4.1 Ionic cross-linking 126
6.4.2 Ion exchange 129
6.4.3 Hydrophilicity 132
6.5 Ionomers based on polyethylene 135
6.6 Elastomeric ionomers 138
6.7 Ionomers based on polystyrene 142
6.8 Ionomers based on polytetrafluoroethylene 144
6.9 Ionomer with polyaromatic backbones 148
6.10 Polyelectrolytes for ion exchange 149
6.11 Polyelectrolytes based on carboxylates 152
6.12 Polymers with integral ions 153
6.12.1 Halato-telechetic polymers 153
6.12.2 Ionenes 154
6.12.3 Polyethylenimine 155
6.13 Polyelectrolyte complexes 156
6.14 Blends of polymers and salts 157
6.15 Biological and inorganic ionic polymers 157
6.16 Conclusions 158
References 159

7 Liquid crystal polymers **161**
D. COATES and M. VERRALL

7.1 Introduction to thermotropic liquid crystals 161
7.1.1 The nematic phase 163
7.1.2 The chiral nematic phase 163
7.1.3 The smectic phase 164
7.1.4 Anisotropic properties 166
7.1.5 Lyotropic liquid crystals 168
7.2 Liquid crystal polymers 168
7.2.1 Main-chain liquid crystal polymers 168
7.2.2 Side-group liquid crystal polymers 169
7.3 The structure of side-group liquid crystal polymers 170
7.3.1 General structural features 170
7.3.2 Liquid crystal poly(acrylate)s 171
7.3.3 Liquid crystal poly(siloxanes)s 172
7.3.4 Liquid crystal poly(styrene)s 173
7.3.5 Liquid crystal poly(ester)s 174
7.3.6 Liquid crystal poly(vinyl ether)s 174
7.3.7 Other classes of liquid crystal side-group polymer 175
7.3.8 The mesogenic core group 176
7.3.9 Glass transitions in liquid crystal polymers 178
7.3.10 Molecular weight of liquid crystal polymers 178
7.4 Elastomeric liquid crystal polymers 179
7.5 Laterally attached liquid crystal side-group polymers 180
7.6 Cyclic polysiloxanes 181

7.7 Applications of side-group liquid polymers 183
7.7.1 Display applications 183
7.7.2 Optical data storage 186
7.7.3 Chromatography 188
7.7.4 Non-linear optical applications 189
7.8 *In situ* photopolymerization 190
7.8.1 Monoreactive liquid crystals 191
7.8.2 Direactive liquid crystals – anisotropic networks 192
7.8.3 Applications of anisotropic networks 196
References 197

8 Structural polymers **200**
I. HAMERTON and J.N. HAY

8.1 Introduction 200
8.2 Introduction to composites 200
8.2.1 Background 200
8.2.2 Processing methods 201
8.2.3 Carbon fibre 203
8.3 Thermosets 203
8.3.1 Unsaturated polyesters and vinyl esters 203
8.3.2 Epoxy resins 206
8.3.3 Cyanate esters 211
8.3.4 Isocyanate polymers 216
8.3.5 Bismaleimides 219
8.3.6 PMR resins 222
8.3.7 Benzocyclobutene resins 225
8.3.8 Toughened thermosets 229
8.3.9 Other thermosets 230
8.4 Thermoplastics 233
8.4.1 Polyamides 233
8.4.2 Polyimides 236
8.4.3 Polysulphones and polyketones 239
8.4.4 Polyphenylene sulphide 241
8.4.5 Polybenzimidazoles 244
8.5 Recent and future directions 247
8.5.1 Molecular composites 247
8.5.2 Stereocontrolled polyolefins 248
Further reading 249

Index **251**

Contributors

M.S.M. Alger	The School of Polymer Technology, University of North London, Holloway Road, London N7 8DB, UK
D. Coates	Merck Ltd, West Quay Road, Poole, Dorset BH15 1HX, UK
R.W. Dyson	The School of Polymer Technology, University of North London, Holloway Road, London N7 8DB, UK
C.A. Finch	Pentafin Associates, 18–20 West End, Weston Turville, Aylesbury, Bucks HP22 5TT, UK
M.T. Goosey	Shipley Europe Ltd, Herald Way, Coventry CV3 2RQ, UK
I. Hamerton	Department of Chemistry, University of Surrey, Guildford, Surrey GU2 5XH, UK
J.N. Hay	Department of Chemistry, University of Surrey, Guildford, Surrey GU2 5XH, UK
B.N. Hendy	New Science Group, ICI plc, PO Box 90, Wilton, Middlesbrough, Cleveland TS6 8JE, UK
M. Verrall	Merck Ltd, West Quay Road, Poole, Dorset BH15 1HX, UK

Abbreviations

ABS	acrylonitrile-butadiene-styrene
AMPS	2-acrylamido-2-methylpropane sulphonic acid
ASIC	application-specific integrated circuit
AT	acetylene-terminated
ATH	alumina trihydrate
BBL	poly-(*bis*benzimidazobenzophenanthroline)
BCB	benzocyclobutene
BDTA	3,3′,4,4′-benzophenone tetracarboxylic dianhydride
BMC	bulk moulding compound
BMI	bismaleimide
BTDE	...
CAD	computer-aided design
CAM	computer-aided manufacturing
CE	cyanate ester
CFCs	chlorofluorocarbons
C-PS	carboxylated polystyrene
CTBN	carboxy-terminated butadiene-acrylonitrile
DABA	*o,o′*-dialkylbisphenol-A
DGEBA	diglycidyl ether of bisphenol-A
DMAc	*N,N*-dimethylacetamide
DMC	dough moulding compounds
DSC	differential scanning calorimeter (calorimetry)
DVB	di-vynil benzene
ENB	ethylene norbornene
EPDM	ethylene–propylene–diene monomer terpolymer
EPIC resin	resin based on catalysed reaction of epoxy resins with isocyanates
EPR	ethylene–propylene rubber
HEMA	hydroxyethyl methacrylate
HDT	heat distortion temperature
HIPS	high-impact polystyrene
HTP	halato-telechelic polymer
ILT	interlaminar toughening (interviewing)
IPN	interpenetrating network
LCP	liquid crystal polymer
LOI	limiting oxygen index
MAO	methylaluminoxane
MDA	methyl dianiline
MDI	4,4 diphenylmethane di-isocyanate
MOCA	4,4 methylene-bis-2-chloroaniline
MOSFET	metal oxide semiconductor field effect transistor

MPD-I	poly-*m*-phenylene isophthalamide
NBR	nitrile rubber (acrylonitrile–butadiene copolymer)
NDI	naphthalene di-isocyanate
NE	norborene acid ester
NMP	*N*-methylpyrrolidinone
OHC	open-hole compression
PA	polyamide (nylons)
PAA	polyamic acid
PAI	polyamideimide
PAN	polyacrylonitrile
PAQR	poly (acene-quinone radical) polymer
PAr	polyarlate
PAS	polyarylene sulphide
PBI	polybenzimidazole
PBOX	phenylene bisoxazolines
PBT	polybutylene terephthalate
PC	polycarbonate
PE	polyethylene
PEC	polyelectrolyte complex
PEEK	polyetheretherketone
PEI	polyethylenimine
PEK	polyether ketone
PEN	polyethylene-naphthalene
PES	polyether sulphone (victrex type)
PESu/PSu	polyether sulphone (udel type)
PET	polyethylene terephthalate
P-F	phenol formaldehyde resin
PBA	polyhydroxy-benzoic acid
PI	polyimide
PMMA	polymethyl methacrylate
PMR	polymerization of monomeric reactants
POM	polyoxy-methylene (acetals)
POSFET	piezoelectric oxide semiconductor field effect transistor
PP	polypropylene
PPA	polyphthalamide
PPD	*p*-phenylene diamine
PPD-T	poly-*p*-phenylene diamine
PPE	polyphenylene ether
PPO	poly dimethyl-phenylene oxide
PPP	poly*para*phenylene
PPS	polyphenylene sulphide
PPSe	polyphenylene selenide
PPTe	polyphenylene telluride
PPTS	polypyrrole toluenesulphonate

PPV	poly-*p*-phenylene vinylene
PS	polystyrene
PSU	polysulphone
PTFE	polytetrafluoroethylene
PTrFE	polytrifluoroethylene
PU	polyurethane
PVA	polyvinyl alcohol
PVC	polyvinyl chloride
PVDF	polyvivnylidene fluoride
PVK	poly-*n*-vinyl carbazole
PVP	polyvinyl pyrrolidone
PVPy	poly(4-vinyl pyridine)
PZT	lead zirconate titanate
RIM	reaction injection moulding
RRIM	reinforced reaction injection moulding
RTM	resin transfer moulding
SAP	superabsorbant polymer
SAXS	small angle X-ray scattering
SBR	styrene butadiene rubber
SBS	styrene–butadiene rubber
SCMC	sodium carboxymethyl cellulose
SIN	simultaneously inter-penetrating network
SMC	sheet moulding compounds
S-PS	sulphonated polystyrene
SRIM	structural reaction injection moulding
STN	supertwisted nematic
TCl	terethaloyl chloride
TDI	tolylene (or toluene) diisocyanate
TFE	tetrafluoroethylene
TGBAP	*N,N,N″N″*-tetraglycidyl-α,α′-*bis*(aminophenyl)-*p*-diisopropylbenzene
TGIC	triglycidyl isocyanurate
TGDDM	*N,N,N″,N″*-tetraglycidyl-4,4′-diaminodiphenylmethane
TGMBAP	*N,N,N″,N″*-tetraglycidyl-α,α′-*bis*(3,5-dimethyl-4-aminophenyl)-*p*-di-isopropylbenzene
THF	tetrahydrofuran
TLV	threshold limit value
TMEDA	tetramethylethylenediamine
TNF	trinitrofluorenone
TP	thermoplastic
TrFE	trifluoroethylene
UPE	unsaturated polyester
UV	ultraviolet
VDF	vinylidene fluoride

VE	vinyl ester
6FCoPI	copolyimide
6FDA	4,4′-hexafluoroisopropylidene diphthalic anhydride

Note: This list includes standard abbreviations for polymers mentioned in the text, even though the abbreviation itself may not be mentioned.

Trade names

Aclar	fluoropolymer	Du Pont
Adiprene	polyurethane	Uniroyal
Algoflon	fluoropolymer	Montedison
Amberlite IR-120	ion exchange resin	Rohm and Haas
Amberlite IRA-93	ion-exchange resin	Rohm and Haas
Amberlite IRC-84	ion-exchange resin	Rohm and Haas
Amberlyst A26	polyelectrolyte	Rohm and Haas
Amberlyst 15	polyelectrolyte	Rohm and Haas
Amodel	...	...
Araldite PT-810	triglycidyl isocyanurate	Ciba–Geigy
Araldite MY 0500	low-viscosity liquid triglycidyl resin of *p*-aminophenol	Ciba–Geigy
Araldite MT 0510	low-viscosity liquid triglycidyl resin of *p*-aminophenol	Ciba–Geigy
Araldite MY 720	epoxy resin based on TGDDM	Ciba–Geigy
Araldite MY 721	epoxy resin based on TGDDM	Ciba–Geigy
Arcton	fluorotrichloromethane	ICI
Ardel	aromatic polyester	Union Carbide
AroCy B	bisphenol-A	Ciba–Geigy
AroCy F	hexafluorobisphenol-A	Ciba–Geigy
AroCy L	bisphenol-E	Ciba–Geigy
AroCy M	tetramethylbisphenol-F	Ciba–Geigy
Arotech 6000	Bisoxazoline-phenolic resin	Ashland
Arylef	aromatic polyester	Solvay
Arylon	aromatic polyester	Du Pont
Arylox	polyphenylene oxide	USSR
Astrel	polyether sulphone	3M
Biapen	polyphenylene oxide	Chemolimpex
Blendure	EPIC resin	Bayer
BT-2000	bisphenol-A	Mitsubishi
Carbopol	acrylic polyelectrolyte	B.F. Goodrich
Carboset	acrylic polyelectrolyte	B.F. Goodrich
Celazole	polybenzimidazole fibre	Hoechst-Celanese
Compimide 183	...	Shell-Technochemie
Compimide 353	...	Shell-Technochemie
Compimide 796	...	Shell-Technochemie
Dowex SBR	aminated styrene copolymer	Dow Chemical
Duolite ES 467	polyelectrolyte	Rohm and Haas
Ekcel	aromatic polyester	Carborundum
Ekonol	aromatic polyester	Carborundum
EPON HPT 1071	epoxy resin	Shell

EPON HPT 1072	epoxy resin	Shell
EPON 825	...	Shell
Flemion	carboxylated fluoropolymer	Asahi Glass
Fluon	fluoropolymer	ICI
Fluon FEP	fluoropolymer	ICI
Fluorel	fluoropolymer	3M
Fluothane	fluoropolymer	USI
Fortron	polyphenylene sulphide	Hoechst
Freon	fluorotrichloromethane	Du Pont
Gantrez AN	polyelectrolyte	GAF
Grilamid TR	aromatic polyamide	Emser
Halon	fluoropolymer	Du Pont
H resin	aromatic hydrocarbon	Hercules
Hostaflon	fluoropolymer	Hoechst
Hostaflon EFE	fluoropolymer	Hoechst
Hostaflon FEP	fluoropolymer	Hoechst
Hostaflon TFA	fluoropolymer	Hoechst
Hostaflor C	fluoropolymer	Hoechst
Hostatec	polyetheretherketone ketone	Hoechst
Hycar	polycarboxyllic acid	B.F. Goodrich
Hycar 1300XB	...	B.F. Goodrich
Hypalon	chloro-sulphonated polyethylene	Du Pont
Ixef	aromatic polyamide	Solvay
Kadel	polyetheretherketone	Amoco
Kalrez	fluoropolymer	Du Pont
Kapton	polyimide	Du Pont
Kel F	fluoropolymer	3M
Kerimid	polybismaleimide	Rhone-Poulenc
Kerimid 601	...	Rhone-Poulenc
Kevlar	aromatic polyamide	Du Pont
Kinel	polybismaleimide	Rhone-Poulenc
Krynac	carboxylated nitrile rubber	Doverstrand
Luranyl	polyphenylene oxide (blend)	BASF
Lycra	polyurethane	Du Pont
Nafion	persulphonated fluoropolymer	Du Pont
Neoflon	fluoropolymer	Daikin Kogyp
Neosepta F	ionic membrane	Tokoyama Soda
Nomex	aromatic polyamide	Du Pont
Noryl	polyphenylene oxide (blend)	General Electric
Paralene C	chlorinated aromatic hydrocarbon	Union Carbide
Parylene D	chlorinated aromatic hydrocarbon polymer	Union Carbide
Parylene N	aromatic hydrocarbon	Union Carbide
Polymin	polyethylenimine	BASF

Prevex	polyphenylene oxide (blend)	Borg Warner
Primaset PT	Novolac resin	Allied Signal
Primef	polyphenylene sulphide	Solvay
Pyralin	polyimide	Du Pont
Pyre ML	polyimide	Du Pont
Radel A	aromatic polysulphone	Amoco
Revinex	carboxylated rubber	Doverstrand
REX-371	Novolac resin	Ciba–Geigy
RTX-366	bisphenol-M	Ciba–Geigy
Ryton	polyphenylene sulphide	Phillips Petroleum
Sirotherm	ampholytic polyelectrolyte	ICI
Skybond-700	polyimide	Monsanto
Sokalan CP2	polyelectrolyte	BASF
Soreflon	fluoropolymer	PCUK/ATO
Stanyl	aliphatic polyamide	DSM
Supec	polyphenylene sulphide	General Electric
Surlyn	polyethylene monomer	Du Pont
Sycar	silicon–carbon thermoset	Hercules
Technoflon	fluoropolymer	Montedison
Tedur	polyphenylene sulphide	Bayer
Teflex	fluoropolymer	Du Pont
Teflon	fluoropolymer	Du Pont
Teflon EFE	fluoropolymer	Du Pont
Teflon FEP	fluoropolymer	Du Pont
Teflon PFA	fluoropolymer	Du Pont
Texigel	polyelectrolyte	Scott Bader
Texipol	polyelectrolyte	Scott Bader
Thermid MC-600	acetylene-terminated oligoimide	National Starch and Chemical
Torlon	polyamide-imide	Amoco
Trogamid T	aromatic polyamide	Dynamit Nobel
Twaron	polyamide	Akzo
Udel	aromatic polysulphone	Union Carbide
Ultem	polyether imide	General Electric
UltemSiltem	polysiloxane-etherimide	General Electric
Ultrapek	polyether ketone	BASF
Ultrason P	aromatic polysulphone	BASF
Urepan	polyurethane	Bayer
Vectra	aromatic copolyester	Celanese
Vectron	liquid crystalline aromatic copolyester fibre	Hoechst-Celanese
Versicol	polyelectrolyte	Allied Colloids
Vespel	polyimide	Du Pont
Victrex	aromatic polyether sulphone	ICI

Victrex PEEK	polyetherether ketone	ICI
Victrex PEK	polyether ketone	ICI
Victrex PES	aromatic polysulphone	ICI
Viton	fluoropolymer	Du Pont
Voltalef	fluoropolymer	PCUK/ATO
Vulkolan	polyurethane	Bayer
Vyrene	polyurethane	US Rubber
XU-7187	dicyclopentadienyl bisphenol	Dow Chemical
Xydar	liquid crystalline aromatic copolyester	Amoco
Xylok	aromatic hydrocarbon	Albright and Wilson
Zenite	liquid crystalline aromatic copolyester	Du Pont
Zytel HFN	...	Du Pont

Introduction

R.W. DYSON

There will be few readers of this book who are not aware of the contribution that polymers make to modern life. They are to be seen around the home, at work, in transport and in leisure pursuits. They take many forms which include plastic mouldings and extrusions, plastic film and sheet, plastic laminates (fibreglass and formica), rubber gloves, hoses, tyres and sealing rings, fibres for textiles and carpets and so on, cellular products for cushioning and thermal insulation, adhesives and coating materials such as paints and varnishes. The majority of these polymers are synthetic and are derived from oil products. The most important of these in terms of tonnage used are polymers based upon styrene, vinyl chloride, ethylene, propylene and butadiene among plastics and rubber materials, and nylons, polyethyleneterephthalate and polyacrylonitrile among fibres. The total amount of these polymers used each year runs into millions of tonnes.

These polymers are sometimes known as commodity polymers because they are used for everyday artefacts. They are available in many grades and formats to meet a variety of applications and processing techniques. The properties can be adjusted by using additives such as heat and light stabilizers, plasticizers, and reinforcing materials. Often, grades are specially designed and formulated to meet particular requirements and, in a sense, these might be regarded as specialities. Much has been written about these materials elsewhere and they are not the concern of this book. Instead, this book is concerned with a wide range of polymers which have been designed and created to meet special needs which often involve properties not usually associated with polymers.

There has been a growing awareness over the past 25 years or so that there are many advantages to using polymers in applications which previously would not have been considered suitable or feasible. A great deal of research and development has gone into creating polymers to meet the demands of these specialized applications. Much of modern technology would hardly be possible without their use. The electronics industry, for example, could not have developed as it has without the use of polymeric materials which have been created to meet the most exacting requirements. Compared with the commodity polymers, the tonnage production of these polymers is small, being measured in tonnes per year. They are usually costly, partly because the quantities required are small and partly because the processes for making them are themselves expensive. Nevertheless, all of the polymers discussed offer a unique combination of properties for a particular application and it is this which marks them as specialty polymers.

Whilst the reader may be familiar with commodity polymers at least in general terms, many specialty polymers discussed in this book might pass unnoticed because of their specialized nature and applications. If this book serves to introduce the reader to the existence and use of these polymers, it will have served its purpose.

1 Polymer structures and general properties

R.W. DYSON

1.1 Introduction

Polymers are large molecular structures in which many small units are linked together. Typically, the molecular weight of polymers vary from 20 000 to infinity depending on the type and structure. The majority of useful polymers are organic chemicals which is to say that they are based upon the elements carbon and hydrogen. They occur in nature as starches, cellulose and proteins for example and some of these are chemically converted into useful derivative polymers which are known as semi-synthetic polymers. The vast majority of polymers used today as plastics, rubbers, adhesive and paints are synthetic and are made from chemicals derived from oil. The reason for this is that the properties of a polymer depend on the chemical structure and composition. Natural polymers exist with a particular chemical structure and although in some cases this can be chemically modified, the scope is necessarily limited. The synthesis of polymers from simple chemicals derived from oil affords the potential to produce an almost unlimited range of polymers with desirable properties. This potential has not been fully realized and even though the range of synthetic polymers is very wide the range is increasing constantly as new areas of use are developed and exploited.

1.2 Polymer structures

There are two fundamental types of polymer: those which are linear or branched and those which are cross-linked. **Linear polymers** are long-chain molecules in which the units which make up the polymer are joined together end to end like beads in a necklace (Fig. 1.1). An example is polyethylene in which ethylene units are joined together in the polymerization process to form polyethlene:

$$CH_2{=}CH_2 \rightarrow -CH_2-CH_2-CH_2-CH_2-CH_2-CH_2-$$

Sometimes, these chains are branched; that is to say, the chain divides at some point to produce two linked chains. In the majority of branched chain structures, the branches are long and few in number. In the case of polyethlene, however, the branches are short and contain an even number of carbon atoms between two and eight depending on the type of polyethylene. A typical long-chain branched structure is shown in Fig. 1.1 Linear and branched chain structures exhibit similar properties and they will be regarded as being the same from here on.

Linear polymers are thermoplastic. If they are heated, they soften at a temperature which is characteristic of the polymer. This softening behaviour is

Figure 1.1 Polymer chains: (a) linear; (b) branched.

reversible, for if the polymer is cooled it will harden again as the characteristic temperature is traversed. Softening occurs over a range of temperatures although specific temperatures are usually quoted in the literature. This is because softening is usually determined by observing a change in a specific property according to a standard test procedure. Polymers which soften below room temperature are rubbery solids or viscous fluids at room temperature. Polymers which soften above room temperature are hard brittle solid or tough ductile solids at room temperature. Linear polymers are soluble in suitable liquids at room or elevated temperatures.

Cross-linked polymers consist of chains which are joined together by a sequence of covalently bonded atoms which form the cross-links. The cross-links may be long or short but in most commercial polymers the cross-links are short. Examples of cross-linked polymers are vulcanized rubbers and thermosetting plastics such as urea formaldehyde. A diagrammatic representation of a cross-linked system is shown in Fig. 1.2. The properties of a cross-linked polymer will depend upon the degree of cross-linking. If the degree of cross-linking is low, there will be relatively long sequences of the main polymer chain between cross-links. In such cases, the polymer will behave in some respects like an uncross-linked polymer. For example, such polymers show softening

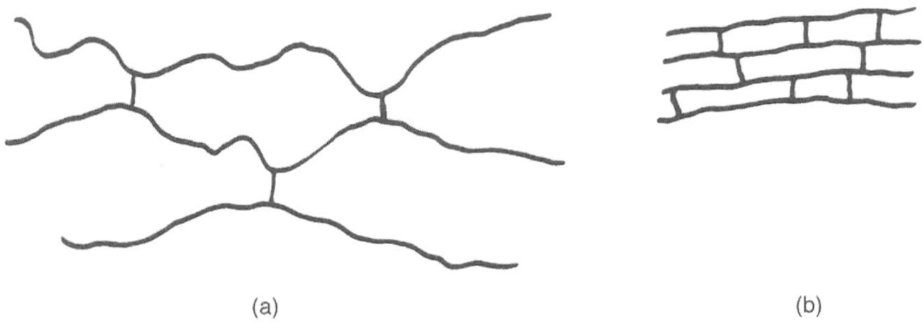

Figure 1.2 Cross-linked polymers: (a) with low degree of cross-linking; (b) with high degree of cross-linking.

behaviour. Vulcanized rubbers usually have a low degree of cross-linking and the cross-links are present to confer the properties of high elasticity and creep resistance on the polymer. These polymers show reversible hardening if cooled sufficiently below room temperature. If the degree of cross-linking is high, the chain sequences between cross-links are very short and softening behaviour is lost. The traditional thermoset polymers such as urea formaldehyde are so called because they set permanently during cross-linking to form hard brittle materials which are not softened by heat.

Cross-linking makes polymers insoluble, although lightly cross-linked polymers will swell in liquids that are solvents for the non-cross-linked polymer. The amount of swelling decreases with increasing cross-link density, and highly cross-linked polymers hardly swell at all.

1.3 Morphology and softening behaviour

Linear polymers can be divided into those which can crystallize and those which cannot. Those which cannot crystallize form an amorphous solid which can be regarded as a frozen liquid structure. Polymers which can crystallize never crystallize completely and are referred to as semi-crystalline even though the amount of crystallinity may vary between 25% and 80%. Whether a polymer can crystallize or not depends upon the regularity of the chain structure.

All linear polymer chains can be regarded as being derived from polyethylene, which is the simplest of chains consisting as it does of a long sequence of covalently bonded carbon atoms (Section 1.2). Other polymer chains can be regarded as being derived structurally from polyethylene by placing other atoms or groups of atoms at regular intervals either in the chain itself or as attachments to the chain. For example, placing an oxygen atom between each carbon atom produces polyoxymethylene (acetals), interspersing amide links produces polyamides (nylons) and ester links produce polyesters. When atoms or groups are attached to the chain, the attachments usually occur on every other main chain carbon atom. Examples include polyvinyl chloride. Polystyrene, polyvinyl alcohol and polymethylmethacrylate among common polymers. Table 1.1 gives the structures of some common polymers to illustrate this.

Polymers will crystallize if they are able to pack together in a highly ordered way. Although there are exceptions, notably polyvinyl alcohol, this implies that the polymer chains themselves should have regular structures. The polyethlene chain is highly regular and therefore readily crystallizes. It should be noted, however, that the highly branched low-density polyethylene is less crystalline than the more linear high-density polyethylene.

In substituted hydrocarbon chains, the attached atom or group (X) is attached to alternate carbon atoms:

$$\begin{array}{cccccccccc} -CH_2- & CH & -CH_2- & CH & -CH_2- & CH- \\ & | & & | & & | \\ & X & & X & & X \end{array}$$

Table 1.1 Polymer structure and glass transition temperature

Polymer	Structural repeat unit	Stiffening agent	T_g (°C)
Polyethylene	$-CH_2-CH_2-$	–	−120
Polypropylene	$-CH_2-CH(CH_3)-$	$-CH_3$ side group	−18
Polystyrene	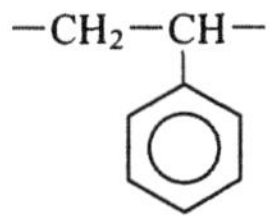	Phenyl ring side group	100
Polyvinylchloride	$-CH_2-CH(Cl)-$	Polar side group	80
Polymethylmethacrylate	$-CH_2-C(CH_3)(COOCH_3)-$	Di-substituted carbon	110
Polycarbonate	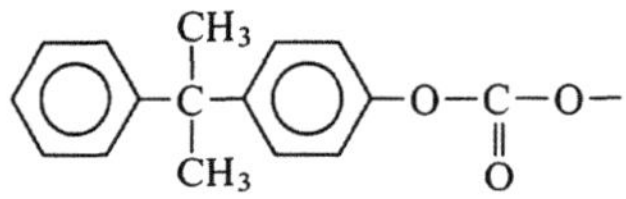	Benzene rings in chain	145
Polyethsulphone	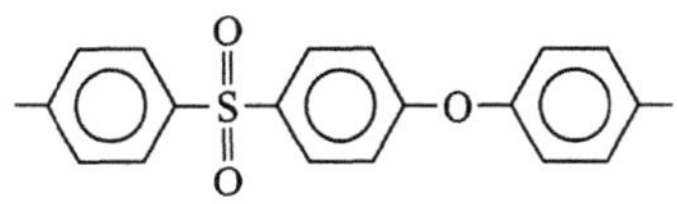	Benzene rings in chain	210
Polyphenylene oxide	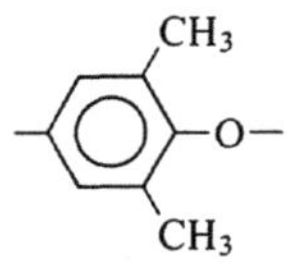	Substituted benzene ring in chain	220

The substituted carbon atom is asymmetric and this leads to three possible chain structures, which are shown in Fig. 1.3. Only the isotactic and syndiotactic forms are regular and these are the form that will crystallize. The atactic form is too irregular to crystallize and atactic polymers are mostly amorphous. Polyvinyl alcohol is atactic but crystallizes because the $-OH$ group is small enough to be accommodated in a close-packed arrangement of the polymer chains.

Interpolation of groups in the polymer chain usually permits crystallization if the interpolation is on a regular basis. A consideration of the polymers of this type in Table 1.2 will show that this is so for common polymers.

An important factor in determining whether a polymer is found in the semi-crystalline state, even if it is crystallizable, is the rate of crystallization. If the rate is very slow, crystallinity may not be observed. An example is natural rubber which crystallizes very slowly at room temperature but appreciably faster at about −23°C. Other examples include many engineering thermoplastics such as polycarbonates which only crystallize at appreciable rates at temperatures

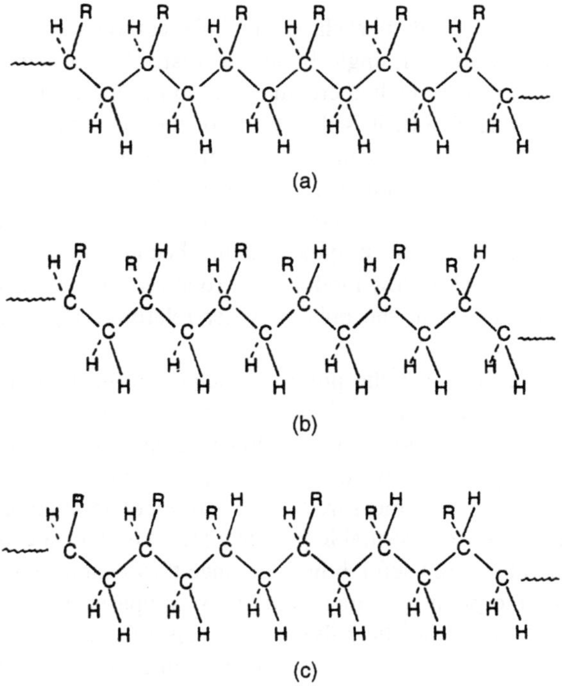

Figure 1.3 Monosubstituted polymer tacticity: (a) isotactic; (b) syndiotactic; (c) atactic.

well above room temperature. Consequently these polymers, although crystallizable, are usually amorphous.

1.4 Amorphous polymers and softening behaviour

The carbon–carbon single bond has the capability of rotating provided that there is sufficient energy available. The same is true of any single bond, and other

Table 1.2 Polymer structure and melting point

Polymer	Structural repeat unit	Interchain bonding	T_m
Polyethylene	$-CH_2-CH_2-$	dispersion forces	100–130
Polyoxymethylene	$-CH_2-O-$		165°C
Polyethylene_ terephthalate	$-CH_2-CH_2-O-C(=O)-C_6H_4-C(=O)-O-$	polar	265
Nylon 66	$-HN(CH_2)_6NH-CO-(CH_2)_4-CO-$	hydrogen	260
Nylon 6	$-NH-(CH_2)_5-CO-$	hydrogen	235

single bonds found in polymer chains include carbon–oxygen and carbon–nitrogen. It is this ability of single bonds to rotate that produces softening in amorphous polymer. Although there may be some contribution to softening from rotations in side chains, it is the ability of the main chain bonds to rotate that determines the softening point of a polymer. Rotation in the main chain, sometimes referred to as the polymer backbone, will occur when the polymer is at a temperature which is high enough to provide sufficient thermal energy. When rotation can occur the polymer chains become flexible and the bulk polymer is relatively soft (a low modulus). If rotation cannot occur, the polymer chains will be rigid, giving the bulk polymer relatively high rigidity (a high modulus).

The temperature at which the polymer changes from a rigid polymer to a flexible polymer is known as the glass transition temperature (T_g). The glass transition temperature is not a well-defined temperature because softening occurs over a range of temperatures (10–20°C). The values quoted in the literature for T_g are values measured when sufficient rotation is occurring to enable a change in some measurable property (e.g. volume, specific heat) to be detected. These values are useful, however, since they characterize the softening temperatures of amorphous polymers and allow comparisons to be made.

The value of T_g for a specific polymer will depend upon the structure of the polymer. Side groups attached to the polymer main chain will in general hinder rotation in the polymer backbone, necessitating higher temperatures to give enough energy to enable rotation to occur. The amount of extra energy will depend upon the size, stiffness, polarity and frequency of the attached groups. An increase in any of these factors will increase T_g (Table 1.1). Likewise, if groups are introduced into the polymer chain T_g will be affected. If rigid groups are introduced, the polymer chain becomes inherently stiffer and will require high temperatures to make the chain flexible. The benzene ring is a common stiffening unit in polymers designed to have a high softening point, and if these are present at sufficient frequency in the chain the glass transition temperature can be well above 200°C. Amorphous polymers retaining glass-like properties to high temperatures are almost always aromatic in character. A consequence of creating stiff chains to produce polymers with high glass transition temperatures is that the melt viscosity also becomes very high. Melt processing therefore becomes very difficult and sometimes impossible. It would appear that a T_g of around 250°C is the highest temperature obtainable that is compatible with melt processing.

1.5 Semi-crystalline polymers and softening behaviour

In order to obtain a high softening temperatures, semi-crystalline polymers should have a high melting point and a high degree of crystallinity. The melting point is the temperature at which sufficient thermal energy is available to disrupt the ordered structure of the crystalline polymer. Since the melting point is higher than the glass transition temperature, the polymer becomes an amorphous liquid above the melting point.

The melting point is given by

$$T_m = \frac{\Delta H}{\Delta S}, \tag{1.1}$$

where ΔH is the heat (enthalpy) of melting and ΔS is the entropy change which accompanies melting. The enthalpy is essentially the energy required to overcome the interchain forces which bind the polymer chains together in the crystalline regions. The energy required will depend upon the strength and frequency of the interchain forces. Interchain forces commonly found in polymers in decreasing order of strength are: hydrogen bonds (e.g. polyamides), polar or dipole bonds (e.g. polyesters) and dispersion forces (e.g. polyethylene). The entropy term is a measure of the change in chain flexibility as the polymer melts. Chains which are inherently flexible in the melt will have a high entropy change whereas polymer chains which are inherently stiff in the melt will have a small entropy change. Flexible chains will be characterized by a low glass transition temperature and a low melt viscosity whereas rigid chains will be characterized by a high glass transition temperature and a high melt viscosity.

For most polymers it is the enthalpy term which determines the melting point, but in the case of polytetrafluoroethylene (PTFE) it is the entropy term which determines the very high melting point (327°C) of the polymer. The interchain forces in PTFE are weak, but the fluorine atoms confer exceptional rigidity on the polymer chain to the extent that even in the liquid state above T_m the polymer chains are virtually inflexible. The melt viscosity is so high that PTFE cannot be processed by normal melt processing techniques.

It follows from the above that a rigid chain structure should lead to a high melting point and to a high glass transition temperature. This is found to be so in practice and there is a rule of thumb derived from experimental observations which relates T_g and T_m (both in kelvin, K), namely:

$$\frac{T_g}{T_m} \approx \tfrac{2}{3}.$$

At temperatures between the melting point and glass transition temperature the amorphous content of the polymer will be soft and rubbery while the crystalline content will retain its integrity until the melting point is reached. Since the crystalline polymer confers rigidity on the polymer as a whole, it is clear that the softening temperature of the polymer is controlled by the amount of crystallinity. For a high softening point it is desirable to have as much crystallinity as possible, but most commercial processes for shaping polymers do not allow significant control of the amount of crystallinity with the majority of semi-crystalline polymers.

1.6 Molecular weight

The molecular weight of linear polymers affects their properties. The molecular weight is related to the number of repeat units in the chain by the equation

$$M = nM_0, \qquad (1.2)$$

where n is the number of repeat units in the chain and M_0 is the molecular weight of the repeat unit. In practice, it is impossible to manufacture polymers in which the chains all have the same length. A polymer sample therefore contains a distribution of molecular weights (Fig. 1.4). Equation (1.2) should be modified to

$$\bar{M} = \bar{n}M_0, \qquad (1.3)$$

where $\bar{M}$ is the average molecular weight and $\bar{n}$ is the average number of units per chain. It is relatively easy to control the molecular weight average during polmerization but not so the distribution. The breadth of the distribution will depend upon the method of polymerization used to manufacture the polymer. In broad distributions the breadth of the distribution can be important because of the presence of low molecular weight and high molecular weight polymer of significantly different molecular weight.

In general, strength properties increase with increasing molecular weight in a non-linear fashion (Fig. 1.5). It is found that there is a minimum molecular

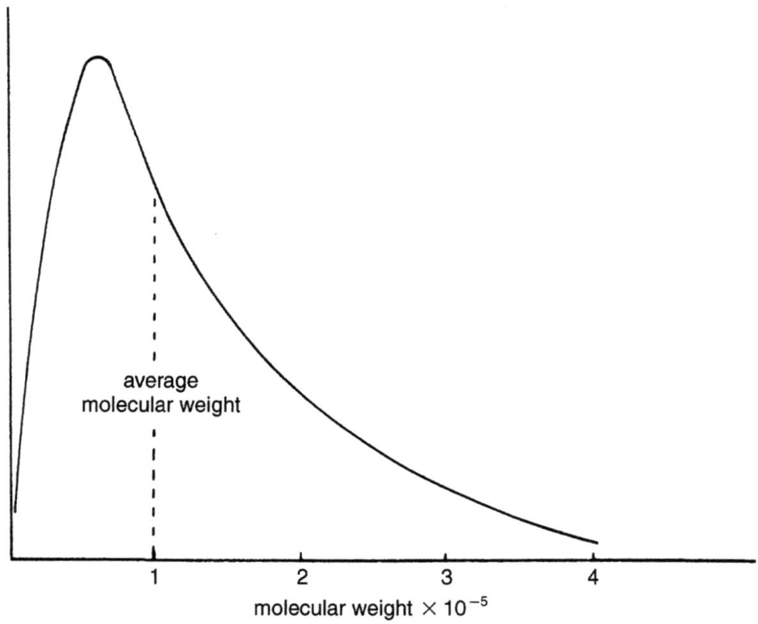

Figure 1.4 Calculated molecular weight distribution curve for polystyrene of average molecular weight 1×10^5.

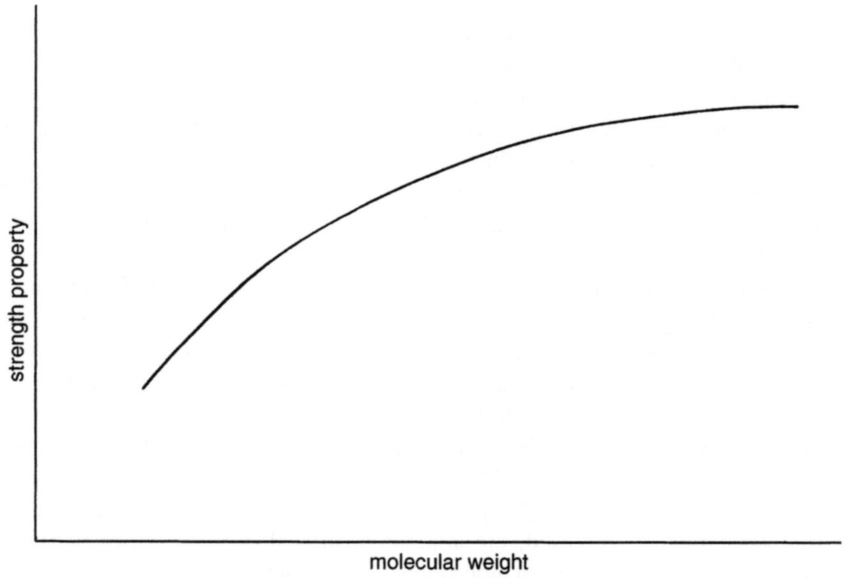

Figure 1.5 Effect of molecular weight (chain length) on strength properties.

weight before desirable strength properties are obtained. This minimum is usually about 20 000. Increasing the molecular weight above this minimum is desirable for increased strength, but increasing molecular weight has other, sometimes undesirable, effects. The melt viscosity for example increases and makes melt processing more difficult, solubility decreases and crystallization becomes more difficult because of the increase in chain entanglements in the melt. If the molecular weight distribution is broad, the presence of the high and low molecular weight fractions in the distribution will affect the properties of the polymer and it will behave differently from a narrow molecular weight distribution of the same polymer having the same average molecular weight. The presence of low molecular weight polymer will, for example, enhance melt flow, but the presence of high molecular weight polymer may lead to incomplete solubility when the polymer is placed in suitable solvents.

1.7 Copolymers

Copolymers are polymer chains which consists of two types of repeat units. A typical example is the styrene–butadiene copolymers which are the most important synthetic rubbers in terms of tonnage used. The name 'homopolymer' is sometimes used for polymers consisting of a single repeat unit. Polymers with three repeat units forming the chain are called terpolymers.

The properties of copolymers depend upon the chemical nature of the constituent units, the relative amounts of these units in the copolymer chain and the arrangement of the units in the drain. There are four types of arrangement in principle, and these are called block copolymers, graft copolymers, alternating copolymers and random copolymers. These are illustrated in Fig. 1.6 where A and B denote the two different repeat units. Random copolymers are the most important commercially, although block copolymers are being increasingly exploited.

Random copolymers tend to be amorphous polymers, especially if both units are present in significant amounts. This is because the random nature of the chain does not allow regular alignment of the polymer chains in crystalline structures. However, if one unit predominates, and this unit is capable of producing a crystalline homopolymer, then the copolymer is likely to be crystalline also, but with reduced crystallinity. For example, the copolymer comprising 90% propylene and 10% ethylene units is a semi-crystalline thermoplastic polymer which resembles polypropylene itself. As the ethylene content of the copolymer is increased, the amount of crystallinity decreases until the copolymer is obtainable only in amorphous form. Commercially, such materials are available as rubbers (EPR, EPDM).

```
AAAAAAA----------AAAAAAABBBBBBB-----------BBBBB
```

Block copolymer

```
AAAAA----------------AAAAAAA
                B
                B
                B
                |
                |
                `------BBBBB
```

Graft copolymer

```
---ABABABABABABABABABABABABABABABABABABABABAB---
```

Alternating copolymer

```
----ABAABAAABBBAABABABBBBBAAABBAAAABAABBBBAB-----
```

Random copolymer

Figure 1.6 Copolymer types.

The glass transition temperature of a random copolymer lies between the glass transition temperatures of the homopolymers made from the constituent repeat units. A convenient equation to estimate the copolymer glass transition temperature (T_{g_c}) is

$$\frac{1}{T_{g_c}} = \frac{w_a}{T_{g_a}} + \frac{w_b}{T_{g_b}}, \tag{1.4}$$

where w_a and w_b are the weight fractions of the two units in the copolymer chain and T_{ga} and T_{gb} are the glass transition temperatures of the respective homopolymers. All temperatures in this equation are in kelvin (K).

Random copolymers which are made from units that are isomorphous may crystallize.

Block and graft copolymers tend to behave as a mixture of two homopolymers, especially if the sequence lengths are long. In other words, they show a combination of two sets of properties rather than a resultant of two sets of properties. Block copolymers show two glass transition temperatures which correspond to the glass transition temperatures of the two homopolymers. Crystallization may occur if either or both homopolymers are crystallizable. If the two sections of a block copolymer are chemically dissimilar the two sections of the chain may be incompatible, leading to phase separation. Careful design of the block copolymer can lead to useful new materials. For example, the SBS triblock copolymer made from styrene and butadiene is a thermoplastic rubber if the butadiene centre block is long and the styrene end blocks are short. The polystyrene and polybutadiene sequences are incompatible and separate into two phases, the polystyrene ends forming small domains in a polybutadiene matrix. At room temperature, the polystyrene phase is hard and rigid and the domains act as cross-links to the flexible rubber phase. On heating to a temperature above 100°C, the polystyrene phase softens and a copolymer melt is produced. On cooling, the effective cross-links are reformed as the polystyrene domains harden again. The copolymer behaves as a cross-linked rubber at room temperature but is thermoplastic in its response to heat.

If SBS block copolymers are made with short polybutadiene sequences and long polystyrene, the resultant copolymer is a rigid thermoplastic material that is tough and transparent.

1.8 Chemical behaviour

Polymers, in common with other organic chemicals, are subject to attack by various agencies, which include heat, chemicals and ultraviolet (UV) light. In general the effect of these agencies is to change the chemical or physical nature of the polymer. If the effect is undesirable, the term 'degradation' is commonly used. The changes brought about include chain scission leading to lower molecular weight polymer, cross-linking leading to increased molecular weight and insolubility, and modification of the chemical structure of the chain. The last mentioned can be a useful method of making new polymers.

The ability to withstand or undergo chemical attack depends on the chemical constituents of the polymer chain. Polymers consisting of C—C, C—H or C—F bonds are resistant to chemical attack as are their simple low molecular weight analogues. Polyolefins are resistant to chemical attack at room temperature except by concentrated oxidizing acids, and fluorinated polymers show a greater resistance that increase with fluorination. Polytetrafluoroethylene is virtually inert over a wide temperature range. The C—Cl bond is also fairly resistant to chemicals; chlorinated hydrocarbon polymers such as polyvinyl chloride and polyvinylidene chloride show good chemical resistance. Other bonds and groups of atoms are more reactive, and polymers containing these reactive entities are themselves susceptible to chemical attack by specific types of chemicals. Two examples must suffice to illustrate this. The C=C double bond is attacked by oxygen and ozone. As a consequence, diene rubbers undergo progressive loss of elasticity and embrittlement on exposure to the atmosphere unless stabilized. The ester and amide linkages in polyesters and polymides are susceptible to hydrolysis and are therefore degraded in the presence of concentrated alkalis.

Not all chemical reactivity of groups in polymer structures is undesirable. Considerable use is made of the reactivity of side groups to modify or completely change polymers into different polymers. Polyvinyl acetate, for example, is readily converted into polyvinyl alcohol by alcoholysis of the acetate group by treatment with ethanol and alkali. Even relatively unreactive polymers such as polyethylene can be converted into other polymers. When polyethylene is treated with chlorine and sulphur in chlorinated hydrocarbon solution, a useful rubber, chlorosulphonated polyethylene, is produced.

The action of heat on polymers depends largely on whether or not oxygen is present. Heat alone causes bond scission at temperatures which reflect the bond strength of the covalent bonds which bind the atoms of the polymer chain together. Table 1.3 gives bond dissociation energies for some linkages found in polymers. The high the bond dissociation energy, the higher the temperature at which the bond breaks. The weakest bond in a polymer will therefore determine the thermal stability of a polymer chain in the absence of oxygen. In practice, thermal stability is usually less than expected from the chemical structure owing to the accidental inclusion of weak links, notably the peroxide link (—O—O—) during polymerization. In presence of oxygen, it is the ability of bonds to withstand reaction with oxygen at elevated temperatures that is important. The bonds with the lowest bond dissociation energy (Table 1.3) are the most susceptible to oxygen attack at elevated temperatures. Consequently, polymers containing —CN, —C=O, —C=C—, —OH and —Cl are all readily degraded by heat in the presence of oxygen. The —CF bond, although not having the highest bond dissociation energy, has remarkable resistance to oxygen and so polytetrafluoroethylene has excellent heat stability in the presence of oxygen (up

Table 1.3 Bond dissociation energies

Bond	Dissociation energy kJ mole^{-1}
C≡N	890
C=O	740
C=C	610
O—H	464
C—F	*c*.500
C—H	414
N—H	389
C—O	359
C—C	347
C—Cl	343
C—N	305
O—O	145

to around 400°C). Polymers with aromatic backbones also show excellent thermal stability owing to the relative inertness of the phenylene ring. The thermal stability of these polymers is determined by the nature of the groups which link the phenylene ring together in the chain.

Electromagnetic radiation can produce a reaction, usually degradative if the radiation has sufficient energy. The energy of the radiation is related to wavelength of the radiation:

$$\text{Energy} = \frac{hc}{\lambda}, \tag{1.5}$$

where h is Plank's constant, λ is the wavelength and c is the speed of light. It therefore increases with decreasing wavelength. Sufficient energy for bond scission is found in the UV region and electromagnetic radiation of shorter wavelength. It is not simply that sufficient energy is available. The energy must be capable of being absorbed. Bonds have characteristic absorption frequencies and only radiation of these frequencies can be absorbed. On the earth's surface, UV light occurs in the range 300–400 nm owing to filtration of smaller wavelengths by the atmosphere. Most bonds in polymers do not absorb in this region, a notable exception being the carbonyl group —C=O. Most polymers should therefore be stable in sunlight. However polymers usually contain UV reactive groups which have been incorporated accidentally during polymerization or processing. Polymers are therefore often less stable than their structures would suggest.

Radiation with energy higher than that of UV light (gamma rays and electron beams) generally have a degradative effect on polymers. In some cases, notably polyethlene, high-energy radiation produces significant amounts of crosslinking. High-energy radiation is exploited commercially to produce crosslinked forms of a polymer to improve mechanical properties, especially at higher temperatures, or to increase insolubility. Alternatively, it is used to produce degradation and hence increased solubility.

1.9 Solubility

A polymer solution consists of polymer chains dispersed as discrete but entangled chains in a solvent medium. The process of dissolution involves three fundamental steps: the penetration of the solvent, the separation of the chains and the solvation of the chains.

For separation of the chains to occur, the solvent liquid must be able to overcome the interchain forces, which vary in strength from weak dispersion forces to strong hydrogen bonds. It follows therefore that since the interchain forces are determined by the chemical nature of the polymer, the chemical nature of the solvent in relation to the polymer is of importance. In addition to separating the chains, the liquid must be able to solvate the polymer chains in order to reduce the free energy of the separated chains so that dispersion of the chains in the liquid can take place.

In general, non-polar liquids are solvents for non-polar polymers, and polar liquids are solvents for polar polymers. Although various approaches, notably the use of the solubility parameter, are used to predict the suitability of a liquid as a solvent for a polymer, no approach is universally applicable. In many cases, one has to resort to the practical method of trial and error and suitable solvents can be found in most cases. Notable exceptions among common polymers are PTFE for which there is no known solvent and polyethylene for which there is no solvent at room temperature. In the latter case, the polymer is soluble in aromatic hydrocarbons at elevated temperatures where the thermal energy helps to overcome the interchain forces.

There are potentially many uses for polymers in aqueous solution. The high polarity of water militates against its use as a solvent unless either the polymer can undergo a specific interaction with the water, usually hydrogen bonding, or the polymer chain is made with a high polarity even to the extent of being ionic. Many natural polymers are water-soluble because they contain hydroxyl groups in sufficient concentration or are able to participate in hydrogen bonding with water. Hydrophilicity in synthetic polymers has been obtained by emulating nature. For example, acrylic polymers based upon methymethacrylate have been made hydrophilic by copolymerizing the methymethacrylate with hydroxyacrylate monomers.

The rate of dissolution depends primarily upon the rate at which the solvent can penetrate the solid polymer structure. The rate of diffusion is determined by the interchain forces and by the polymer morphology. Amorphous polymer will tend to dissolve more quickly than will crystalline polymer because the closer packing of the crystalline chains hinders diffusion. In some cases, the rate of diffusion is so slow that the polymer is effectively insoluble in the solvent. In such cases, raising the temperature often produces dissolution. Polyvinyl alcohol does not dissolve at an appreciable rate in cold water but rapidly dissolves in hot water.

Molecular weight can also significantly affect the solubility of polymer chains. Low molecular weight polymer dissolves more readily than high molecular weight polymer. In less than good solvents, the high molecular weight polymer may well be insoluble. Where a polymer has a wide molecular weight distribution, this has led to the lower molecular weight fractions dissolving but not the higher molecular weight fractions. This partial solubility can often be converted into complete solubility by heating the solution.

Cross-linked polymers will not dissolve even in ideal liquids. The cross-links consist of covalently linked atoms and consequently the polymer chains are permanently tied together. Although penetration of the liquid may be possible, dispersion of the chains is not. The result is that cross-linked polymer swells as the liquid is absorbed but it will not dissolve. Highly cross-linked polymers will swell very little but lightly cross-linked polymers may absorb several times their own weight of liquid to produce highly swollen gels.

Highly swollen gels may also be obtained with linear polymers if the solvent is capable of undergoing specific reactions with the polymer to the extent that the solvent molecules effectively form cross-links. This is the case with starch in water where, provided the starch concentration is high enough, the water molecules from hydrogen bonded bridges between the chains. These bridges are strong enough to prevent true dispersion and a gel results. These types of polymers are useful as thickening agents.

1.10 Electrical properties

Polymers in common use are non-conducting materials. Conductivities vary from 10^{-10} to 10^{-18} S cm^{-1} and consequently, polymers are widely used as insulating materials. Good conductors such as copper (conductivity 6×10^5 S cm^{-1}) conduct because there are electrons available to transmit the current when an electrical potential is applied. This is not the case with most common polymers since the electrons are tightly bound in covalent bonds. Polar polymers such as polyvinyl chloride have higher conductivities than do non-polar polymers because of the electric dipole which is present. Nevertheless, polar polymers are essentially non-conductors and are widely used in low-voltage insulation. They also make useful dielectric materials where their polarity and insulating character enables small capacitors of high storage capability to be manufactured. Especially useful in this area are polyvinyl fluoride and polyvinylidene fluoride in film form.

The most common method of making polymers electrically conductive is to incorporate a conductive material in the polymer. These materials include carbon and metal powders. However, polymers that are intrinsically conductive are feasible provided that electrons are available to transmit current. Polymers with conjugated double bonds are found to be more conductive than non-conjugated polymer structures. The conduction is due to the delocalization of the π electrons of the double bonds. The simplest polymer of this type is polyacetylene:

$$-CH{=}CH-CH{=}CH-CH{=}CH-CH{=}CH-$$

and, among the commercially available polymers, polyphenylene sulphide

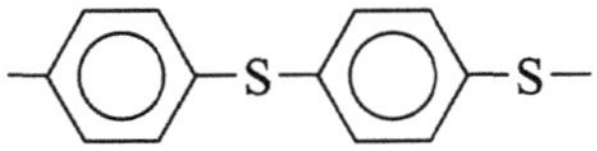

shows conductive capability as a result of delocalization involving the aromatic ring and the sulphur atom.

In itself, delocalization does not significantly improve the conductive nature of the polymer, but if the polymer is doped with additives (electron donors or acceptors) conductivity values of up to 10 S cm^{-1} can be achieved depending on the additive, its concentration and the polymer. In effect, the dopant is making electrons or positive holes (owing to the removal of electrons) available for conduction. Other methods of releasing electrons include the use of heat, light and mechanical stress.

1.11 Conclusion

The content of this chapter has of necessity been introductory but it is hoped that sufficient information has been given to indicate the importance of the chemical structure in determining the properties of a polymer chain. In order to achieve a desired property, a polymer chain of specific character and structure is necessary and, today, methods can usually be found to create the required polymer. Methods of making polymers are reviewed in Chapter 2. It must be remembered that in striving to achieve a particular property, other properties are inevitably affected and often adversely. In most cases, a compromise has therefore to be accepted. The reader is referred to the reading list for further information on polymer structures and properties.

Further reading

Billmeyer, F.W. (1962, 1971) *Textbook of Polymer Science*, John Wiley, New York.
Cowie, J.M.G. (1973) *Polymers: Chemistry and Physics of Modern Materials*, Intertext [The Blackie Publishing Group], Glasgow and London.
Flory, P.J. (1953) *Principles of Polymer Chemistry*, Cornell University Press, Ithaca, NY.
Hiemenz, P.C. (1984) *Polymer Chemistry*, Marcel Dekker, New York.
Meares, P. (1965) *Polymers: Structure and Bulk Properties*, Van Nostrand, New York.
Odian, P.J. (1970) *Principles of Polymerization*, McGraw-Hill, New York.
Ritchie, P.D. (1965) *Physics of Polymers*, Iliffe Books [Butterworth], London.
Rodrigues, F. (1985) *Principles of Polymer Systems*, McGraw-Hill, New York.
Saunders, K.J. (1973) *Organic Polymer Chemistry*, Chapman & Hall, London.

2 Polymerization

R.W. DYSON

2.1 Introduction

Almost all polymers in use as plastics, rubbers and fibres are synthesized from chemicals derived from oil. The chief exceptions are natural rubber and related polymers, and polymers derived from cellulose such as cellulose acetate. This chapter is principally concerned with the methods of polymerization, the salient features of which are outlined.

There are two principal methods of synthesis each with many variations. These give rise to what are known as **addition polymers** and **step-growth polymers** (also known as condensation polymers).

2.2 Addition polymerization

2.2.1 Monomers

The synthesis involves taking a simple chemical, called the monomer, and causing the molecules to join together to form long chains, for example:

$$\begin{array}{c} \mathrm{M\quad M\quad M\quad M\quad M\quad M\quad M} \\ \downarrow \\ \mathrm{-M-M-M-M-M-M-M-} \end{array}$$

The polymerization is brought about by the presence of initiators in trace quantities (0.1–1.0%).

The links between the monomers do not occur instantly but in rapid succession. A polymer chain of 1000 units can be formed in about one second. However, the conversion of the monomer to polymer usually takes between several minutes and several hours, because in most polymerizations of this type the chains are not all initiated at the same time.

The monomers contain at least one carbon–carbon (or carbon–oxygen) double bond. Ethylene is the simplest monomer:

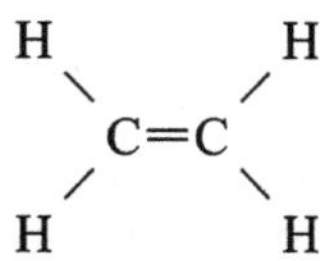

and other monomers may be considered to be derivatives of ethylene. Most of these monomers are monosubtituted:

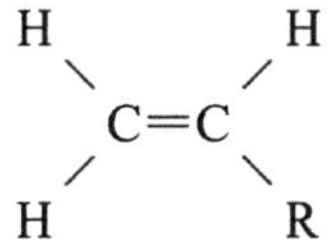

For example propylene ($R = CH_3$), styrene (R = phenyl) and vinyl chloride (R = Cl), but some are disubstituted:

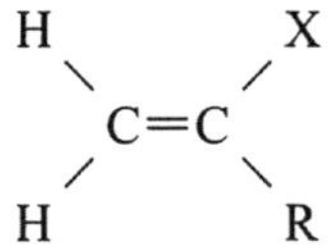

For example methyl methacrylate ($R = CH_3$, $X = COOCH_3$) and vinylidene chloride (R = Cl, X = Cl). Tetrafluoroethylene is a tetrasubstituted form of ethylene:

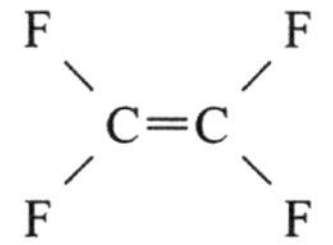

Monomers containing two C=C bonds are termed dienes and are used for producing rubbers, for example butadiene:

$$H_2C{=}CH{-}CH{=}CH_2$$

In all cases, the function of the initiator is to open up the double bond causing the monomers to link together in rapid succession. This can be represented by:

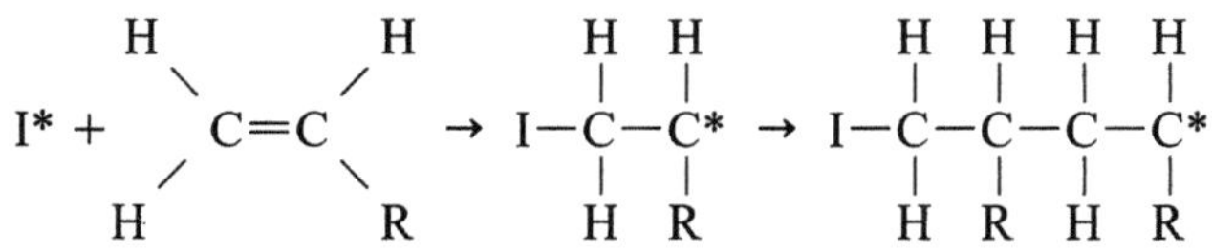

and so on. I* is a fragment derived from the initiator and may be a free radical, a cation or an anion. In practice, only a few monomers can be polymerized by all three types of initiator.

2.2.2 Free radical polymerization

Most free radical polymerizations are brought about by the decomposition of organic peroxides, peresters or azo compounds at elevated temperatures. Some use is made of ultraviolet light, which produces the decomposition of certain initiators (e.g. azo compounds) at room temperatures.

The polymer produced is essentially linear but with some long-chain branching. Polyethylene (low-density) is a notable exception, being highly branched with short side chains.

The polymer chains grow to different sizes, usually measured in terms of molecular weight, which results in a distribution of molecular weights in a given polymer sample. The average molecular weight can be controlled by reaction conditions and by the use of trace additives which are sometimes known as molecular weight modifiers. Chemically, they are transfer agents which compete with the monomer in reacting with the growing chain. Instead of adding on to the growing chain, however, a radical (usually H•) is transferred to the end of the growing chain and stops further growth of that chain. The transfer agent, having transferred a radical, itself becomes a free radical and a new chain polymerizes from the residue of the transfer agent. The transfer process and its effect is shown schematically in Fig. 2.1. The average molecular weight of the

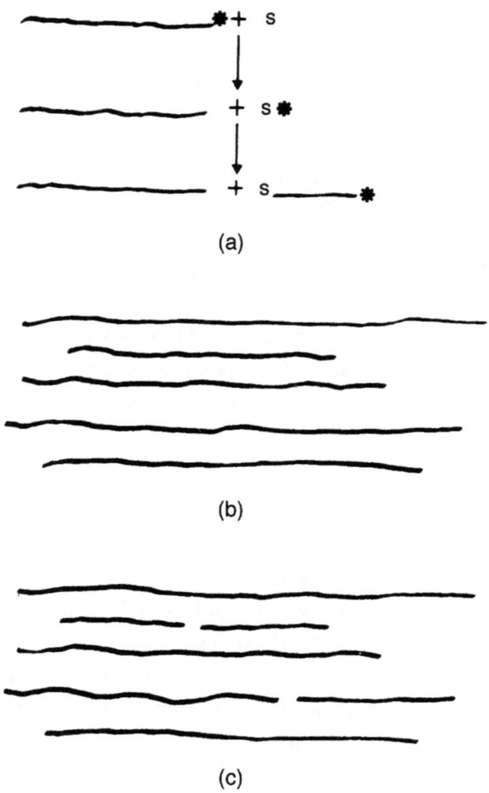

Figure 2.1 (a) Transfer process in which the activity of a growing chain is transferred to a small molecule s (chain modifier). (b) Polymer chains produced in the absence of chain modifier. (c) The same polymer produced in the presence of chain modifier.

polymer formed in the presence of the transfer agent ($\bar{M}$) is related to the average molecular weight of the polymer formed in the absence of the transfer agent ($\bar{M}_0$) by the equation

$$\frac{1}{\bar{M}} = \frac{1}{\bar{M}_0} + C_S \frac{[A]}{[M]M_m}, \tag{2.1}$$

where $[A]$ is the concentration of transfer agent, $[M]$ is the monomer concentration, C_S is a constant for the system (the transfer constant) and M_m is the molecular weight of the monomer. The value of C_S is usually very large so that only trace amounts of transfer agent are needed to control the molecular weight average of the polymer being formed. The molecular weight distribution cannot be controlled so readily.

There is little stereoregulation of the way in which the monomers add to the growing chain in free radical polymerization. Consequently, the polymer produced from monosubstituted monomers is essentially atactic although lower polymerization temperatures favour an increase in syndiotactic content compared with higher polymerization temperatures.

Some polymers are manufactured commercially by heating the monomer together with the initiator, a process known as mass (or bulk) polymerization. There are considerable heat problems, however, since polymerization is highly exothermic. Commonly, the monomer containing the initiator is dispersed as droplets in water (suspension polymerization), the water acting as a heat sink. Other methods involve dissolving the monomer in a suitable solvent (solution polymerization) or emulsifying the monomer with a non-solvent (emulsion polymerization). These methods are used mainly when the polymer is to be used as a solution or a latex to avoid extraction problems. An advantage of emulsion polymerization is that high molecular weight polymer can be produced more easily than with the other methods. Gaseous monomers are usually polymerized at high pressure, otherwise atmospheric pressure is used.

2.2.3 Ionic polymerization

Ionic polymerization is always performed in solution since the ions require a solvent for their existence. The choice of solvent is important and polymerization will not occur if the solvent polarity is too high or too low. Polymerization temperatures are much lower than those used for free radical polymerization, ranging from just above room temperature to well below it, depending upon the polymerization system. In most anionic polymerizations the rate of polymerization increases with decreasing temperature and conversion of the monomer to polymer is often very rapid, taking a few minutes rather than hours.

The molecular weight distribution is much narrower than in polymers produced by free radical polymerization, and in some anionic polymerizations it is very narrow. Chain branching in non-existent in many instances so that the polymer produced is truly linear. A high degree of stereoregulation is possible using certain initiating systems, solvents and temperatures.

From a practical point of view ionic polymerizations must be performed in uncontaminated conditions since merely *traces* of contaminants (such as water) will prevent the polymerization. Ionic polymerizations are generally expensive and are carried out either when a polymer cannot be made by free radical techniques (e.g. polyacetals) or when use can be made of the unique properties of some ionic polymerizations to produce polymers of desired structure.

2.2.4 Catalytic surface polymerizations

These polymerizations have many features in common with ionic polymerizations and are generally used for the polymerization of ethylene and propylene, although their use is potentially wider. They involve the use of organometallic complexes, the best known of which are the Ziegler–Natta type. These involve titanium chlorides and aluminium alkyls. Another similar type of system involves chromium oxide. The great advantage of these systems is that they are highly stereoregulating and are therefore used for producing isotactic polypropylene. They also produce linear polymer (e.g. high-density polyethylene).

2.3 Step-growth polymerization

Step-growth polymerization makes use of the fact that certain chemical groups will react with other chemical groups under appropriate conditions to link the molecules to which the groups are attached. For example in

$$\mathrm{R{-}OH + R'{-}COOH \rightarrow R{-}O{-}OC{-}R' + H_2O}$$

R and R′ are linked by an ester group and water is condensed out, hence the term condensation reaction. R and R′ may be any of a wide range of organic species giving rise to a whole family of esters.

If the reacting species are difunctional, for example HO—R—OH and HOOCR′COOH, the reaction will lead ultimately to a long linear chain of R and R′ entities linked by ester groups, that is a polyester, namely ···R′COOROOCR′COOROOCR′···.

If at least one of the reacting species is trifunctional, as in

$$\begin{array}{c}\mathrm{H{-}R{-}OH} \\ | \\ \mathrm{OH}\end{array}$$

a branched polyester will be formed:

$$\sim R'-COO-R-OOC-R'\sim$$
$$\begin{array}{c} O \\ O \\ C \\ R' \\ \wr \end{array}$$

in the first instance, and ultimately a cross-linked system will be produced.

If both the reacting species are trifunctional or if the reacting species have higher functionalities then a more highly branched chain will be formed initially and, ultimately, a more highly cross-linked end product will result.

The above examples illustrate the general approach. Groups other than —OH and —COOH can be made to react together to form useful polymers which are usually named after the predominant linking group. Examples are polyamides, polycarbonates, polyethers and polyurethanes.

Since R and R′ groups can be chosen from an enormous range of chemicals there are potentially vast families of each of these polymers. In practice, only a few are exploited commercially.

The chemistry of step-growth polymers is often complex. For example, polyurethanes are widely used as foams, and one way of producing the foamed polymer is to incorporate traces of water, which leads to the formation of carbon dioxide which is the blowing agent. This in turn leads to several other linking groups in addition to the urethane link:

$$\begin{array}{c} -N-C- \\ \;\;|\quad\;\, \| \\ \;\;H\quad O \end{array}$$

which is produced by reacting isocyanate groups (—NCO) with the hydroxyl group (—OH).

The formation of polymer can be brought about by heating a mixture of the required starting materials in solution with appropriate catalysts. There are a number of problems if this simple approach is taken. The reaction is relatively slow and gets slower as the reactants are consumed. It is not commercially viable to make high molecular weight polymers by this approach and it is therefore restricted to the production of low molecular weight chains. Such chains must be cross-linked to make them into useful products. Once cross-linked, any shaping process involving flow cannot be used. Therefore, this technique is used for making low molecular weight polymer which will be cross-linked during (or after) the shaping or application process.

The low molecular weight chains need not be polyfunctional (i.e. made from polyfunctional starting materials). If they are, heat and an appropriate catalyst will usually complete the cross-linking. Linear chains (made from difunctional starting materials) can subsequently be cross-linked in a number of ways. For example, cross-linking agents (polyfunctional molecules) can be added which react with the ends of the short polymer chains to join them together to form a cross-linked system. In the case of polyesters used in glass fibre composites the polyester is made to contain C=C bonds in the chain so that it can be cross-linked by free radical polymerization of, usually, styrene. In some cases, cross-linking is brought about by the presence of moisture in the atmosphere, for example, in the case of silicone rubbers.

It may be noted here that cross-linking is not restricted to short chains. There are advantages to cross-linking long polymer chains. This is essential in the case of rubbers and, in the majority of cases, sulphur is used as the cross-linking agent, reacting with the residual C=C in the long rubber molecular which is made from the polymerization of a diene. Polyethylene is sometimes cross-linked. This has been done for many years by free radical means or by high-energy radiation techniques. Both of these have limitations. In the former cross-linking occurs as a result of heat produced during processing but is often be affected by other additives which may present, especially carbon black. In the latter case, cross-linking occurs after moulding but is generally restricted to small articles, and the hazards of high-energy radiation are obvious. A more recent technique involves treating polyethylene with silicone compounds such that the polymer contains attached silicone groups which are reactive to water. The polymer is processed in the normal way and subsequent cross-linking occurs through the ingress of atmospheric moisture. In practice, the product is often 'steamed' immediately after processing to speed up the cross-linking process. Potentially, this type of cross-linking can be applied to many polymers but commercial use is largely confined to polyolefin polymers.

There is significant use of high molecular weight, linear condensation polymers as thermoplastics, including the polyamides (known as nylons), polycarbonates, polysulphones and thermoplastic polyesters.

Since the process of simply reacting starting materials together in solution produces only low molecular weight polymer economically, the thermoplastic condensation polymers are produced by a variety of other techniques which are often quite devious. The main reason why relatively few of these polymers are used compared with thermoplastic polymers produced by addition polymerization is that the processes for making such polymers are often uneconomic. However, this method of polymerization produces thermoplastic polymer with the best heat stability coupled with good all-round mechanical performance.

2.4 Copolymer formation

The production of new polymers by addition polymerization depends on the development of new polymerization techniques of finding new monomers. Both seem unlikely. As an alternative, copolymerization of mixtures of existing monomers using existing techniques has led to a wide range of new polymers–copolymers. Sometimes these copolymers have been produced to modify properties of polymers; for example polypropylene copolymer, comprising about

90% propylene and 10% ethylene, retains the essential features of polypropylene, but the ethylene content gives the polymer superior toughness and reduces the brittle temperature to about –20°C. At other times, completely new polymers have been produced: for example a random copolymer containing approximately equal amounts of ethylene and propylene is not a plastic but a rubber. These examples illustrate the concept that if one monomer constitutes the majority (about 90% or more) of the copolymer then the polymer properties are essentially those of the polymer made from that monomer. However, if both monomers are present in more equal proportions then new polymers are obtained with often quite different properties from polymers prepared from either of the individual monomers. The arrangement of the monomers in the copolymer is also significant (Chapter 1). The majority of commercial copolymers are made by addition polymerization.

2.4.1 Step-growth polymerization

This offers, potentially, more scope for the production of new polymers without resorting to copolymerization. It is still restricted by the need to find new starting materials and commercially viable methods of converting them into useful polymers. Copolymers are produced to a limited extent by condensation polymerization of mixed starting materials – for example Nylon 66/610.

Conventionally, the term copolymer is used for linear or branched polymer chains, for example thermoplastic materials. However, many polymers prepared for thermosetting materials are in fact copolymers since they are made from mixed starting materials. The polyesters used in glass fibre composites come under this heading, being made not from a diol and a diacid but from *two* diols and *two* diacids. The ability to use mixtures of reactants has enabled many developments in polyurethanes, for example, to take place.

2.4.2 Random copolymers

A random copolymer is one in which the two repeat units in the chain are arranged to a greater or lesser extent in a random fashion. The most common way of producing these is by addition polymerization, the majority of methods being by a free radical process.

Addition polymerization of a mixture of two monomers by any of the above methods will in general produce a random copolymer. The composition of the copolymer can be adjusted by choosing an appropriate monomer feed composition. In the vast majority of cases, the copolymer composition does not have the same composition as the monomer feed. This is because the monomers have different reactivities towards the growing chain radicals, of which there are two types. Usually, one monomer is preferentially incorporated into the copolymer so that the copolymer composition depends not only upon the relative amounts of the monomer in the feed but also upon the relative reactivities of the monomers with each type of chain end.

There are thus four possible addition reactions:

$$\sim\!\sim A\bullet + A$$

$$\sim\!\sim A\bullet + R$$

$$\sim\!\sim B\bullet + A$$

$$\sim\!\sim B\bullet + B$$

where A and B are the monomers and $\sim\!\sim A\bullet$ and $\sim\!\sim B\bullet$ are the two types of growing chain. In simple terms, it is the relative rates of these four reactions that determines the copolymer composition. An equation can be derived:

$$F_{\mathrm{A}} = \frac{r_{\mathrm{A}} f_{\mathrm{A}}^2 + f_{\mathrm{A}} f_{\mathrm{B}}}{r_{\mathrm{A}} f_{\mathrm{A}}^2 + r_{\mathrm{B}} f_{\mathrm{B}}^2 + 2 f_{\mathrm{A}} f_{\mathrm{B}}}, \tag{2.2}$$

relating the copolymer composition (F) to the feed composition (f) and to the relative reactivities of the monomers to a given radical, expressed as the reactivity ratio (r). F and f are mole fractions of monomers A and B, as denoted by the subscripts. The values of r may be obtained in most cases from the wealth of experimental values available. The fact that the copolymer composition is not the same as the monomer feed composition means that there will be a drift in the composition of both the polymer and the monomer feed. This is remedied by monitoring the feed and keeping its composition constant by constantly topping up with an appropriate monomer mixture.

The arrangement of the monomers in the copolymer is a function of the same factors that govern the composition, namely reactivity ratios and mole fractions of the monomers in the feed. Composition and monomer arrangements cannot therefore be varied independently. The monomers occur in a distribution of sequence lengths which is characteristic of the feed ratio and reactivity ratios. Figure 2.2 shows the sequence length distribution for a truly random copolymer. The average sequence length ($\bar{s}$) is two for both monomers. In general, the average sequence length is given by:

$$\bar{s} = 1 + r_{\mathrm{A}} \frac{f_{\mathrm{A}}}{f_{\mathrm{B}}} \quad \text{and} \quad \bar{s}_{\mathrm{B}} = 1 + r_{\mathrm{B}} \frac{f_{\mathrm{B}}}{f_{\mathrm{A}}}. \tag{2.3}$$

A special case is when r_{A} and r_{B} both equal zero because this always produces an alternating copolymer $\sim\!\sim$ABABABABABABABABABABABABABAB irrespective of the feed composition.

Mixtures of suitable monomers can also be addition polymerized by ionic methods and indeed some are (e.g. acetal copolymer). The principles outlined above apply although the r values for a pair of monomers will be different from those for free radical polymerization. They will be dependent upon reaction conditions and upon the solvent in particular. Ionic polymerization is used principally for making block copolymers (Section 2.4.3) where use can be made of certain unique features of some ionic systems.

Random copolymers can be produced by step-growth reactions. The procedure is usually relatively simple; an appropriate mixture of the reactants is mixed in the required proportions and reacted. Since the reacting species generally have similar reactivities, the copolymer composition is essentially that of the feed, and the copolymer produced is random.

As an example, let us consider nylon 66/6 10 copolymer. Nylon 66 is obtained by polymerizing the salt $H_2N{-}(CH_2)_6{-}NH_3^+ \ {}^-OOC{-}(CH_2)_4COOH$. Likewise, $H_2N{-}(CH_2)_6{-}NH_3^+ \ {}^-OOC{-}(CH_2)_8COOH$ is used to prepare nylon 6 10. If the two salts are mixed, the nylon 66/6 10 copolymer is produced with the general formula

$$\sim\!NH{-}(CH_2)_6{-}NH{-}CO{-}(CH_2)_x{-}CO\!\sim$$

in which x may be 4 or 8 in random fashion. Another example is the formation of unsaturated polyesters used as thermosetting resins in glass fibre composites. These resins are made from *two* diacids and a diol or two diols. General purpose resins can be made from maleic acid and phthalic acid (in equal proportions) and

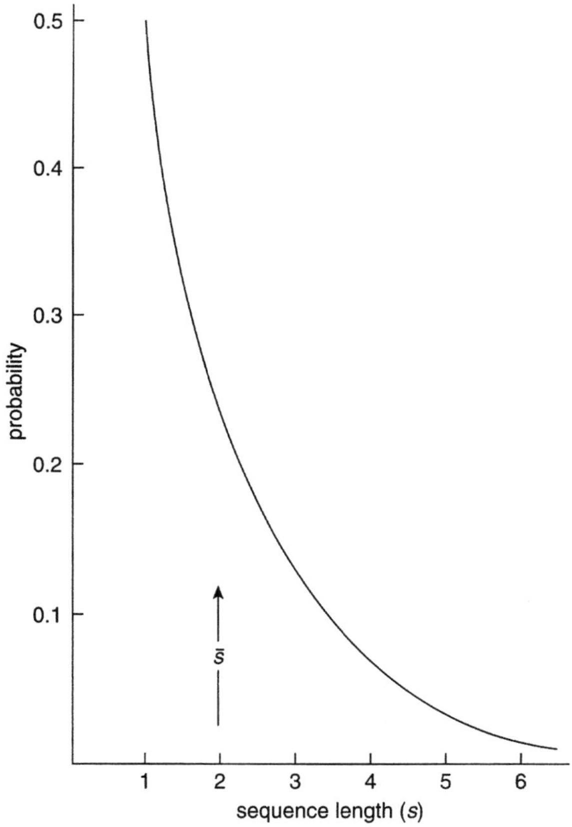

Figure 2.2 Sequence length distribution for a truly random copolymer.

equivalent amounts of ethylene glycol. If the reactants are mixed initially, the resultant polymer comprises random maleic (M) and phthalic (P) residues interspersed with glycol (G) residues, for example GPGPGPGPGMG or GMGPGPGMG. If the copolymerization is carried out in two stages a more regular structure is produced. In stage one, the phthalic acid is reacted with all of the glycol to give GPG Stage two involves reacting the maleic acid with the product of stage one to give GPGMGPGMGPG. This illustrates that some control over the copolymer structure is possible with step-growth reactions.

2.4.3 Block copolymers

Block copolymers consist of two units (A and B) which occur in long sequences called blocks. Several types of block arrangements are possible, for example:

A B block	A——AB——B
A B A block	A——AB——BA——A
alternating block	A—AB—BA—AB—BA—A

Block copolymers are not made by copolymerizing a monomer mixture but either by sequential polymerization or by linking prepolymers together.

Sequential polymerization is an addition polymerization involving the use of **living polymer** and introducing the monomers to the reaction vessel in sequence so that only one monomer is polymerizing at a time. As noted in Section 2.2.1, addition polymerization involves rapid growth of a polymer chain by the addition of monomer units to an active chain, the activity being free radical or ionic. In most cases the chain is terminated after about a second by a mechanism involving the collision of two growing chain ends. In ionic polymerizations no such mechanism is possible and in certain anionic polymerizations there is no alternative mechanism to terminate the growing chain.

Once the monomer is used up, the chains do not die but remain active and are called living polymers. They do not grow, however, because there is no monomer. However, if more monomer is added, growth begins again until all of the added monomer is used up. This process of successive addition of monomer can in theory be continued indefinitely. In practice this is not so. The anionic activity remains only so long as a high degree of purity is maintained in the reaction vessel. Sooner or later, air leaks in and moisture or other contaminants kill the chains. However, the reactivity can last long enough to make block copolymers.

The process involves introducing a known quantity of monomer A to the reaction vessel which contains a known number of ions. Polymerization of A ensues and the chain length (block sequence length) is controlled by the ratio of monomer to ions. The result is living polymer.

$$A\text{———}A^-$$

If a regulated amount of monomer B is introduced, this will polymerize from the ends of the living polymer A chains:

A——AB——B⁻

If an AB block is required, a **chain stopper** is added to kill the polymer, otherwise more monomer A is added to give an ABA block

A———AB———BA———A⁻

This can be killed if an ABA block is required or polymerization can be continued. There is no reason why a third, or fourth, monomer should not be introduced at any stage if required.

A special case of this is the use of **coupling monomers** at some stage to form radial or star multiblock copolymers. These are copolymers comprising three or more block sequences radiating from a single point, for instance:

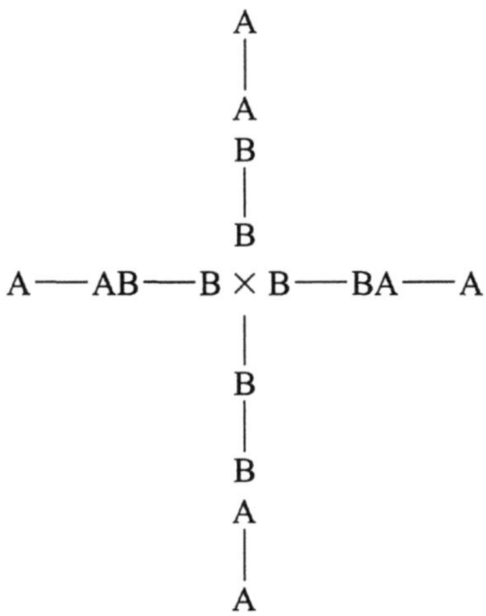

The polymerization is similar to that of the formation of the linear ABA block copolymer describes above. In the above example, monomers A and B are sequentially polymerized as above. Then instead of adding more A, a polyfunctional coupling monomer (X) is added so that one coupling monomer adds to each chain. The polymerization is then completed by sequentially adding monomers B and A before terminating the reaction.

Zeigler–Natta type polymerizations are also used for making block copolymers since they are also capable of producing living polymers.

The great advantage of living polymerization systems is that the block sequences can be closely controlled.

The linking of prepolymers involves preparing prepolymers of A and B such that each has a reactive group at each end. These reactive groups (say X and Y) are chosen such that they can be made to react with each other. The final stage involves mixing the two prepolymers in appropriate amounts and allowing the linking reaction to occur. For example, if the ratio of prepolymer A to prepolymer B is 2 : 1, an ABA block is formed:

$$\mathrm{XA{-\!\!-}A{-}X \qquad YB{-\!\!-}B{-}Y \qquad X{-}A{-\!\!-}A{-}X}$$
$$\downarrow$$
$$\mathrm{X{-}A{-\!\!-}AXYB{-\!\!-}BYXA{-\!\!-}A{-}X}$$

These block copolymers may be further coupled if required. The chain lengths of both prepolymers can be controlled during their polymerization. Use is made of this type of copolymerization in forming polyurethanes, for example.

Since the linking reaction is a condensation type of reaction, it is easier to make the prepolymers by a condensation method as the prepolymers will naturally have reactive groups at the end. However, it is possible to prepare addition polymers with reactive groups at each end by using suitable initiators. An initiator fragment remains attached to the end of the chain and it is the initiator fragment that has the reactive group attached. A reactive group results at both ends of the chain if during the chain termination reaction the two growing chains combine to form one long chain. Most addition polymers formed by free radical polymerization terminate in this way.

2.4.4 Graft copolymers

These are generally made either by polymerizing one monomer in the presence of a prepolymer or by linking reaction which joins two prepolymers.

When a monomer is polymerized in the presence of another polymer by free radical means, a small amount of chain transfer to polymer occurs. This occurs during the polymerization of a single monomer in the later stages of polymerization when there is a significant amount of polymer present. The mechanism can be described pictorially as in Fig. 2.3. A growing chain collides with a polymer chain and a component, often H, is transferred to the end of the chain, terminating its growth. The polymer chain now has an active centre from which monomer can polymerize to form a side chain. This leads to branched polymer in homopolymerization.

If polymer A is dissolved in monomer B, the polymerization of B will involve to a greater or lesser extent (depending on the chemical nature of polymer A) the formation of side branches of B attached to polymer A (Fig. 2.4). Some use is made of this commercially (in toughened polystyrene and ABS) but it is difficult to control the structure of the graft copolymer.

As an alternative, prepolymer A can be made with reactive side groups which may take part in the initiation of the polymerization of monomer B. For example, if certain chlorine-containing groups are present in the prepolymer A, these will react with metal carbonyls to produce a free radical following the abstraction of a chlorine atom. Polymerization of B is initiated at this side, leading to the attachment of a B side chain. A little more control is possible since the placing of the chlorine-containing groups can be engineered to some extent by copolymerization.

Use can be made of linking reactions to produce copolymers if one prepolymer has reactive groups at its ends while the other has reactive groups along the chain:

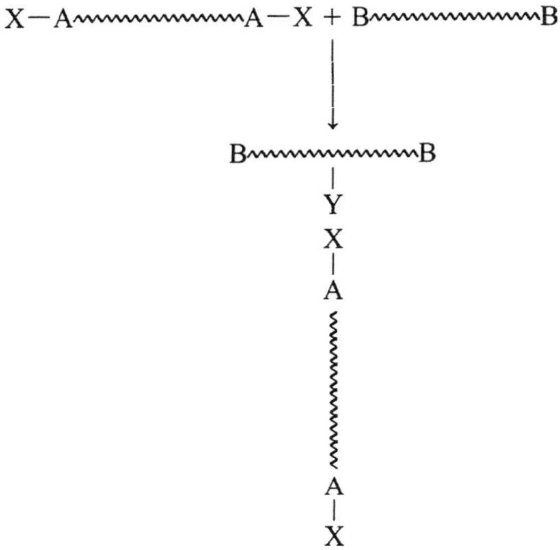

The situation is similar to the formation of block copolymers by this method (Section 2.4.3).

2.5 Chemical conversion

New polymers can be made by chemically converting polymers by suitable reactions. Although chemical conversion is not a polymerization process, it will be mentioned here because it is used commercially to produce polymers with quite different properties from the starting polymer and in some cases provides the only practical method of obtaining a desired polymer. The following examples will illustrate the principle of the technique.

Polyvinyl alcohol cannot be polymerized form vinyl alcohol because the monomer is unstable. The polymer is therefore obtained by the reaction of ethanol on polyvinyl acetate under alkaline conditions:

$$\sim\!CH_2-\underset{\displaystyle OOCCH_3}{\underset{|}{CH}}\!\sim + CH_3CH_2OH \rightarrow \sim\!CH_2-\underset{\displaystyle CH}{\underset{|}{CH}}\!\sim + CH_3COOCH_2CH_3$$

The process is called alcoholysis. The amount of conversion can be controlled so that a range of polymers with slightly different properties can be obtained. For example, polymers with 88% of the acetate groups converted to hydroxyl groups are readily soluble in cold water whereas those with a higher hydroxyl

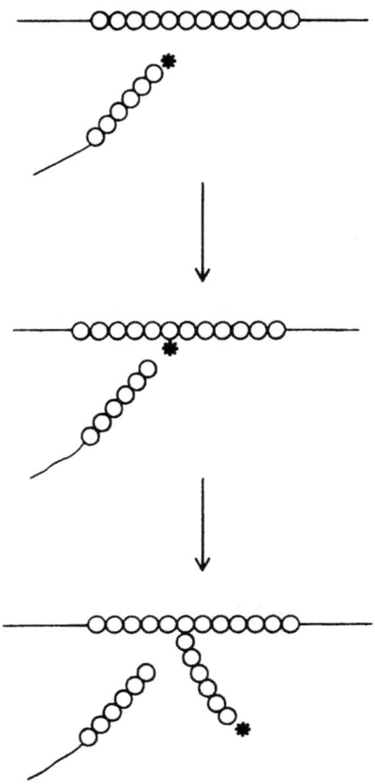

Figure 2.3 Transfer to polymer producing a side branch.

content are less readily soluble, and if the hydroxyl content is above about 98% heat is required to produce a solution. Technically, all polyvinylalcohol polymers are copolymers of vinyl alcohol and vinyl acetate unless alcoholysis is complete (100% conversion).

Cellulose is converted into cellulose acetate polymers by treatment with acetic acid or acetic anhydride. The acetic acid reacts with the hydroxyl groups of the cellulose repeat unit:

$$-OH + CH_3COOH \rightarrow -OOCCH_3 + H_2O$$

In practice, the acetate content of these polymers is usually between 2.2% and 3% per ring depending on the ultimate use of the polymer.

Even chemically unreactive polymers such as polyethylene can be modified. Treatment with chlorine and sulphur dioxide produces chlorosulphonated polyethylene which is a useful rubber commonly known by its trade name Hypalon^{TM}.* The silane treatment of polyethylene (Section 2.3) is a chemical modification of the polymer to produce a useful cross-linkable form of polyethylene.

* Names and manufacturers of trademarked chemicals are listed at the beginning of this book.

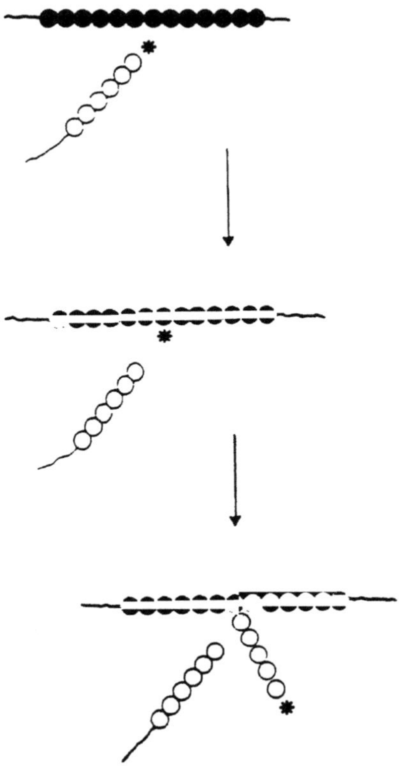

Figure 2.4 Polymerization of monomer B (○) in the presence performed polymer A–A (●) leads to polymer B chains being grafted onto polymer A.

2.6 Polyblends

A polyblend is a mixture of two (or more) polymers (including copolymers) which is brought about by physically mixing the polymers in the required amounts. The production of polyblends would therefore appear not to be part of a chapter on polymerization. However, the production of many polyblends does in fact involve polymerization in order to achieve the appropriate dispersion of the polymers.

The degree of intimacy of mixing depends upon the compatibility of the polymers and the method of mixing. In most cases, it is desirable to have a blend in which there are two phases (at least) in order to obtain the required properties. This implies a degree of incompatibility of the polymers. One can therefore envisage a blend consisting of at least two phases interspersed with each other. If one component is in excess, this can be more simply envisaged as domains of the minor component dispersed in a matrix of the major component. Of considerable importance is the degree of dispersion, that is the size of the domains. This is controlled to a large extent by the method of mixing.

In order for mixing to take place at all, the polymers must be in a liquid or semi-liquid state. The most intimate mixing is obtained when the two polymers are dissolved in the same solvent and the blend is obtained either by removing the solvent or by precipitating the blend from solution. Solvent techniques are not convenient on a large scale.

As an alternative, the polymers may be softened by heat and mechanically mixed using mills (heated rollers), internal mixers or extruders. Since polymer melts are very viscous liquids, the mixing is not very efficient in terms of dispersion but it is often adequate for many uses. Problems arise, for example, when polymers have very different softening temperatures.

The use of heat introduces a chemical factor which may be beneficial. Heat tends to promote the generation of free radicals which in any case result from the mechanical shear forces of mixing. This often leads to some grafting of the polymers, especially at the phase interface. In extreme cases, cross-linking can occur.

Where more control over the dispersion or where a smaller domain size is required than can be obtained by mechanical mixing, other techniques must be used. These usually involve solutions or latices. The production of toughened polystyrene will serve as an example. Toughened polystyrene contains about 10% of a rubber dispersed in the polystyrene. Mechanical mixing produces a modest improvement in impact strength. The improvement is limited by the relative inefficiency of the mixing process which produces rubber domains which are too large for the rubber to be efficient at stopping the propagation of cracks through the blend as a result of an impact. If the rubber is dissolved in styrene monomer which is then polymerized, it is found that the toughening effect of the rubber is much greater. This is because the rubber domains are more finely dispersed in the polystyrene and it has been shown that a rubber domain size of about 30 μm gives the optimum toughening effect. What is more, the polymerization of the styrene in the presence of the rubber leads to grafting of the polystyrene to the rubber which further improves the ability of the rubber to resist crack propagation.

The choice of mixing method will ultimately depend upon the convenience of the method and the ability to achieve an adequate dispersion to obtain the properties desired of the blend.

It should be noted that polymer blends are sometimes called polymer alloys by analogy with metals. The name alloy is often given specifically to a blend which produces a single phase.

2.7 Conclusion

This chapter has, it is hoped, given an indication of the many ways in which polymers can be made, often with engineered structures and therefore properties. Mention must be made of the fact that although many polymers can be made with specified desirable properties they often have less desirable characteristics as well. These can be ameliorated by the use of additives or by blending two or more polymers together. Sometimes, these techniques are used to enhance the required property of the polymer. For example, glass fibres are often added to improve strength and stiffness, especially at elevated temperatures, and rubbers are blended with thermoplastics to improve toughness, especially at low temperatures. Thus not only can polymers be tailor-made to meet specific requirements but also their range of usefulness can be enhanced by blending and the judicious use of additives. Further information can be obtained from the reading list given at the end of Chapter 1.

3 High-temperature and fire resistant polymers

M.S.M. ALGER

3.1 Introduction

Organic polymers provide one of the most versatile groups of materials and find widespread use as plastics, rubbers, fibres, adhesives and coatings. The advantages of these materials, especially their ease of fabrication and low density, for use in an ever widening range of products have slowly become more and more appreciated by designers and engineers. However, some inherent deficiencies in these materials, which have been apparent even from the earliest days of their use, have restricted their use in certain areas of application. Compared with traditional materials, especially metals, organic polymers show extreme sensitivity to temperature. Most importantly, owing to the intrinsic flexibility of their molecular chains, they exhibit very low softening temperatures. Thus whereas most metals do not soften appreciably below their melting points (which may be 1000°C or higher) many common polymers used as plastics, including polystyrene, polyethylene and polyvinyl chloride (PVC), soften sufficiently by about 100°C to be of no use in any loadbearing applications. As explained in Chapter 1, a measured softening point is usually just below the T_g for an amorphous polymer or between T_g and T_m for a partially crystalline polymer. Therefore polymers will have high softening points if they have a high T_g and/or T_m.

A further restriction to their use at high temperature is their susceptibility to atmospheric oxidation when used at elevated temperatures. This oxidation usually leads to polymer chain scission by way of a free radical chain reaction (Fig. 3.1). A deterioration in mechanical behaviour, most notably resulting in embrittlement, is the usual result. Many common polymers oxidize at appreciable rates at temperatures as low as 100°C, so long-term use at these temperatures or above can lead to premature product failure. The susceptibility of a polymer to thermal oxidation is largely dependent on the strengths of the C—H bonds from which hydrogen atoms are abstracted by the propagating free radicals. Table 3.1 lists some values of C—H bond dissociation energies from which it can be seen that aromatic C—H bonds are much more stable than aliphatic bonds and that, among the latter, tertiary, benzylic and allylic hydrogens are particularly reactive.

In the absence of atmospheric oxygen, however, most polymers do not undergo break-up of their molecular structure, that is, degradation, until much higher temperatures, usually in excess of 300°C. Whilst this purely thermal degradation resistance is not relevant to most polymer applications (since polymers are usually used in air) purely thermal degradation is the first stage in

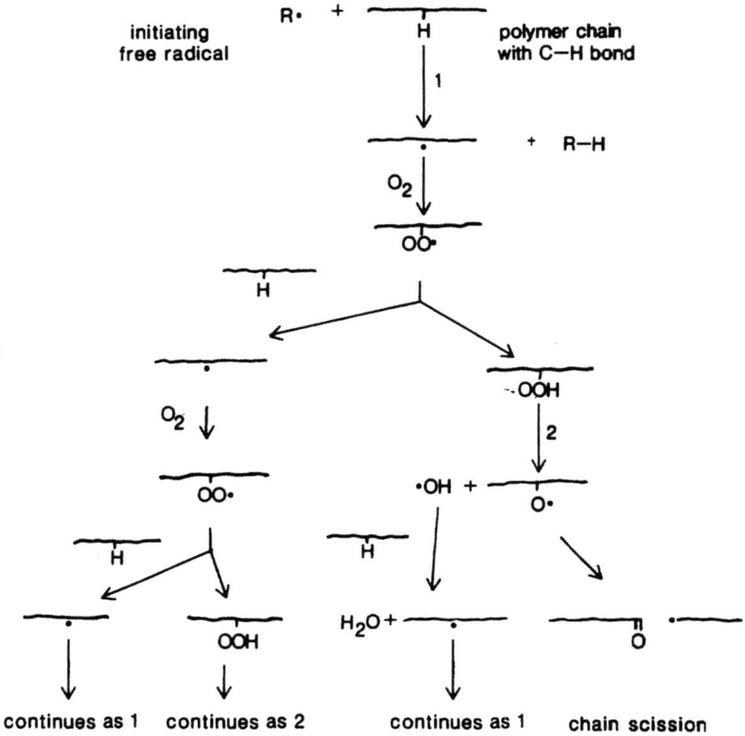

Figure 3.1 Mechanism of thermal oxidation.

Table 3.1 Carbon–hydrogen bond dissociation energies

Bond		Dissociation energy (kJ mole^{-1})
Primary	$-CH_2-H$	409
Secondary	$>CH-H$	397
Tertiary	$\geq C-H$	380
Allylic	$CH=CH-\overset{\vert}{\underset{\vert}{C}}-H$	355
Benzylic	$C_6H_5-\overset{\vert}{\underset{\vert}{C}}-H$	355
Aromatic	C_6H_5-H	455

the burning of polymers. In general, polymers with good thermal stability also offer good fire resistance.

3.2 The need for thermally resistant polymers

The wide spectrum of properties available with polymers makes them particularly suitable to a range of application areas but, unfortunately some of these applications are very restricted because of the poor thermal resistance discussed above. In two particular application areas this restriction has long been particularly evident. Owing to their uniquely good electrical insulation properties, polymers are very widely used in many electrical products. However, many electrical components are required to operate at high temperatures, for example electric motors and some domestic appliances. There has therefore been a long-standing demand for higher temperature resistance in organic polymers for use as insulating materials.

Another important characteristic property of polymers, compared with other structural materials, is their low density and hence high specific stiffness and strength, especially when used as the basis for fibre reinforced composite materials. This has led to their use in transport applications, especially in aerospace and particularly in military and deep-space vehicles, where weight saving is of vital importance and materials cost is secondary. However, here again, many applications also demand high temperature resistance. Indeed, in the late 1950s and early 1960s, the US and Russian space programmes provided the major impetus for the development of new high-temperature resistant plastics. More recently road vehicle manufacturers have been attempting to save weight (and hence improve fuel consumption) by replacing heavy metal components with light plastic ones. In many under-the-bonnet products, plastics are used to advantage owing to their ease of moulding into complex-shaped parts. Here again resistance to elevated temperatures is often also needed, so high-temperature resistant materials may be called for. Although electrical and transport applications have perhaps provided the biggest demand for such materials, increasingly the more exotic high-temperature resistant polymers are being used in more mundane consumer goods, especially appliances where exposure to elevated temperatures can occur, such as hairdryers, toasters and microwave ovens.

3.3 Improving low-performance polymers for high-temperature use

The deficiencies of polymers concerning heat resistance described above apply to all the large tonnage commodity plastics – polyethylene (PE), polypropylene (PP), polyvinylchloride (PVC) and polystyrene polymers (PS, HIPS and ABS) to such an extent that their maximum temperature of use is restricted to below

100°C, or slightly higher in the case of PP. Furthermore, even among the five main types of higher performance engineering polymers used as plastics – nylons (PA), acetals (POM), polycarbonate (PC), thermoplastic polyesters [polyethylene terephthalate (PET) and polybutylene terephthalate (PBT)] and polyphenyleneoxide/polystyrene blend (PPO) – only PC offers use temperatures approaching 150°C. An attractive approach to producing improved materials is to modify suitably these existing, and relatively cheap, polymers without going to the trouble and expense (with consequent commercial risks) of developing new polymers.

Much may be achieved by compounding, and thus modifying, lower performance materials with appropriate additives to upgrade the material. This approach has been widely used to improve mechanical performance (stiffness and strength properties) by the use of appropriate fillers, especially glass and carbon fibres and particulate mineral fillers. In addition, incorporation of such fillers usefully increases the softening point and hence can raise temperature of use. Table 3.2 provides some examples of the effects obtainable with glass fibres with the five main engineering plastics (PA, POM, PC, PET/PBT, PPO). A disadvantage is that the incorporation of fillers increases processing problems due to increased melt viscosity and wear on machinery. With fibrous fillers, orientation during

Table 3.2 Typical mechanical properties of unfilled and glass-fibre filled engineering plastics[a]

	Acetal homopolymer		PPO/PS blend		Polycarbonate	
	Unfilled	20% glass-filled	Unfilled	30% glass-filled	Unfilled	20% glass-filled
Heat distortion temperature (°C)	110	150	131	151	135	140
Yield strength (MPa)	70	60	67	119	63	112
Flexural modulus (GPa)	2.9	6.2	2.5	7.7	2.4	5.6
Izod impact strength ($J\ m^{-1}$)	74	43	267	123	757	106

	Nylon 66 (wet)[b]		PBT	
	Unfilled	30% glass-filled	Unfilled	30% glass-filled
Heat distortion temperature °C	75	254	59	228
Yield strength (MPa)	60	154	56	119
Flexural modulus (GPa)	1.3	6.3	2.3	7.7
Izod impact strength ($J\ m^{-1}$)	106	139	54	69

[a] Property data in this and the following tables are merely indicative values. A measured value depends on grade of polymer, method of specimen preparation and precise test conditions. In this and the following tables mechanical property data are quoted for 23°C unless otherwise noted. The values for heat distortion temperature (HDT) are for the 1.82 MPa loading condition. HDT values are particularly sensitive to the stress applied. The values of some mechanical properties are included to provide a comparison with the polymers to be discussed later.

[b] Some properties of nylons and other polyamides are very dependent on the amount of water absorbed, which can be considerable, typically several per cent. Standard test conditions usually refer to samples that have been conditioned in an atmosphere of relative humidity (RH) of 50%. Dry samples are often referred to as DAM (dry as moulded).

moulding introduces anisotropy of properties, which may be undesirable. However, high temperature performance is ultimately restricted by the polymer $T_g$value and/or thermo-oxidative resistance.

An alternative is to modify the chemical structure of the polymer chain either by cross-linking, which necessarily restricts molecular motion and hence raises softening point (an example is the radiation cross-linking of polyethylene for high-temperature cable use) or by some other chemical reaction (for example, the chlorination of PVC to raise its softening point by about 30°C). However, such practices can considerably add to the cost of the material, and what can be achieved is limited.

Similarly, high-temperature oxidation resistance may be improved by incorporating antioxidant additives. A problem for long-term use is lack of permenance of the antioxidant through its volatility at the higher temperatures involved. Although improvements have been made over the years it is still not possible to provide good long-term protection for polymers that are sensitive to atmospheric oxidation, especially the polyolefins, for continuous use at temperatures much above about 100°C.

In recent years a further way of upgrading a cheaper polymer that has been increasingly used is to combine it with a proportion of a second higher performance polymer to give a polymer blend. Alternatively, this may be seen as using a low-performance polymer to reduce the cost, with some sacrifice to properties, of the higher performance polymer. Blends such as ABS–PC, PBT–PC and blends with even higher performance polymers, such as polysulphones, have been used. In general, upgrading performance by any of these methods is limited by the intrinsic properties of the basic polymer or by penalties that have to be paid in other properties, such as processability. In order to produce materials with really useful high-temperature properties it is necessary to design the polymer molecular structure with this end in mind

3.4 The need for fire resistant polymers

All organic materials will burn under most fire conditions, and the poor fire resistance of organic polymers used as plastics, fibres and rubbers has long been apparent. It has become of increasing public concern when polymers are used in building and transport applications. The problem has been highlighted by certain fires where polymers have contributed to the rapid growth of the fire, resulting in many deaths.

The burning of polymers may be considered to consist of three stages. First, heat from the fire causes thermal (non-oxidative) decomposition of the polymer, producing the expected thermal degradation products. These may consist of non-volatile char and volatile non-flammable or flammable products. Second, evolution of the volatiles occurs from the burning surface, thus feeding fuel to the flame, where they mix with atmospheric oxygen. Third, in the flame itself, high-temperature (600–900°C) oxidation takes place by a complex series of free radical chain reactions, but involving different free radicals such as •OH and •CHO, from those found in low-temperature thermal oxidation. The flame reactions are highly exothermic and the heat evolved causes further thermal degradation of the polymer. If insufficient heat reaches the polymer to allow it to continue to burn when heat from other sources is removed, then the polymer is said to be self-extinguishing. The overall process is represented by:

$$\text{Polymer} + \text{Heat} \longrightarrow \begin{bmatrix}\text{Non-volatile}\\ \text{residue}\\ \text{(carbon char)}\end{bmatrix} + \begin{matrix}\text{Flammable}\\ \text{volatiles}\end{matrix} + \begin{bmatrix}\text{Non-flammable}\\ \text{volatiles}\end{bmatrix}$$

$$\text{Flammable volatiles} \xrightarrow{O_2\text{(air)}} [\text{Heat}] + CO + CO_2 + H_2O$$

(with the [Heat] returned to the Heat input of the first step)

The situation at and above a burning surface is depicted in Fig 3.2.

Interruption of the processes occurring at any of the above three stages will reduce polymer flammability. There are many different aspects to the burning of polymers which must be taken into account in assessing completely the hazards presented by any particular material. These include ease of ignition, rate of spread of flame, and the production of smoke and toxic gas. In recent years the

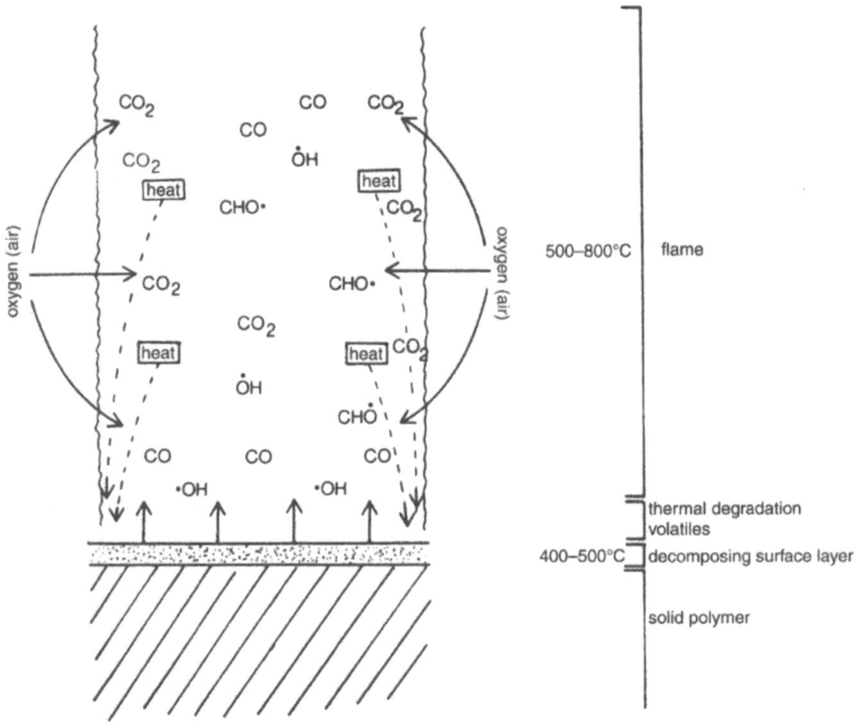

Figure 3.2 A burning polymer surface.

Table 3.3 Smoke emission on burning (NBS smoke chamber, flaming condition)

Polymer	Maximum smoke density
ABS	800
Polyvinyl chloride	520
Polystyrene	475
Polysulphone	230
Polycarbonate	215
Polyamid-imide	169
Polyarylate	109
Polytetrafluoroethylene	95
Phenolic	75
Polyethersulphone	50
Polyetheretherketone	35

latter hazards have become much more widely recognized. They arise because flames involving large amounts of polymer are fuel rich and result in incomplete combustion. This produces large amounts of carbon monoxide, which is the single largest cause of death, and carbon-rich black smoke, which inhibits escape from the burning environment. Table 3.3 compares smoke production from burning different polymers.

Although no single test can give an adequate picture of the fire hazards because so many factors are involved, the limiting oxygen index (LOI) test is the single most widely used test of polymer flammability. The test measures the minimum amount of oxygen, in an oxygen–nitrogen mixtures, required to just sustain burning under specified standard conditions. The test is simple to carry out and, unlike many fire tests, is very reproducible. LOI values for some representative polymers are shown in Table 3.4.

Table 3.4 Limiting oxygen indices (LOI) of polymers[a]

Non self-extinguishing polymer	LOI	Self-extinguishing polymer	LOI
Acetal (polyoxymethylene)	16	Polycarbonate	27
Polyethylene	18	Polyphenylene oxide	32
Polypropylene	18	Polysulphone	33
Polystyrene	18	Liquid crystal polymer	35
Polyisoprene (natural rubber)	18	Polyarylate	34
Polymethylmethacrylate	18	Polyetheretherketone	35
ABS	19	Polyethersulphone	38
Polybutylene terephthalate	22	Polyvinyl chloride	42
Nylon 66	24	Polyamide-imide	43
Polyethylene terephthalate	25	Polyphenylene sulphide	44
Polychloroprene	26	Polyvinylidene fluoride	44
		Polyether-imide	47
		Polybenzimidazole	48
		Polyimide (pyromellitimide)	50
		Polyvinylidene chloride	60
		Polytetrafluoroethylene	95

[a] The limit of self-extinguishment is often taken as an LOI of 27%, not 21% (which is the volume concentration of oxygen in the air), to correct for a lack of convective heating in the LOI test.

3.5 Polymers for low fire-hazards

As with other properties, such as high-temperature resistance discussed above, good fire resistance can be achieved in a polymer-based material by one of two approaches. Either the base polymer is designed to have inherently good fire resistance by virtue of its chemical structure, or the desired level of resistance is achieved by incorporation of a suitable additive–in this case a fire retardant.

Several general conclusions regarding the flammability of individual polymers may be drawn from Table 3.4. Aliphatic hydrocarbon polymers are all very flammable but, as the heteroatom content increases, so flammability decreases. Aromatic polymers have low flammability, which reflects their much greater thermal stability compared with aliphatic polymers. Although aromatic polymers are useful mostly for their high-temperature resistance, they are increasingly specified for fire sensitive applications, such as transport and electrical insulation or for fire resistant textiles, specifically because of their low flammability. A further advantage of aromatic polymers is that they produce much lower smoke levels than do aliphatic polymers, as shown by the data in Table 3.3. The partially aromatic structure of phenol formaldehyde (PF) materials, as well as their cross-linked nature, imparts low flammability and smoke production on burning. This has led recently to an increased use of PF in transport applications where fire performance is important. The presence of halogen in a polymer produces the most dramatic increase in LOI and hence decrease in flammability. Rigid PVC, with an LOI value of 42, has the lowest flammability of any common, and most not so common, polymers. Unfortunately, this is offset by the large amount of smoke and toxic gas (hydrochloric acid) produced when PVC, and other chloropolymers, burn. Furthermore, plasticization of PVC considerably reduces the LOI: for example, it is reduced to 25 when 50 phr dioctyl phthalate is used as a plasticizer. Fluoropolymers also have low flammability: the fully fluorinated polytetrafluoroethylene burns only in 95% oxygen under the LOI test conditions! However, although smoke production is not as great as with chlorinated polymers, hydrogen fluoride is produced as a highly toxic and corrosive gas. Furthermore, fluoropolymers rarely provide a cost-effective solution to flammability problems.

One method of increasing the fire resistance of a polymer is to incorporate a halogen-containing comonomer (as a so-called 'reactive' fire retardant) into the polymer chain of a non-halogen polymer during its polymerization. This is most frequently done with certain thermosetting plastics. Examples include the replacement of a proportion of maleic anhydride in unsaturated polyester resins by chlorendic acid:

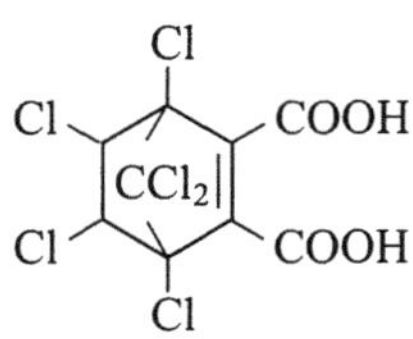

and the replacement of some bisphenol A in epoxy resins by tetrabromobisphenol A. As with additive halogen fire retardants effectiveness is much improved by the use of antimony oxide as well, which acts synergistically with organic halogen. However, for many polymers such an approach is not possible owing to non-availability or cost of suitable comonomers. Furthermore, incorporation of halogen, especially bromine, lowers both the thermal stability and the photostability of the polymer.

It will be apparent that, owing to the cost, processing or property limitations of the low-flammability polymers, it is often not possible to achieve a desired low-flammability level by choice of polymer alone. Much more frequently, low flammability is achieved by the use of fire retardant additives. Unfortunately, the use of many fire retardants results in undesirable side effects. For example, the most widely used system, an organohalogen compound with antimony oxide, although considerably reducing flammability leads to the production of additional smoke and toxic products. On the other hand additives which do not increase smoke and toxic gas production, such as aluminium and magnesium hydroxides (the use of which has recently increased considerably for this very reason) suffer from other disadvantages. They have considerable processing limitations since they decompose at relatively low temperatures (especially the aluminium compound) and they are relatively ineffective, so that large quantities (100 phr is not unusual) are needed. Such large 'filler' quantities lead to undesirable reductions in the strength properties of the material. Smoke suppressants, mostly based on molybdenum compounds (such as molybdenum oxide) and boron compounds (such as zinc borate) are increasingly being used specifically to reduce smoke production. However, detailed discussion of the use of additives to achieve high-performance properties in polymers is not appropriate here.

3.6 Polymers for high-temperature resistance

For use at temperatures much above 100°C, and certainly above 150°C, it is necessary to choose a polymer that has been specially produced for its high temperature resistance rather than to rely on upgrading the more general purpose polymers by the use of additives. As mentioned above, one of the earliest demands for such polymers came from the needs of aerospace applications, especially the US Apollo programme. As a result, in the late 1950s and early 1960s many exotic polymers with high thermal stability were synthesized. However, their mechanical behaviour or thermo-oxidative stability were rarely characterized, and hardly any of these materials were commercialized. Today it is more the consumer goods industries that are demanding high-temperature resistance in polymers and in recent years a bewildering variety of polymers have been offered as commercial products. Nearly all of these materials are aromatic polymers with benzene rings as part of the repeat unit in the polymer

Table 3.5 Relative temperature index (RTI) for 10 000 h, loss of 50% tensile strength

Main engineering polymers[a]	RTI (°C)	High-performance polymers[a]	RTI (°C)
PBT	80	PSU	160
PPO	80	PES	180
POM	110	PEI	170
PA 66	110	PPS	200
PC	120	LCP	210
PET	120	PAI	240
		PEEK	250
		PI	270

[a] Abbreviations are given in full at the beginning of this book.

chain. Not only do such structures impart good high-temperature resistance, but they also lead to stiff and often strong materials. Thus these polymers are often also known as much for their engineering properties as for their high-temperature resistance. The majority of these polymers are thermoplastics, and hence can easily be injection moulded into complex components.

Table 3.5 compares the long-term high-temperature resistance, usually governed by oxidation resistance, of a variety of engineering polymers according to the widely used US Underwriters Laboratory (UL) relative temperature index (RTI) test. This is a demanding test since the result is the temperature at which there is a 50% retention in the property of interest, for example tensile strength (the most commonly used criterion) after exposure to the temperature concerned for a long time (usually 10 000 h). For many applications this temperature is a conservative estimate of the maximum service-life temperature, especially for short-term exposure.

Short-time heat resistance is mostly limited by the softening behaviour and may be indicated by the heat distortion temperature (HDT) [also called the deflection temperature under load (DTUL)]. This behaviour is, of course, dependent on the loading conditions. For the standard HDT test a high load (1.82 MPa) or low load (0.45 MPa) is used. Quite different results can be obtained with the different loads. Reinforcing and other fillers can also give very different results at a particular load, especially with crystalline polymers, as shown in Table 3.6.

The main requirements for a high-temperature resistant polymer are, as we have seen, a high softening point and resistance to thermal oxidation. In terms of molecular structure these demands may be met in several ways. In the earlier days of interest in such materials, much emphasis was placed on thermal stability and hence on the strengths of chemical bonds in the polymer molecules. Bond strengths (Table 1.2) suggest that inorganic polymers could provide the potential for good high-temperature resistance owing to the high bond strength between atoms other than those involving carbon. The bond strengths for Si—O and B—N, for example, are 445 kJ mol^{-1} and 440 kJ mol^{-1}, respectively.

Table 3.6 Heat distortion temperature (°C) values at 1.82 MPa

Polymer[a]	Unreinforced	30% glass reinforced
Crystalline polymers:		
PTFE	55	–
PBT	59	228
PA 66 (wet)	75	254
PET	90	224
POM	110	160
PPS	135	>260
PEEK	152	315
LCP	180	271
Amorphous polymers:		
PPO	131	151
PC	135	145
PSU	174	181
PES	203	216
PAI	278	282

[a] Abbreviations are given in full at the beginning of this book.

Much effort was devoted to synthesizing inorganic polymers. However, with the single exception of the silicone polymers, that is, polysiloxanes consisting of

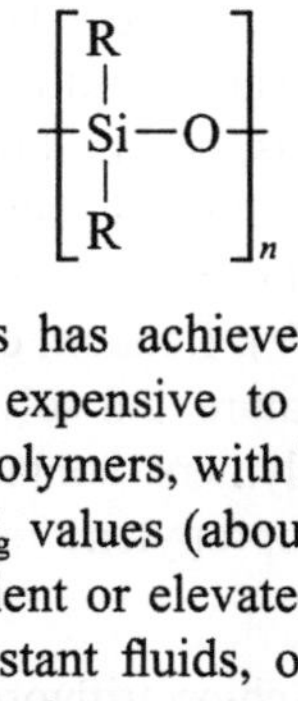

chains, none of these materials has achieved commercial significance. The polymers proved difficult and expensive to make and often suffered poor stability to hydrolysis. Silicone polymers, with R either methyl or mixed methyl-phenyl groups, have very low T_g values (about -100°C) and hence, if of high molar mass, are rubbery at ambient or elevated temperatures. Low molar mass silicones are useful as heat resistant fluids, oils and greases. The high molar mass silicone rubbers offer not only good high-temperature resistance but also good low-temperature flexibility. The fluorosilicone elastomers based on the polymer

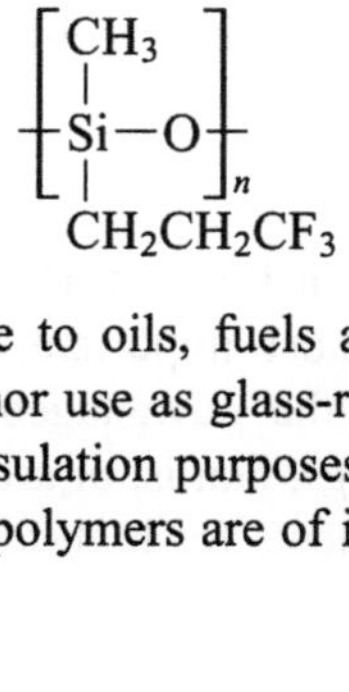

have much improved resistance to oils, fuels and to solvent swelling. Cross-linked silicones have found minor use as glass-reinforced thermosetting plastics for high-temperature electric insulation purposes and other special applications. A few other silicon-containing polymers are of interest as composite laminating resins, such as a silicon–carbon chain thermoset polymer Sycar and a polysiloxane–etherimide Ultem Siltem, the latter having a T_g of about 220°C.

3.6.1 Fluoropolymers

Carbon–fluorine bonds have higher bond strengths (typically 485 kJ mol^{-1}) than carbon–hydrogen (typically 350–435 kJ mol^{-1}) or carbon–carbon (typically 350–410 kJ mol^{-1}) bonds. Fluoropolymers are therefore extremely resistant to attack by the free radicals involved in oxidative degradation and hence to thermo-oxidative degradation. In addition they have excellent resistance to chemical and solvent attack. The C—C backbone chains have high flexibility, so the polymers have low T_g values. Polymers with structural regularity are crystalline and are therefore plastic materials, whereas non-crystalline polymers are rubbery. The most important commercial polymer is the fluoro-analogue of ethylene–polytetrafluoroethylene (PTFE) of repeat unit $-[CF_2-CF_2]-$ (products such as Algoflon™, Fluon™, Hostaflon™, Soreflon™, Teflon™). It is a relatively soft plastic (tensile modulus of about 0.5 GPa) so is not suitable for most load-bearing engineering uses.

The polymer is produced in a very linear high molar mass form and has an unexpectedly high melting temperature of 327°C. Even above its T_m the melt viscosity is so high that it cannot be melt processed. Products are made by sintering or special extrusion techniques. However, despite its good high-temperature resistance most applications utilize its extreme chemical resistance (these include pump parts, gaskets, seals, valves) or its well-known non-stick properties (in coatings). Similar melt processable polymers are available, such as polychlorotrifluoroethylene, $-[CFC-CF_2]_n-$ (Aclar™, Fluothane™, Halon™, Hostaflor C™, Kel F™, Voltalef™). Random copolymers, being non-crystalline provide the basis of high-temperature and solvent resistant rubbers: the fluoroelastomers. Examples are copolymers containing vinylidene fluoride units, $-CH_2-CF_2-$, with hexafluoropropylene units, $-CF_2-\underset{\displaystyle CF_3}{\underset{|}{C}}F-$ (Fluorel™, Technoflon™, Viton A™), with chlorotrifluoroethylene units (Voltalef™, Kel F™), or with tetrafluoroethylene (Fluon FEP™, Neoflon™, Teflex™, Teflon FEP™, Hostaflon FEP™), or of tetrafluoroethylene with fluoroalkyl aryl ether. The latter (Kalrez™, Hostaflon TFA™, Teflon PFA™, Teflon EFE™) is fully fluorinated and therefore offers the best swelling and thermal resistance.

3.6.2 Aromatic polymers

By far the most successful approach to the design of high-temperature resistant polymers, especially for the most important areas of application as plastic materials, has been to utilize the advantageous features of polymer chains composed of linked benzene rings, that is, of aromatic polymers. Such polymers combine the two main requirements for high-temperature resistance. First, they have good oxidation resistance, since the C—H bonds of the benzene ring resist abstraction by the chain-propagating oxidation free radicals, having typical bond strengths of 435 kJ mol^{-1}, compared with alpihatic C—H bond strengths of 350–400 kJ mol^{-1}. It is roughly true that a polymer's susceptibility to thermal oxidation correlates with its aliphatic C—H bond content. Second, the incorporation of the benzene ring structures into the polymer chain stiffens the chain and hence raises the T_g (and T_m if the polymer is crystalline). However, the stiffening effect may be overdone. The polymer consisting entirely of benzene rings linked together, that is, poly-*p*-phenylene,

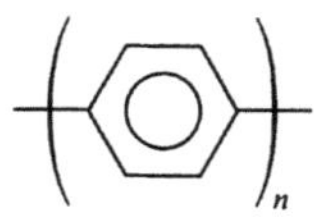

although superbly resistant to oxidation, has a softening point of over 500°C and so is not melt processable. This example illustrates the need to achieve a compromise between high softening point (for end-use requirements) and adequate processability, preferably by conventional melt processing techniques such as extrusion and injection moulding. However, for polymers whose end use requires the highest softening points, special processing techniques more akin to metallurgical techniques may be needed.

Therefore commercial polymers that achieve the required compromise are mostly of the type

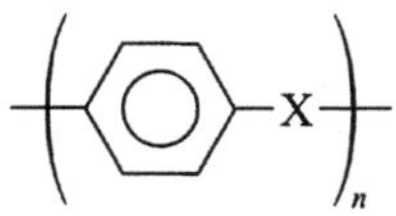

where X is an atom or group of atoms, giving the polymer chains a certain degree of flexibility and hence melt processability. In most commercial polymers the benzene rings are joined through the 1,4- or *p*-positions. This provides the most symmetrical type of structure and hence the best possibility for crystallization, which of course increases the softening point. However, many of these polymers do not crystallize or do so only reluctantly, and hence rely for the high softening point entirely on a high T_g.

A great variety of flexibilizing atoms or groups (X of the above structure) are to be found in commercial polymers. X is chosen not only to give the desired degree of flexibility (and hence appropriately high, but not too high, T_g and/or T_m), but must also be oxidatively inert. In general, therefore, X should not contain aliphatic C—H groups since these reduce the oxidative stability. Most simply X is a single atom, such as S in polyphenylene sulphide or O in polyphenylene oxide. More commonly it is some simple functional group such as $-SO_2-$ in polysulphones, —NHCO— in polyamides, $-\underset{\underset{O}{\|}}{C}O-$ in polyesters,

$-\underset{\underset{\displaystyle O}{\|}}{C}-$ in polyketones. Such aromatic polymers which contain no aliphatic linking atoms in the polymer chain are termed wholly aromatic polymers. Partially aromatic polymers naturally have properties intermediate between those of their wholly aliphatic and wholly aromatic counterparts. Although some important commercial polymers are partially aromatic, such as the thermoplastic polyesters polyethylene terephthalate (PET)

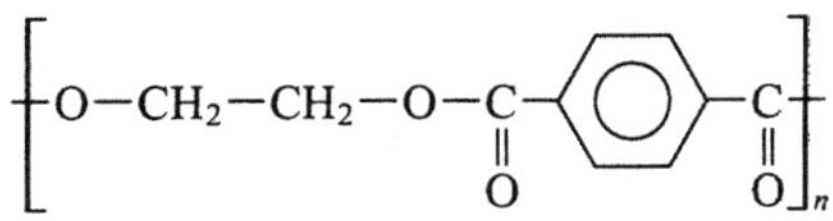

polybutylene terephthalate (PBT)

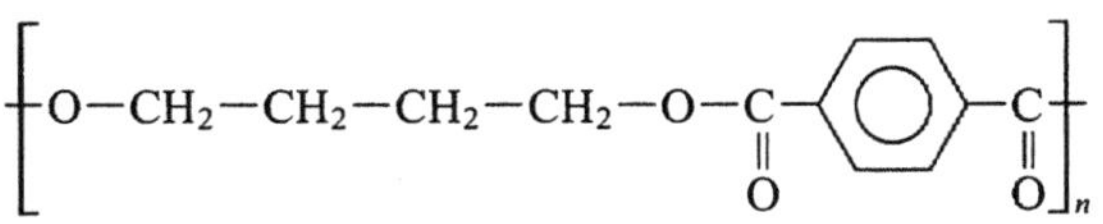

and bisphenol A polycarbonate (PC)

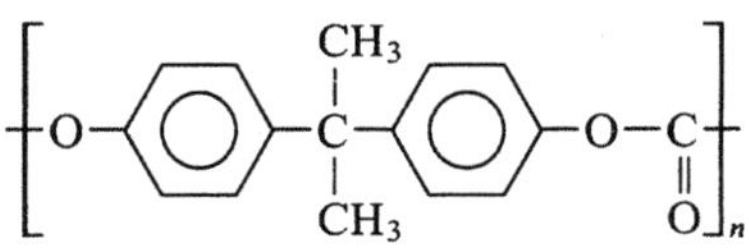

they do not have sufficiently high softening points to be considered as truly high-temperature resistant polymers. However, special copolymer grades of these homopolymers in which additional benzene rings are incorporated into the chain have higher softening points and use temperatures. Examples are polyethylene-naphthalene–polyethylene terepthalate (PEN–PET) copolymers and polyester carbonates with terephthalate units replacing carbonate units.

Sometimes, the desired balance of properties is achieved in copolymer structures especially by incorporating the relatively flexible ether links to improve processability, as in polyethersulphones and polyethereketone. Some commercial materials are also copolymers containing both aromatic and heterocyclic ring structures, as in polyether imides and polyamide-imides; these are considered in a later section.

Structures of the type

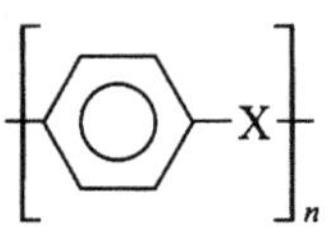

therefore can provide a variety of polymers with a useful balance of high-temperature properties. Because of their stiff chains they also have enhanced mechanical properties, especially stiffness, compared with the five main engineering polymer types, particularly if the chains become highly oriented, as with LCPs. Furthermore, they have thermal (as well as thermo-oxidative) stability and have low flammability and low smoke production as previously mentioned. They are therefore sometimes selected more for these features than simply for high-temperature resistance for use as plastics or fibres. In addition, some aromatic polymers are electrically conducting, especially when doped, and also of interest for this reason. Furthermore, such structures are relatively easy to synthesize compared with other types of high-temperature resistant polymers as the monomers are reasonably accessible and polymerization procedures are often fairly straightforward. Nevertheless the monomers needed are not cheap large-tonnage materials such as those used for commodity and general purpose engineering polymers. So, in general, aromatic high-temperature resistant polymers are relatively expensive materials, ranging from about 5 to over 100 times the cost of commodity plastics and rubbers.

3.6.3 Hydrocarbon polymers

As previously mentioned, high molecular weight polyphenylene is a highly intractable material, so despite its superb high-temperature resistance to oxidation it is not commercially useful. However, low molecular weight branched polymers with mixed isomeric linking (H-resin™) have been used as soluble, flexible precursor polymers for fabrication of high-temperature resistant composites with glass fibre, or asbestos, which form cross-linked networks on curing at high temperatures (up to 250°C). Maximum use temperature in air is 215°C (continuous) or 350°C for short times. Typical of highly aromatic polymers, they have low flammability, with a limiting oxygen index of 55. More recently, high molecular mass polymers have been produced by bacterial oxidation of benzene. Polyphenylenes are also of interest as electrically conducting polymers especially when doped, for example with AsF_5. The related polyphenylene-vinylene (PPV)

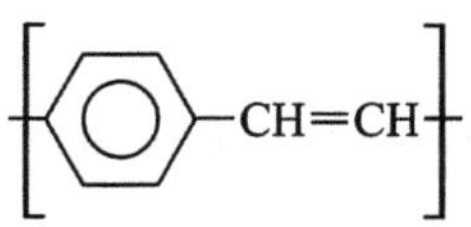

has a similar high conductivity (about $1\ S\ cm^{-1}$) when doped and is electrochromic.

Linking the benzene rings through aliphatic methylene ($-CH_2-$) groups is one way of introducing chain flexibility and hence improving processibility. However, the presence of aliphatic groups reduces resistance to thermal oxidation. The simplest such polymer

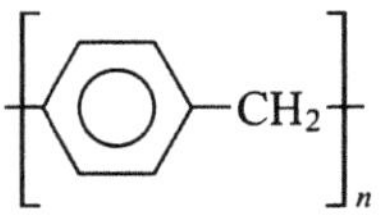

has not been synthesized, but the early commercial Xylok™ resins of typical structure

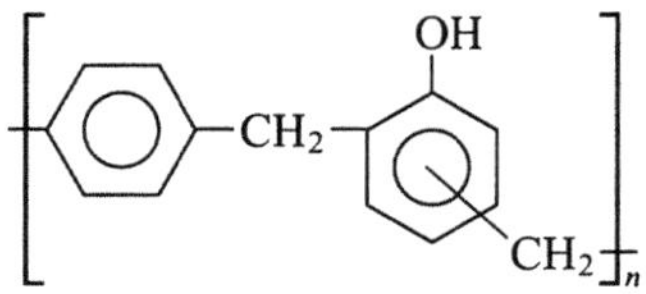

have been produced as low molecular weight prepolymers which may be cured to thermoset products by heating with hexamine or epoxy compounds. The products are intermediate in structure between a novolak phenolic resin and a *p*-phenylene, so have properties expected of a high-temperature resistant phenolic resin. For example, strength retention at elevated temperatures (e.g. 250°C) is eight times as long as for an epoxy-cured novolak. In addition, the product has low flammability and low smoke production on burning.

Polymers of the type

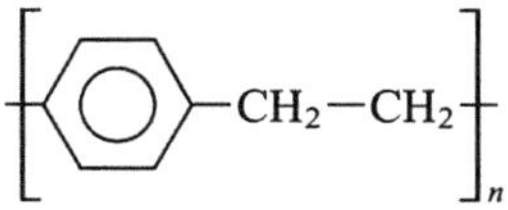

the poly-*p*-xylylenes, are also commercial products (Parylene N™) which are useful as electrically insulating coatings. They are produced by the vacuum pyrolysis of the cyclic dimer of *p*-xylene, the volatile pyrolysis products of which spontaneously polymerize on the cool surface to be coated. The polymer has a T_m of about 400°C and a T_g of 80°C, but is susceptible to oxidation above 100°C. Chlorinated versions (Parylene C™ and Parylene D™) have better oxidation resistance.

3.6.4 Polyethers

Ether links provide a very good choice of flexibilizing group since the —C—O—C— link is itself very flexible and is also highly resistant to thermal oxidation. The polymer

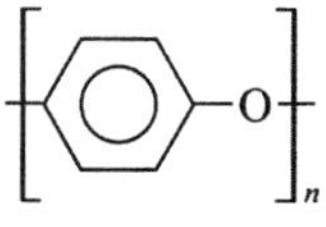

has a T_g of only about 90°C, but commercial polyphenylene oxide (PPO) is based on the 2,6-dimethyl substituted polymer

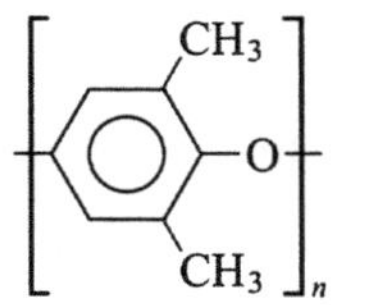

with a T_g of 208°C. It does not crystallize from the melt owing to the proximity of the T_g to the T_m (about 250°C). It has been offered as a high-temperature resistant polymer in unmodified form (Biapen™, Arylox™) but the main long-established commercial products (Noryl™) are blends with polystyrene, high-impact polystyrene or ABS. Although the blends have higher softening points (heat-distortion temperatures up to 150°C) than commodity polymers, they are not high enough for the materials to be considered as high-temperature resistant polymers. Recently, other companies have marketed similar polymer blends (Prevex™, Luranyl™), which are often referred to as polyphenylene ether (PPE) polymers.

Ether links are also employed in some aromatic copolymers, such as polyethersulphones and polyetherketones, to introduce flexibility into otherwise rather intractable homopolymers.

3.6.5 Polyphenylene sulphide

The polymer

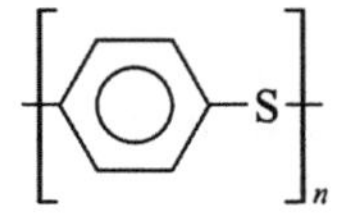

(PPS) has been available commercially (Ryton™) for many years. It is normally crystalline (T_m 285°C, T_g 85°C) and shows such good thermal and thermo-oxidative stability that it may be used in air at temperatures above 200°C for long periods. Several different types of PPS have evolved. The earlier polymer was produced by post-polymerization, heat treating it in air, the initial low molar mass polymer chain extending and branching, improving mechanical behaviour and producing a highly insoluble and chemically resistant material. Later, high molar mass branched polymer was introduced by using a single-stage polymerization process (modified PPS). More recently, linear polymers have been produced in a single-stage polymerization process and have better strength properties. Its limiting oxygen index is 44, so it has low flammability. Reinforced polymer is used for injection-moulded products, which include electrical connectors, high-temperature lamp holders and reflectors, pump parts, valves and, especially when filled for example with PTFE or graphite, as bearings. Products may be amorphous, but usually a high mould temperature is used to induce high (typically about 65%) crystallinity. Unfilled polymer is also a useful high-temperature surface-coating material and is used for encapsulation of

Table 3.7 Properties of polyphenylene sulphide

		40% glass-filled		
Property	Unfilled	Early branched	New linear	Later modified
Tensile strength (MPa)	66	150	185	115
Elongation at break (%)	1.6	1.1	1.6	1.2
Flexural modulus (GPa)	3.8	12.0	13.0	12.5
Heat distortion temperature (°C)	135	>260	>260	>260
Limiting oxygen index (%)	44	47	47	47

electronic components and as a textile fibre. Recently, other manufacturers have introduced PPS (e.g. Tedur™, Fortron™, Supec™, Primef™). Typical properties are shown in Table 3.7

3.6.6 Polysulphones

Poly-*p*-phenylene sulphone itself

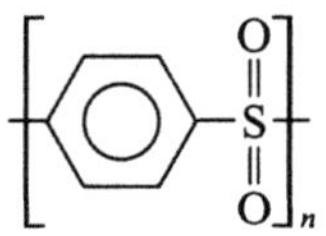

has too high a softening point (T_m about 520°C) to be melt processable. Several commercial aromatic polysulphone copolymers, which contain other flexibilizing groups, are available. Depending on the nature and amount of these groups, the different commercial products have different T_g values. They do not crystallize, so the T_g determines the softening point. The structure of the material referred to as polysulphone (PSU) (Udel™) is

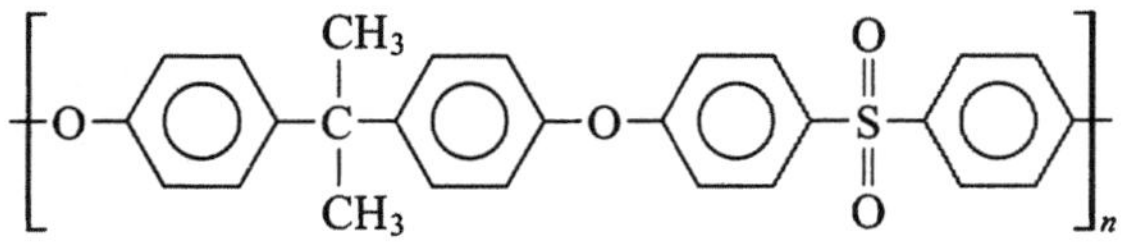

and it has a T_g of about 190°C. Polyethersulphone (PES) (Radel A™, Victrex PES™, Ultrason P™) contains both

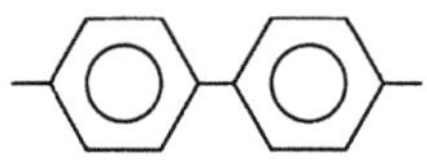

and

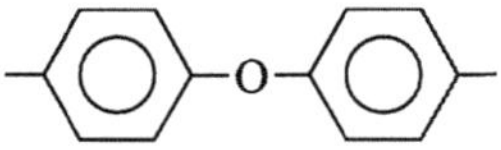

sulphone linked groups and has a T_g of about 200°C. An earlier material (PESU) (Astrel™) is similar to PES but with fewer ether links, and has a T_g of about

Table 3.8 Properties of aromatic polysulphones[a]

Property	PSU	PSU (30% glass)	PES	PPSU
Tensile strength (MPa)	64	108	84	90
Tensile modulus (GPa)	2.5	7..4	2.4	2.6
Elongation at break (%)	50–100	2	40–80	10
Flexural modulus (GPa)	2.1	7.6	2.6	2.8
Heat distortion temperature (°C)	174	181	203	274
Notched Izod impact strength (ASTM, D256) ($J\ m^{-1}$)	69	75	100	–
Limiting oxygen index (%)	–	–	38	–

[a] Abbreviations are given in full at the beginning of this book.

285°C. A further type, polyphenylsulphone (Radel R™) has outstanding impact resistance after hot ageing. PSU is also blended with ABS to lower its cost (and of course with some sacrifice to properties) but still has a HDT of 166°C when glass-filled. Typical properties are shown in Table 3.8. In mechanical behaviour the polymers are like high-temperature versions of polycarbonate, with PES materials capable of operating at up to 180°C. They show outstandingly good creep resistance, especially at somewhat elevated temperatures. Figure 3.3 compares creep behaviour of PES with other engineering polymers.

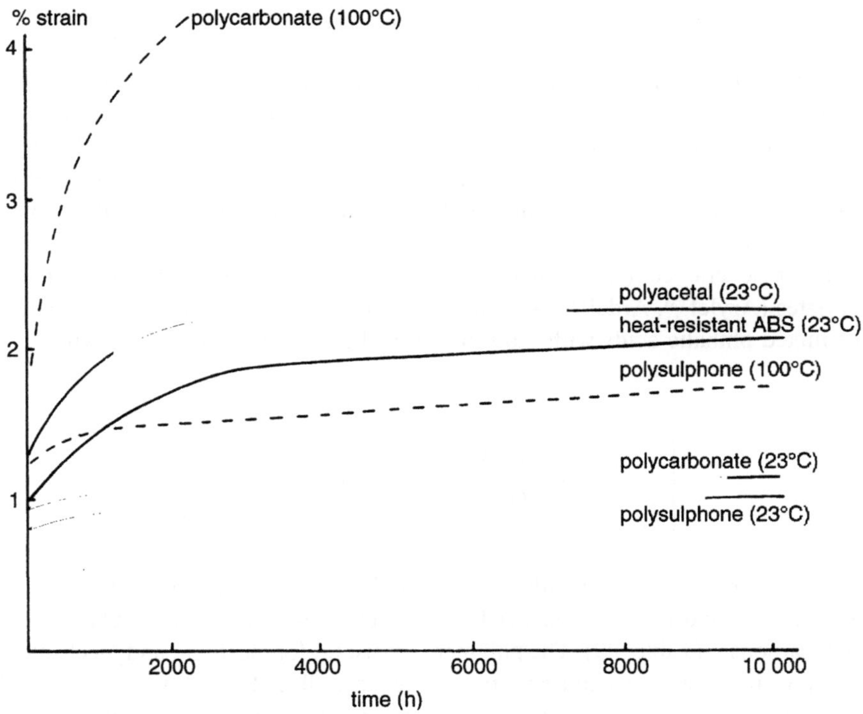

Figure 3.3 Comparative tensile creep behaviour at constant stress of 21 $MN\ m^{-2}$.

The polysulphones have found widespread use as injection-moulding materials, where good dimensional stability at elevated temperatures is required: for example electrical components, including circuit boards, connectors, coil bobbins, and relays; appliances operating at high temperatures, for example fan heaters, hairdryers, microwave ovens; and pumps, valves and lamp housings and bases. The materials are transparent (being amorphous), though often slightly yellow and susceptible to environmental stress cracking in certain fluids. They have low flammability (with a limiting oxygen index of typically 38), and burn with little smoke production.

3.6.7 Polyesters

The partially aromatic polyesters, polyethylene terephthalate and polybutylene terephthalate, are well established as plastics and fibre-forming materials. However, they have limited heat resistance. The earliest commercial wholly aromatic polyester was polyhydroxy-benzoic acid (PHBA) (Ekonol™):

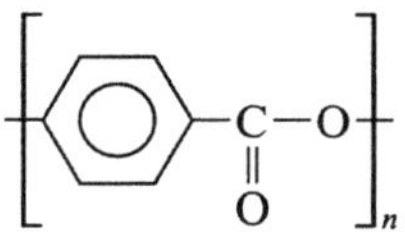

which is crystalline and softens only above 500°C, so must be processed by metallurgical techniques such as sintering and hammering. It may be used at temperatures up to 320°C and has found use as a bearing material. More tractable and melt-processable copolymers, containing ester units from terephthalic or isopthalic acid and aromatic diols (Ekcel™) are also marketed and may be injection moulded.

Later several companies marketed aromatic polyesters described as polyarylates (Arylef™, Ardel™, Arylon™). These are non-crystallizing copolymers of mixed phthalic acids with a bisphenol, so have repeat units of the type

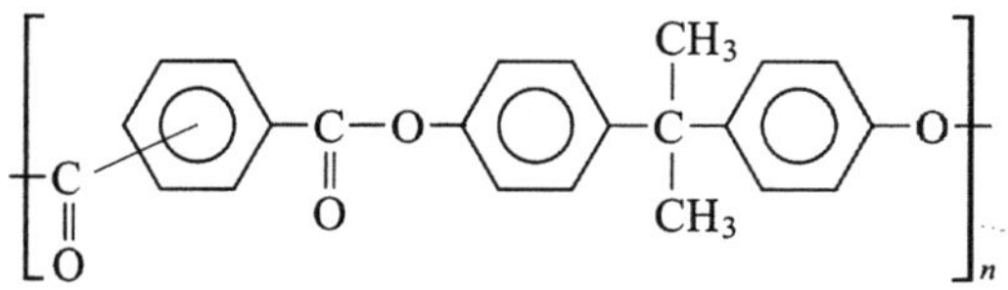

They are melt processable with T_g (and heat distortion temperatures) in the range 150–200°C, and have similar mechanical properties to polyethersulphones and polycarbonate (Table 3.9). The polymers are useful for electrical and mechanical components requiring good heat resistance, as well as for lighting fixtures and consumer goods operating at elevated temperatures such as hairdryers and microwave ovens.

Table 3.9 Properties of aromatic polyesters[a]

Property	Ekonol™	Ekcel™	Arylef™ or Ardel™	Arylon™
Tensile strength (MPa)	74 (flexural)	70–99	70 (yield)	68
Elongation at break (%)	–	–	50	25
Tensile modulus (GPa)	–	1.3–2.4	2.1	2.0
Flexural modulus (GPa)	7.1	3.2–4.9	2.0	2.1
Heat distortion temperature (°C)	>550	300	175	155
Notched Izod impact strength (ASTM, D256) ($J\ m^{-1}$)	–	–	150–250	288
Limiting oxygen index (%)	–	–	34	26

[a] Details regarding trade names are given at the beginning of this book.

A new approach to improving the balance of processability and high softening point is increasingly successfully exploited in another type of aromatic copolyester, the liquid crystalline polymers (LCPs) (Vectra™, Xydar™, Zenite™) in which the retention of crystalline order in the melt gives lower melt viscosities than would otherwise be achieved.

Owing to their unusual morphological structure, these materials possess a unique range of useful properties. The materials have high molecular order and hence have exceptionally high stiffness (tensile modulus about 10 GPa) – higher than any other unreinforced plastic moulding material. They also have very low thermal expansion coefficients, which may even be negative, and high HDTs (180–240°C) characteristic of the other aromatic polyesters. Indeed, PHBA is itself a thermotropic LCP. This exciting combination of properties led many polymer manufacturers to investigate the production of such LCPs during the 1980s. Unfortunately, this early enthusiasm was not maintained when difficulties arose in producing satisfactory mouldings, notably because of poor weld-line strength. Until recently only the Vectra™ and Xydar™ materials reached true commercialization. However, many of these problems are now being solved and the initial disappointment in LCPs is now giving way to a wider interest. The recent introduction of a further material (Zenite™) illustrates this renewed confidence.

All LCPs for use as engineering plastics are aromatic polyester copolymers, Xydar™ is a copolymer of PHBA, dihydroxybiphenyl and terephthalic acid, whilst Vectra™ has 2,6-disubstituted naphtalene units in the chian. One type with a T_m of 280°C is a copolymer of 2,6-dihydroxynaphthoic acid with 2,6-naphthalene dicarboxylic acid and 2-hydroxy-6-naphthoic acid, whilst another (T_m 315°C) is a copolymer of PHBA and 2,6-dihydroxynaphthoic acid. Typical properties are shown in Table 3.10.

Many uses of these material exploit their very good flow properties and they are increasingly being used as a blend component to improve the flow of other polymers, such as polycarbonate. Typical LCP products are electrical and electronic components of complex shape and/or with long flow paths, such as connectors, switches, relays and surface mounted devices (SMDs). Other uses

Table 3.10 Properties of liquid crystalline polymers[a]

Property	Unfilled Vectra™	30% glass-filled Vectra™	40% glass-filled Xydar™	30% glass-filled Zenite™
Tensile strength (MPa)	165	207	146	150
Elongation at break (%)	3.0	2.2	1.5	2.7
Tensile modulus (GPa)	9.7	16.6	19.9	21
Flexural modulus (GPa)	–	–	–	12
Heat distortion temperature (°C)	180	230	241	265
Notched Izod impact strength ($J\ m^{-1}$)	520	150	176	130
Limiting oxygen index (%)	35	50	–	38

[a] Details regarding trade names are given at the beginning of this book.

include electrical brush holders, coil bobbins and high temperature resistant automotive products. A fibre product (Vectron™) has also been manufactured.

3.6.8 Polyamides

Aliphatic polyamides (the nylons) have amongst the highest softening points for wholly aliphatic polymers, owing to their high melting points, generally above 200°C. These high values arise from the complete hydrogen bonding of the amide groups in the crystalline regions. However, even the highest melting polymer, nylon 66 (T_m 264°C) has a maximum continuous-use temperature of only about 120°C even when glass-filled. More recently an even higher melting nylon

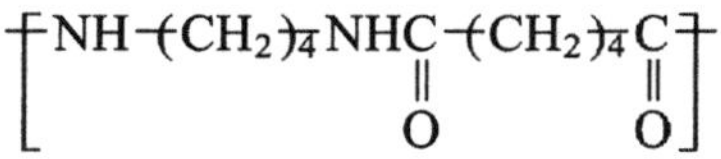

nylon 46 (Stanyl™) with a T_m of 295°C, has become available. The polymer has a claimed maximum working temperature of above 150°C, and a heat distortion temperature of above 150°C when unfilled and of 295°C when 30% glass-filled. Recently, several companies have introduced high-temperature resistant nylons (HTNs), such as Zytel HTN™ with a T_m of 335°C and a T_g of 125°C (Table 3.11).

At the other extreme, wholly aromatic polyamides (also known as aramids) have such high melting points that they are not melt processable, so are not usable as plastics moulding materials. However, two long-established products are significant commercial polymers. The meta-linked polymer, poly-*m*-phenylene isophthalamide, Nomex™,

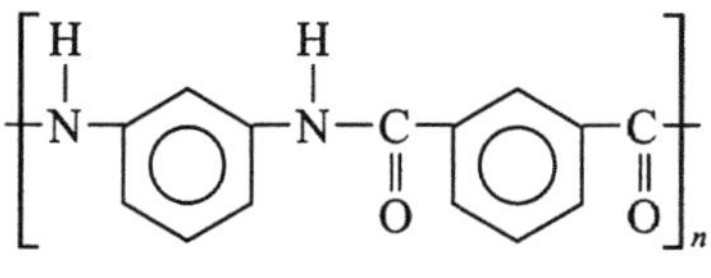

has been available since 1972. It has a T_m of above 400°C and is processed into fibres by wet spinning from solution or to a film by casting from solution. A

Table 3.11 Properties of Polyamides[a]

Property	Nylon 66 (dry)	Stanyl nylon 46 (dry)	Ixef™ (high-impact grade)	Ixef™ (30% glass-filled)	Zytel HTN™ (30% glass-filled) dry	Zytel HTN™ (30% glass-filled) wet
Tensile strength (MPa)	65	100	177	255	220	205
Tensile modulus (GPa)	3.2	–	13.4	20	12.8	13.0
Elongation at break (%)	100	–	2.7	1.9	2.3	2.0
Flexural modulus (GPa)	3.3	3.1	10.2	17.7	–	–
Notched Izod impact strength ($J\ m^{-1}$)	69	140	159	85	11	10.5
Heat distortion temperature (°C)	95	150	217	–	–	–

Property	Amodel™ (30% glass-filled) dry	Amodel™ (30% glass-filled) wet	Trogamid T™	Nomex™ (fibre)	Kevlar™ (fibre)
Tensile strength (MPa)	221	193	60	670	3000
Tensile modulus (GPa)	13.1	13.1	3.0	17.6	134
Elongation at break (%)	2.5	2.1	132	22	2.6
Flexural modulus (GPa)	11.4	11.4	–	–	–
Notched Izod impact strength ($J\ m^{-1}$)	285	–	–	–	–
Heat distortion temperature (°C)	–	–	124	–	–

[a] Details regarding trade names are given at the beginning of this book.

maximum use temperature of 370°C is claimed but, as a fibre, Nomex™ is mostly used for its fire resistance in fire resistant textiles, for example for racing drivers' and astronauts' suits. As a film, Nomex™ paper, owing to its exceptional stiffness, is also useful as a honeycomb layer in high-performance composite laminates especially when bonded to carbon fibre reinforced plastic outer layers, in aircraft and performance speedboats.

Para-linked aromatic polyamides have even higher crystallinities, stiffness and melting points owing to the greater symmetry of the structure. Exceptionally high-stiffness fibres have been available since 1970 (Kevlar™) which are based on poly-*p*-phenylene terephthalamide:

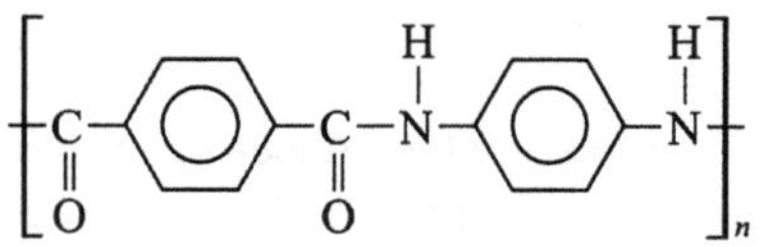

Kevlar™ fibres are produced by wet spinning from solution. Crystallinity in the fibres is enhanced by the liquid crystalline nature of their solutions. The original incentive for the development of Kevlar™ was their use as tyre cord reinforcement. Some fibre is used in high-performance tyres, but the material is better known as a high-modulus reinforcing fibre in plastic composites. It has the advantage over other reinforcing fibres such as carbon and especially glass of having lower density and, more importantly, of being tougher. The fibres are therefore not so liable to fracture even at large strains. This property is also useful in the construction of bulletproof clothing for military use.

Melt processable aromatic, or partially aromatic, polyamides are produced as random copolymers, which do not crystallize (and are therefore transparent) but are still capable of high-temperature use since they have high T_g values. Several commercial polymers of this type have been developed. Trogamid T™ contains repeat units of

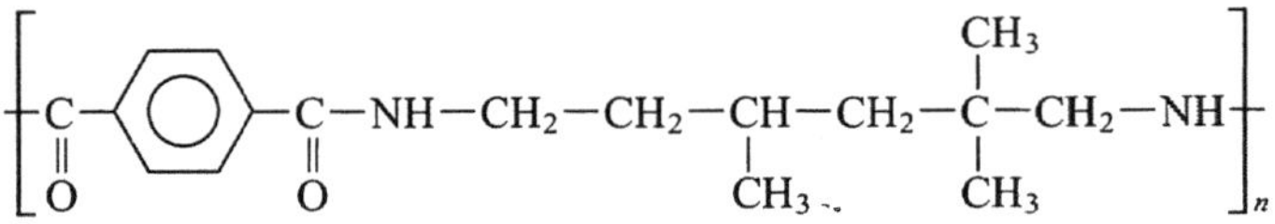

and of the 2,4,4-dimethyl isomer and has a T_g of about 150°C. Grilamid TR™ with a T_g of about 160°C is a copolymer with units of

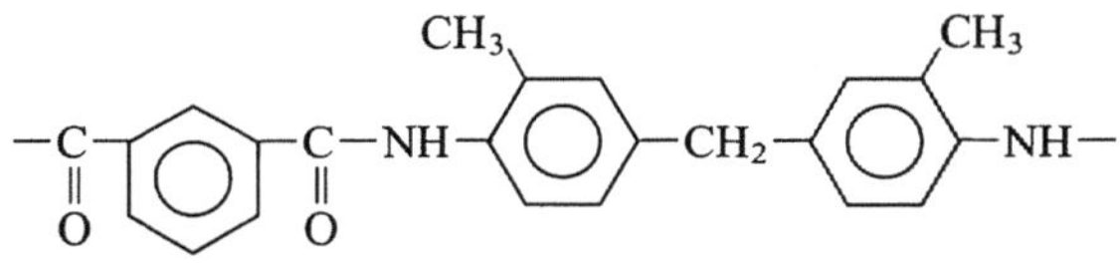

and of

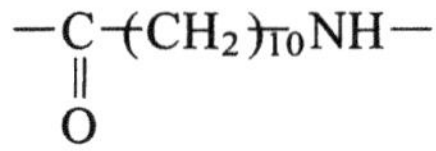

Several other companies have developed similar glass-clear, high softening point, oil and solvent resistant polymers, but these other products do not seem to have achieved much commercial significance.

A crystalline, partially aromatic, polyamide, poly-*m*-xylylene-adipamide, also known as MXD-6 (Ixef™),

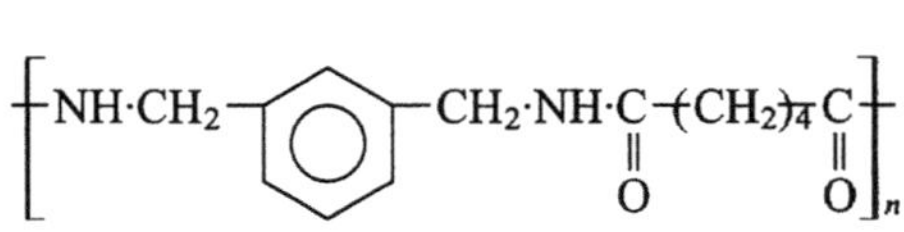

is also available as a heat resistant engineering plastic, generally similar in properties to nylon 66, having a T_m of 243°C but with reduced water absorption and somewhat greater stiffness. Although its heat distortion temperature is only 96°C (its T_g is at about 90°C), when 30% glass-filled this is increased to about 270°C. A further material, polyphthalamide (PPA) (Amodel™) has been recently introduced. Its mechanical properties are between those of the main engineering plastics and the high-performance plastics. Typical properties of some of these polyamides are shown in Table 3.11.

Copolymers containing amide and imide units, the polyamide-imides, are discussed in a later section on polyimides.

3.6.9 Polyketones

Aromatic polyketones are the most recently developed type of aromatic polymer and probably offer the best high-temperature resistance of any melt-processable thermoplastic material. Victrex PEEK™ (polyetherether ketone) has been available since 1978 and has the structure

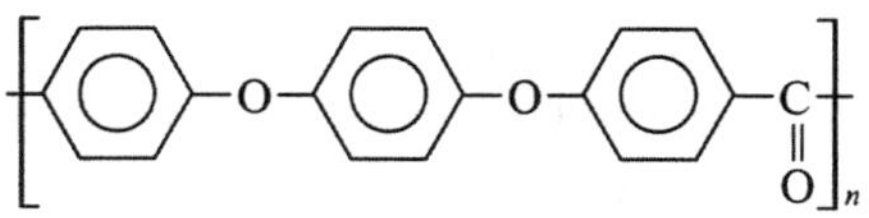

The T_g is 143°C but the polymer crystallizes with a T_m of 334°C. These values give the unreinforced material a heat distortion temperature of 160°C, but this increases to 315°C for 30% glass fibre reinforced grades. Being wholly aromatic the polymer also has excellent thermal oxidation resistance so, overall, the maximum continuous working temperature is 250°C. Its chemical and, especially, its hot-water resistance (no effect after thousands of hours at 250°C) are exceptional. The latter property contrasts favourably to polyimides which are sensitive to hydrolysis. Flammability behaviour is also good with a limiting oxygen index of 40 and very low smoke and toxic gas production, characteristic of many wholly aromatic polymers. Applications include electrical products such as high-temperatures connectors, film wrapping for cables, wire and cable insulation for fire resistant uses such as warships and oil wells, hot-water pumps, valves and engine components. Composite prepregs with carbon fibres have been developed for structural aircraft components. Other types of polymer, so-called polyetherketones (PEK) with fewer flexibilizing ether links, have also been developed (Victrex PEK™, Kadel™, Ultrapek™, and PEEKK Hostatec™). These polymers have even higher T_g values and heat distortion temperatures (up to 340°C for glass-filled grades). Typical properties for PEEK are shown in Table 3.12.

Table 3.12 Properties of polyetherether ketone (PEEK) and polyether ketone (PEK)

Property	PEEK Unfilled	PEEK 30% glass-fibre-filled	PEK (40% glass-fibre-filled)
Tensile strength (MPa)	92	178	172
Elongation at break (%)	>50	1.7	2
Flexural modulus (GPa)	4.1	9.7	12.4
Notched Izod impact strength (ASTM, D256) ($J\ m^{-1}$)	65	92	107
Heat distortion temperature (°C)	152	>300	326
Limiting oxygen index (%)	35	–	–

3.7 Heterocyclic polymers

Heterocyclic structures in the polymer chain provide another type of ring structure to stiffen the chain and raise the T_g, the T_m and the softening point. In the earlier days of the search for high-temperature resistant polymers, many such polymers were synthesized and their thermal degradation behaviour was investigated. However, hardly any of these materials were ever commercialized and, despite the much greater variety of thermally stable heterocyclic polymers potentially available, very few have had any commercial success. By far the most important of these materials are the polyimides, which were developed comparatively early and which have been the only real commercial success. The polybenzimidazoles, also developed in the early 1960s, have now achieved commercial status. Discussion is largely restricted to these two groups of materials.

3.7.1 Polyimides

Dupont offered the first commercial polyimide in the early 1960s, made by condensation between pyromellitic dianhydride and 4,4-diamino diphenyl ether:

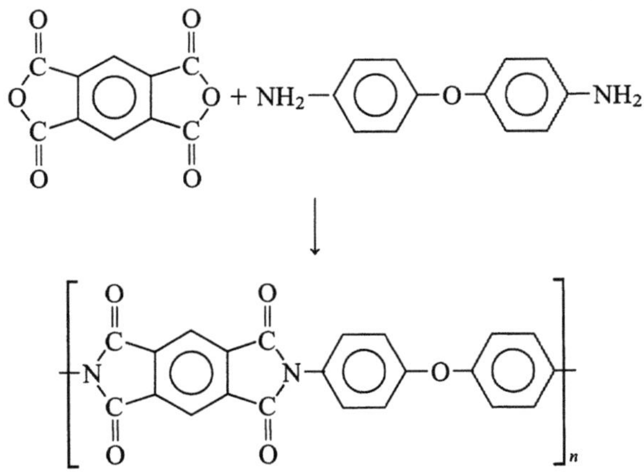

polyimide

The polymer has an exceptionally high heat distortion temperature (357°C) and thermo-oxidative resistance. Indeed for many years there was no other commercial organic polymer which could achieve such good resistance to high temperatures, giving it a continuous-use temperature of about 270°C. However, this high heat resistance is at the expense of mouldability. The polymer must be shaped as the tractable polyamic acid precursor polymer, which is then converted to the final polyimide structure by further heating. As a result only a limited range of product types is possible. Dupont offer insulating wire enamel, varnishes and laminating resins (Pyre ML™); film (Kapton™) cast from solution; simple stock shapes (rod, tube) (Vespel™) for subsequent machining to the required product; and sintered parts (in Vespel™). The varnish and film are mostly used for

Table 3.13 Properties of unfilled polimides[a]

Property	Vespel™	Torion™	Ultem™	Kinel™
Tensile strength (MPa):				
25°C	90	192	9	~40
150°C	67	105	46	–
260°C	58	52	–	~25
Flexural modulus (GPa):				
25°C	3.5	4.6	3.2	3.8
150°C	2.7	3.6	2.5	–
260°C	2.3	3.0	–	2.8
Heat distortion temperature (°C)	357	282	200	–
Limiting oxygen index (%)	35	42	47 –	

[a] Details regarding trade names are given at the beginning of this book.

exceptionally high-temperature electrical insulation applications. The solid resin products are also available in glass filled and graphite-filled grades and are used for seals, bearings, gaskets and piston rings, for example in aero engines and other high-temperature engineering applications. Carbon and glass fibre or fabric and other coated prepreg materials (Pyralin™) are available for moulding, for example by vacuum bag technique, to high-performance laminates. Typical properties of Vespel™ are included in Table 3.13.

Several other companies have offered similar condensation polyimides. An example is Skybond-700™ of structure

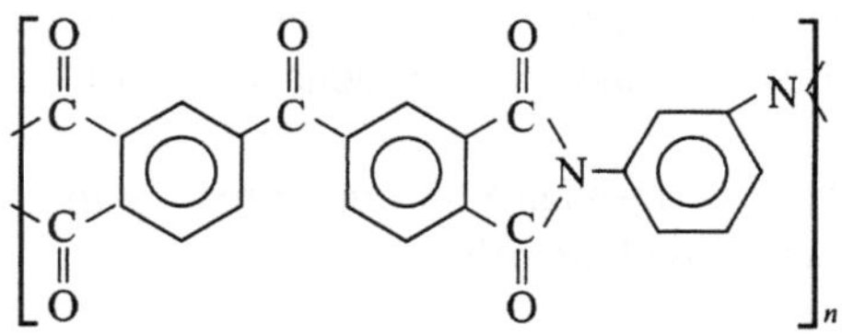

which is supplied as a solution of the monomer reactants, which cure to the polyimide structure. It is used particularly in prepreg form. However, several such products were short-lived and none has been as successful as the original Dupont materials. A fluorinated polyimide (Eymyd L) with a T_g of about 430°C has been developed.

In order to obtain more tractable, and sometimes melt-processable, polyimides, various copolymers have been introduced commercially. Among the better known materials is Torlon™, a polyamide-imide of structure

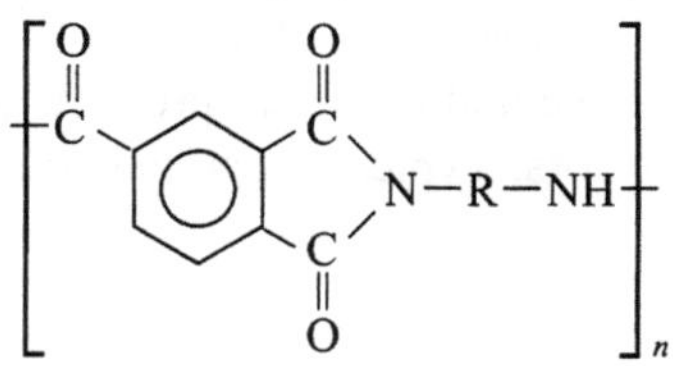

an injection-moulding material giving products with a continuous-use temperature of up to 260°C. The material also has exceptional stiffness and creep resistance. It is moulded at about 353°C with mould temperatures of about 230°C. Mouldings do require extensive annealing at high temperatures to develop the best mechanical performance. Typical properties are compared with other polyimides in Table 3.13. Products include electrical parts, valves, pumps, gears and (when filled with a lubricant such as graphite or PTFE) bearings.

A different type of polyimide, produced by an addition rather than a condensation reaction, is the polybismaleimide (Rhone–Poulenc) of structural type

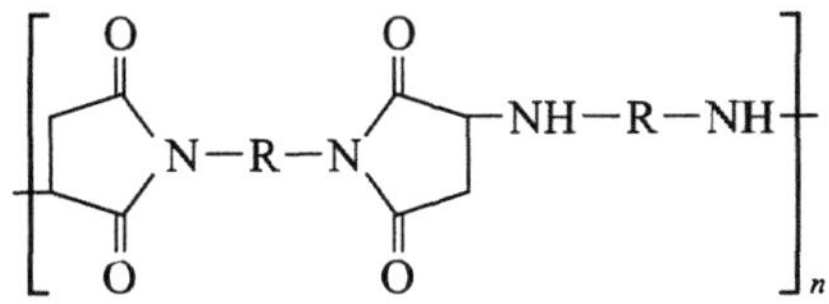

with unsaturated end groups which cross-link the chains giving thermoset products on moulding. The Kerimid™ material is a laminating and filament-winding resin for use with glass and carbon fibres. Kinel™ is a thermosetting moulding material, containing a reinforcing or lubricating filler, for such products as friction pads, gears, cams, bearings and seals. Typical properties of Kinel™ mouldings are shown in Table 3.13. Several similar types of polyimide have been developed, largely for use as matrices for laminates. In one type, a complex monomer with norbornene terminal groups is incorporated with the reinforcement from solution, the solvent evaporated and the monomer cured without any volatiles being produced. It is therefore known as a polymerization by monomeric reactant (PMR) material and gives products for use at temperatures up to 300°C.

A somewhat lower performance, but more readily melt-processable, copolymer is the polyether-imide Ultem™

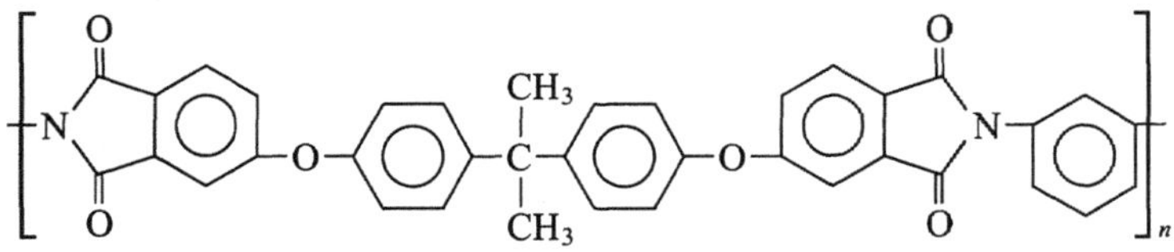

The material has a heat distortion temperature of 200°C with a continuous-use temperature of 170°C and is of low flammability (limiting oxygen index 47%). It is mostly used as an injection-moulding material needing melt temperatures of about 360°C and mould temperatures of 70–150°C. Application areas include under-the-bonnet automotive and electrical products such as microwave ovens, circuit boards and connectors. Typical properties are shown in Table 3.13.

3.7.2 Other polymers

Many other heterocyclic structures offer excellent thermal and thermo-oxidative stability. Among the many linear polymers, polybenzimidazoles (PBI) were developed early and are the only materials to show any real commercial promise, but they have had nothing like the success the polyimides have achieved, despite their comparable high-temperature resistance. The polymer developed commercially is poly-(2,2′-m-phenylene-5,5′-bibenzimidazole)

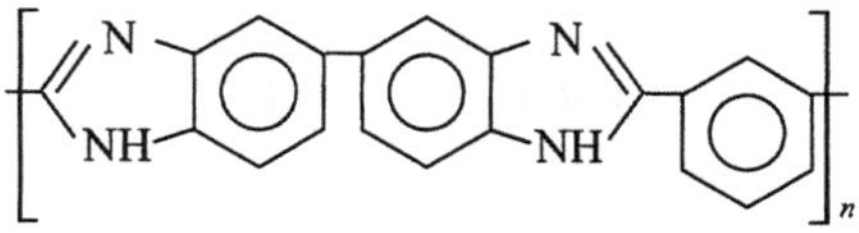

It has been used as a laminating resin and as an adhesive, as a reasonably tractable low molecular weight precursor polymer. The cured product has a T_g above 400°C. Although short-term strength at very high temperatures is better than for polyimides, long-term strengths deteriorate more rapidly. Such factors, together with the high cost and toxicity of the aromatic tetramine monomers used, has limited its commercial development. There is some interest in solution-spun fibres (Celazole™) for use as low-flammability textiles, since the fabric's limiting oxygen index is very high at 48, compared with other organic fabrics, and it has very low smoke and toxic gas generation. Other heterocyclic polymer types have been developed for their exceptionally high stiffness as fibres spun from solutions which are lyptropic liquid crystalline. They are polybenzoxazole

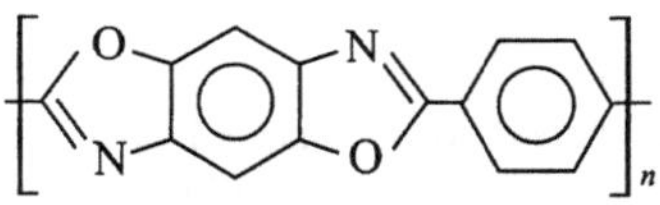

and polybenzothiazole

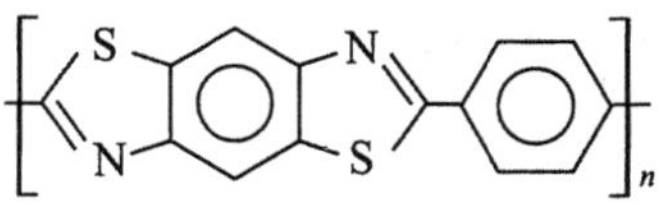

They also have exceptional high-temperature resistance, with T_g values of >400°C and with use temperatures up to 700°C (short-term) and 350°C (long-term). Tensile modulus is about 250 GPa.

In principle, even higher temperature resistance may be obtained with fused ring chain polymers (double strand or ladder polymers) of schematic structure

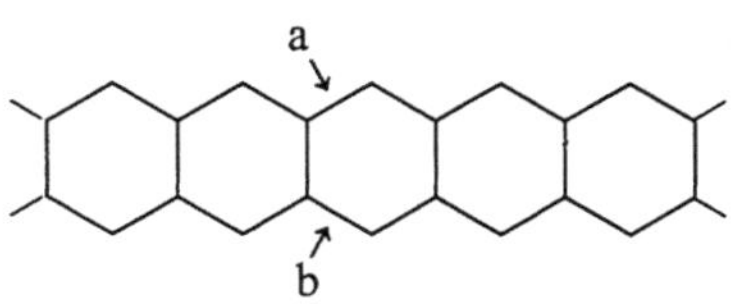

or of heterocyclic analogues. Compared with single-strand analogues, not only would such polymers have exceedingly rigid chains, and hence high softening points, but any chain scission through oxidation would require cleavage to occur twice in the same ring structure (e.g. at points *a* and *b* in the above structure), which is statistically very unlikely until massive oxidation has occurred. Several such polymers have been investigated. For example the polymer BBL [poly-(bisbenzimidazobenzophenanthroline)].

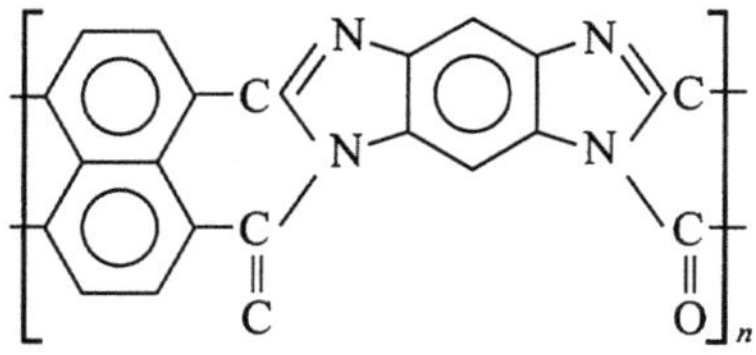

was developed as a fibre material with exceptional high-temperature resistance. However, no such polymers have been commercialized. One of the main problems lies in the difficulty of achieving complete cyclization, and hence ladder formation, during the synthesis of the polymer.

It is perhaps appropriate to end this discussion of high-temperature resistant polymers with these highly aromatic ladder polymers since they seem to offer almost the ultimate that is attainable with organic polymers, with possible use temperatures approaching 500°C.

Further reading

Cassidy, P.E. (1980) *Thermally Stable Polymers*, Marcel Dekker, New York.

Collyer, A.A. (1990) *A Practical guide to Selection of High Temperature Resistant Engineering Plastics*, Elsevier, London.

Cogswell, F.N. (1992) *Thermoplastic Aromatic Polymer Composites*. Butterworth-Heineman, Oxford.

Critchley, J.P., Knight, G.J. and Wright W.W. (1983) *Heat-resistant Polymers*, Plenum Press. New York.

Cullis, C.F. and Hirschler, M.M. (1981) *The Combustion of Organic Polymers*, Oxford University Press, Oxford.

Domininghaus, M. (1992) *Plastics for Engineers*, Hanser, Munich.

Donald, A.H. and Windle, A.H. (1992) *Liquid Crystalline Polymers*, Cambridge University Press, Cambridge.

Frazer, A.H. (1968) *High Temperature Resistant Polymers*, Wiley Interscience, New York and Chichester.

Lewin, M., Atlas, S.M. and Pearce, E.M. (eds) (1975) *Flame Retardant Polymeric Materials* Plenum Press, New York.

Troitzsch, J. (1990) *International Plastics Flammability Handbook*, Hanser, Munich.

Wilson, D., Stenzenberger, H.D. and Hergenrother, P.M. (1990) *Polyimides*, Blackie, London.

4 Hydrophilic polymers

C.A. FINCH

4.1 Introduction

Hydrophilic polymers are those polymers which dissolve in, or are swollen by, water. Many compounds of major technical and economic importance fall within this definition, including many polymers of natural origin. Many foodstuffs – containing substantial amounts of carbohydrate and protein – can be classified as hydrophilic polymers, and some have important technical and industrial uses, apart from their nutritional value. For example, although over 95% of the starches produced from corn (maize), wheat, potato, tapioca and other vegetable sources are used as foods (human or animal), the remaining quantity represents an important part of the technical polymer market. In fact, more than two-thirds of hydrophilic or water-soluble polymers used in industry are derived from polymers of natural origin, so coming from renewable resources (harvested crops, trees, waste animal products and so on) rather than from petrochemical sources of finite availability.

For convenience, hydrophilic and water-soluble polymers can be considered as falling into three principal groups:

- natural polymers, nearly all based on carbohydrates or proteins, usually of complex chemical structure;
- semisynthetic polymers, mainly based on celluloses (from wood pulp, or cotton linters), which are reacted with functional chemicals of petrochemical origin;
- synthetic polymers, prepared by polymerization of monomers of petrochemical origin.

There is no precise definition of 'hydrophilic' or 'water-soluble' polymers, other than the obvious functional statement. The terminology, although useful, is largely a matter of convenience. The borderline between 'water-swellable' and 'water-soluble', in terms of solubility, is less than clear, so both swellable and soluble polymers are discussed in this chapter. However, in terms of molecular structure, all of these polymers are characterized by the presence of polar groups attached to the main polymer 'backbone' of the molecule – notably, hydroxyl (—OH), carboxyl (—COOH) and amino (NH_2) groups. Although the detailed interrelationships between molecular structure and physical properties (including water solubility) are complex, broadly speaking, the solubility behaviour varies with the molecular chain length, amount of intermolecular cross-linking and the number and polarity of the side-chain substituents (see Table 4.1, which relates mainly to synthetic polymers).

Table 4.1 Relation between polymer structure and water sensitivity

Structural feature	Water sensitivity
Chain length increased	decreases
Proportion of polar groups increased	increases
Polarity of polar groups increased	increases
Cross-linking increased	decreases

Applications of hydrophilic and water-soluble industrial polymers are diverse (Fig. 4.1). Over 150 different applications have been identified, but the most important include:

- paper manufacture, where water-soluble polymers are used in the formation of the paper web, in pulp retention and as the binder of coating compositions;
- textile sizing (during spinning of the yarn) and finishing (of the woven fabrics);
- water purification and effluent treatment (in flocculation processes);
- in the formulation and production of many specialist fluids where it is necessary to control viscosity and the flow properties, for example in water-based adhesives and in emulsion paints,
- in film-forming materials, where the flexibility and moisture sensitivity of the film can be modified by the incorporation of plasticizers, humectants or cross-linking agents, for example many water-soluble polymer mixtures such as water-based adhesives and emulsion paints, and coatings for paper.

In this last example the white pigment (a china clay mixture), such as the coating of some book papers, is bonded during manufacture to the cellulosic (wood-pulp-based) substrate by using a mixture of water-soluble polymers – probably using a mixture of starches (for reasons of cost-effectiveness).

Many more specialist applications in which water-soluble polymers are used as part of a mixture include: medical applications, such as surgical dressings and sanitary products and the hydrogel polymers of contact lenses, which are discussed below; personal products (cosmetics, ointments, anti-perspirants, hair-sprays and lotions); domestic products (detergents, cleaners, polishes); latex processing aids (e.g. for carpet backing, where a controlled viscosity is required for coating to an even thickness); and oil-well-drilling fluids and additives.

The pattern of demand and application of hydrophilic polymers has changed over the years, with traditional starch-derived products at a disadvantage compared with the semi-synthetics and the synthetic polymers, which can, to an extent, be tailor-made for particular end uses. However, modified starches and starch derivatives have been developed – the borderline between these and the semi-synthetics is indistinct – and modified starches are, generally, lower in cost than semi-synthetic and synthetic polymers and natural gums.

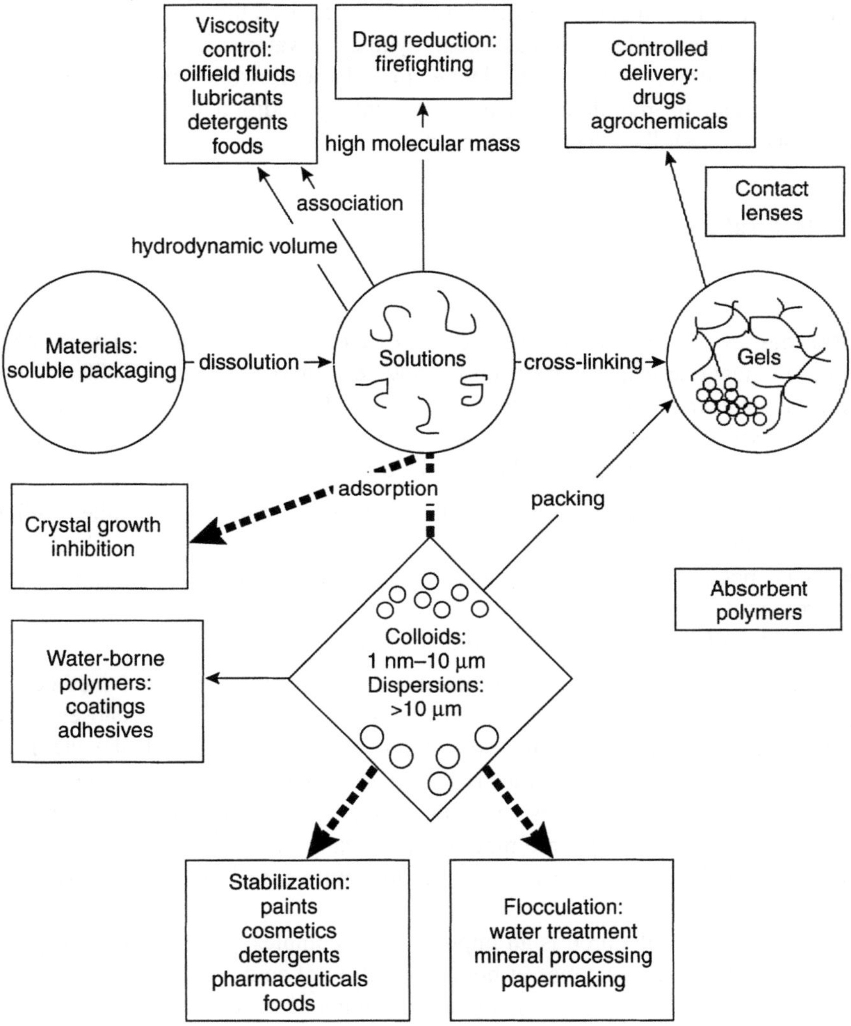

Figure 4.1 Some important properties and applications of water-soluble polymers.

This change in demand pattern is particularly true of natural gums, which are produced mainly in the tropical regions of Africa and Asia (e.g. the carbohydrate gum arabic, more than 80% of which originates in the Sudan). The harvest of these gums are subject to variations in weather and climate, and collection methods are labour intensive (Section 4.2.1).

The biodegradability of water-soluble polymers, which contrasts to water-resistant, organic solvent soluble and solid polymers, can contribute to their industrial value. The rate of biodegradation depends on the molecular structure and molecular size of individual polymers (starches, proteins, carbohydrate gums) and semi-synthetic polymers (such as cellulose ethers) (Section 4.3) are readily biodegradable, whereas most synthetic water-soluble polymers are relatively resistant to biological degradation. However, biological degradation processes can be retarded to a useful extent by addition of relatively small amounts of carefully selected bactericide.

4.2 Natural polymers

4.2.1 Carbohydrates

Starches are the most widely used group of industrial polymers, although their use, in tonnage terms, represents only a few per cent of the total of 2000 million tonnes per year of carbohydrates used for food. Starches are widely distributed as the reserve carbohydrate in the leaves, stems, roots and fruits of most land plants. The major sources are the cultivated plants:

- seeds of cereal grains: corn (maize), wheat, rice, sorghum (also called millet and 'milo');
- roots: potato (also called 'farina'), cassava (also called 'manioc'), tapioca, arrowroot;
- pith: sago (from the trunk of a tropical tree).

The properties of each type of starch vary with the source and method of processing. There is, however, a considerable overlap in the starch chosen for a particular application, since the cost-effectiveness can alter with the yield of harvests in different years. Industrially, most important starches are based on corn (and sorghum) in the western hemisphere, and potato starches in Europe. In many cases, basic starches are modified physically or chemically (or both), by, for example, heat treatment or grinding, to improve performance in particular applications.

The hydrolysis of starch is an important and complex reaction which may be carried out by acids, enzymes or both. The action of acids produces glucose and other products. Oxidation of the hydroxyl groups of starch, with sodium hypochlorite for example, gives aldehydes, ketones or carboxylic acids, depending on the reaction conditions.

Starch gives a well-known blue complex with iodine, which is used as a test for starch: a similar colour reaction is also given by polyvinyl alcohol.

Starches, as polyhydroxy polymers, can undergo conversion to esters and ethers, and the resulting compounds have useful properties. For example hydroxyethyl starches, prepared by reaction with ethylene oxide, are used in paper coating (and in wallpaper paste). The cationic starches (which are the tertiary and quaternary aminoalkyl ethers) are also used widely in papermaking for pigment retention. Other starch esters are used as food emulsion stabilizers.

The four major features which affect the properties of starches are as follows.

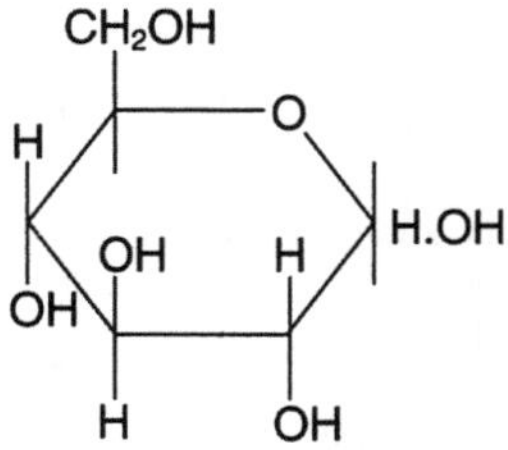

Figure 4.2 D-glucose.

- Starch is a polymer of glucose (Fig. 4.2).
- The polymer is of two types: linear (amylose) and branched-chain (amylopectin) (Figs 4.3 and 4.4). Most starches include about 25% linear molecules and about 75% branched molecules, but some waxy corn starches (from mutant seed varieties, produced by a type of genetic engineering) may include up to 85% of amylose.
- The linear polymers can associate with each other, so giving insolubility in water.
- The granular form of the starch – the size and shape of the granules – is characteristic of each type of starch and markedly affects the properties, since the disruption of the granule is necessary for the polymer to disperse in water.

Detailed design of starch formulations for particular applications considers all of the above factors.

Natural gums are a more specialist group of hydrophilic carbohydrate polymers based on polysaccharides obtained from vegetable sources, including seaweed extracts, plant exudates (rosin, of certain types only) and gums derived from seeds and roots. Several of these natural gums are of scientific and industrial importance, as are other, rather similar, recently established gums

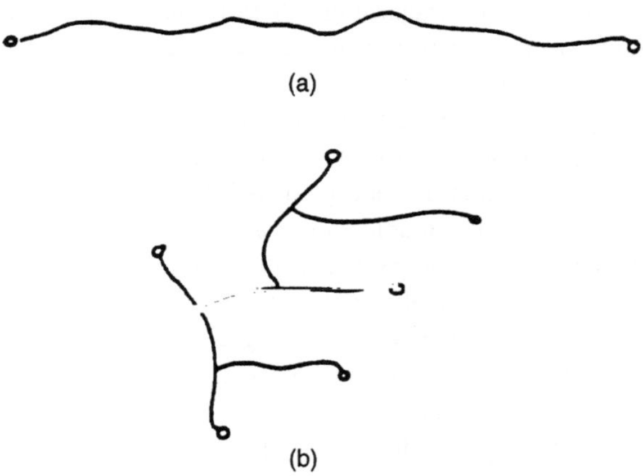

Figure 4.3 Simplified structures of (a) linear amylose; (b) branched amylopectin.

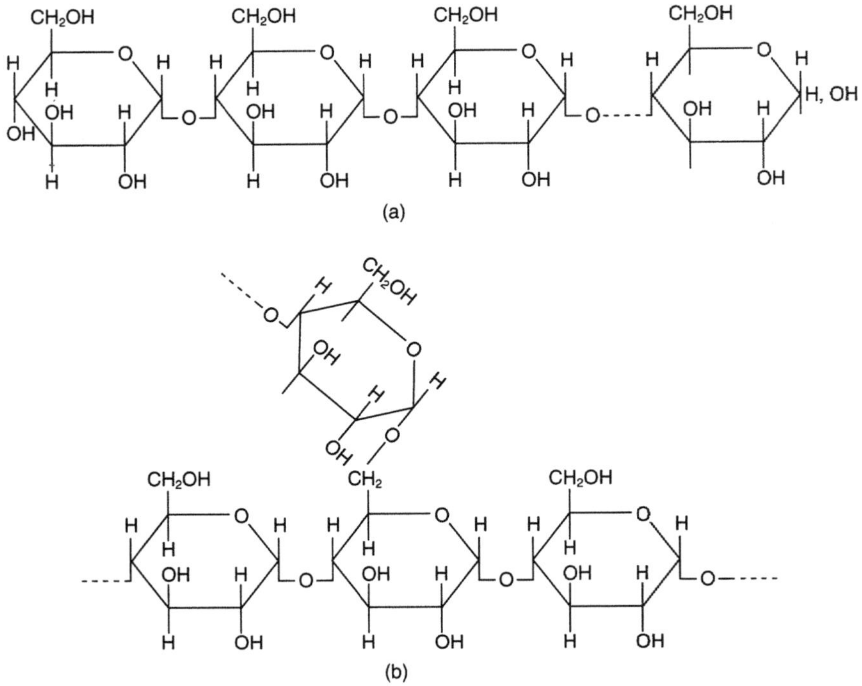

Figure 4.4 Polysaccharide chains of (a) linear amylose; (b) amylopectin (showing detail of branch points).

obtained by microbial fermentation. The properties and applications of these botanical gums are, like starches, determined by the source, climate, harvesting season and methods of collection, extraction and purification.

Many plant gums occur in nature, but only a few, notably gum arabic, gum karaya, gum ghatti and gum tragacanth (all from plant exudates) and guar and locust bean gums (from seeds), are of technical significance. By far the most important are gum arabic (obtained from the acacia, a desert bush growing in sub-Saharan Africa), and guar gum (the only gum obtained in quantity from an annual crop – a leguminous plant grown in the subtropical regions of Asia and North America). Both gums are accepted as edible, so can be used in compounded food products. Gum arabic shows a high protective colloid power for polymer emulsions and is compatible with starches in solution. Guar gum, because of its origin, has relatively stable sources of supply. It is one of the more important seed gums of commerce, although guar is mainly used as a food crop for animals, and in human foods (as a thickener). Much guar gum is also used in derivatives, notably as hydroxypropyl guar, which has wide applications in paper manufacture (to improve dry strength), oil and gas drilling operations (to increase the viscosity of drilling mud) and as a thickener in textile printing and dyeing formulations.

Amongst **seaweed** products, agar is extracted from certain types of red seaweeds, marine algae of the class Rhodophyceae (nearly 100 types have been identified as potential sources). The chemical structure of agar is complex and variable with source, but native agar comprises at least 70% alternating polysaccharide units of D-galactopyranose units and 3,6-anhydro-L-galactopyranose, with a half-ester sulphate on about every tenth D-galactopyranose unit. Agar is insoluble in cold water but soluble in boiling water. A 1.5% solution is clear and, when cooled to 32–39°C, forms a firm resilient gel which does not melt below 85°C. For these unique properties, these gels are used as solid culture media for microbiological testing, since the gel is stable and can be sterile and inert at incubation temperatures up to 65°C.

Alginates are mainly obtained from brown seaweeds, notably *Macrocystis pyrifera*, the giant kelp, which is mechanically harvested from the sea in some parts of the world, notably the Northern Pacific and from various varieties of Laminaria. The alginates appear as structural components of the cell walls, as mixed calcium–magnesium, sodium and potassium salts of alginic acid, which is composed principally of D-mannuronic and D-guluronic acids (both of which are saccharide residues). The composition ratio of the saccharide residues varies with the origin and type of seaweed source. Alginates have many industrial uses, related to their thickening, stabilizing, film-forming and gel-forming properties. Alginate solutions can be spun into fibres which swell in water. These fibres are used in wound dressings.

Carrageenan, also extracted from some red seaweeds, is a mixture of sulphated polysacharides. It produces high-viscosity solutions in water and, unusually, gels that interact with proteins, especially casein, and is used in the preparation of gelling milk-based products for food desserts.

The **carbohydrate gums** – arabic, guar, karaya and tragacanth – have generally similar uses in thickening and stabilizing aqueous solutions, with overlapping applications, mainly in prepared foods, based on regular availability and cost-effectiveness. Some of the less pure grades are also used in the paper and mining industries. Several chemical derivatives of alginates similar to those of cellulose are available, notably propylene glycol alginate (used as a texture modifier in 'soft' ice-cream) and carboxymethyl guar (used in oil-well-drilling muds).

Microbial gums (which are semi-synthetic polysaccharides) have been developed. Dextran, the longest established product, is obtained by fermentation of sucrose by *Leuconostoc* sp. It is used as the gel packing in chromatographic separations and in blood plasmas. Xanthan gum is fermented from many substrates by using *Xanthomonas campestris*. This microbial gum dissolves in hot or cold water to produce high-viscosity solutions at low concentration which have unusual rheological properties – the viscosity decreases sharply as the rate of shear increases. These gums also have complex interactions with galactomannan gums, forming elastic gels, giving unexpected increases in the viscosities of mixed solutions. The gums are salt- and temperature-stable in aqueous solution and are increasingly employed in oil-well drilling and in novel food and pharmaceutical applications.

Welan gum, produced by an *Alcaligenes* sp. in an aerobic submerged fermentation, with broadly similar properties to xanthan gum, has also been introduced commercially. Welan has high thermal stability: solutions are stable at 150° for 1 h, and the solutions, which are pH insensitive, have high viscosity at low polysaccharide concentrations. For this reason their use has been suggested in oil-well-drilling muds, where bottom-hole temperatures often reach 120°C. Other microbial gums, with improved properties, are still being discovered and may be introduced, but the costs of the production by large-scale fermentation in relatively dilute solution under sterile conditions, followed by the high costs of converting to concentrated solution, gelling and drying to remove water, yielding re-dissolvable powders, means that manufacturing costs are relatively high compared with those of starch from annual farmed crops.

4.2.2 Proteins

Some hydrophilic proteins, which have, in some aspects, similar properties to those of the above gelling carbohydrates, should be noted.

Proteins are complex macromolecules which are polymers based on repeating amino-acid units (—CO.NH.CHR—). Of the several hundred amino acids known, only about 20 occur in animal protein. The water-soluble proteins of most importance for industrial purposes are gelatin and casein, which are derived, respectively, from animal connective tissue and from milk. Casein is prepared from skimmed milk by coagulation with rennet or lactic acid, followed by precipitation and drying. It is a phosphoroprotein, which occurs to the extent of about 3% in cow's milk: acid casein is normally used industrially. It is an agglomerate which can be separated by electrophoresis into fractions of different proteins. It can be cross-linked with formaldehyde into an insoluble form once used as a rigid plastic, traditionally for buttons (as the material can be dyed after moulding). It can also be cross-linked with glyoxal or acetaldehyde, and also by chromium compounds. Apart from its food applications it is used in large quantities in the sizing of paper and, to a lesser extent, in adhesives.

Gelatin is obtained from collagen, the multichain protein of skin, bones and connective tissue of animals, by hydrolysis with either acid (normally hydrochloric acid) or alkali (by 'liming' in calcium hydroxide). This process ruptures some of the hydrogen bonds which connect the protein chains and releases the partly hydrolysed protein into aqueous solution, from which the gelatin is extracted by evaporation and air drying. The gelatin is obtained in a range of molecular sizes and shapes and is characterized by its viscosity in hot water and its gelling power, measured under standard conditions. Animal glue is similarly obtained, but the hydrolysis is carried out under the more severe conditions of pressurized steam which produces a lower-quality product (i.e. of lower viscosity and gelling point) which is used in remoistenable adhesives, paper coating (notably that used for currency notes) and as the binder in match-head compositions.

In the last case, the animal glue has a unique combination of properties, namely those of binding power, gelling at a convient temperature and 'foaming' so that the flame is propagated through the microporous head without explosion.

Gelatin is normally produced to food-grade standards of chemical and biological purity and it is not a single chemical substance. The polypeptides of which it is composed have the same amino-acid composition as the parent collagen from which it is derived. It has a broad molecular weight distribution, from 15 000 to 250 000 (average 50 000–70 000). Solutions gel on cooling, but the process is reversible and this gelatin can act as a protective colloid. More than half the gelatin produced is used in food products, but there are other industrially significant applications, which include the production of pharmaceutical capsules, both 'hard' and 'soft'. Administration of medicines in gelatin capsules is a much more effective method of drug delivery to the point of use, and use of gelatin in this application is increasing. Different types of gelatin are used for 'hard' and 'soft' capsules and the latter also include a flexibilizing agent, normally glycerol. Gelatin has been used in the walls of microcapsules used in the coating of carbonless copying paper (where other polymers are now used), and, especially, in the manufacture of photographic emulsions. The latter have an exceptionally complex physicochemical behaviour where gelatin plays an essential part in the sensitization of the active silver halide component from which the photographic image is developed.

4.3 Semi-synthetic polymers

Semi-synthetic polymers are, in practical application, an important group of hydrophilic polymers, prepared by reaction of various natural polymers with reactive compounds of petrochemical origin. They are, therefore, prepared mainly from renewable sources, with a proportion (usually small) of raw materials with finite availability.

Most semi-synthetic polymers are prepared by the conversion of cellulose, by way of alkali cellulose, to a series of derivatives, collectively known as cellulose ethers. These can be considered as two major groups:

- nonionics, including methyl cellulose, hydroxyethyl cellulose hydroxypropyl cellulose, hydroxypropyl methyl cellulose and ethyl hydroxyethyl cellulose;
- anionics, principally sodium carboxymethyl cellulose (SCMC).

Cellulose ethers and derivatives are prepared from purified cellulose from wood pulp or cotton 'linters' (short lengths of cotton too small to process into yarn). The cellulose is first masticated with sodium hydroxide to yield alkali cellulose, which is then reacted with:

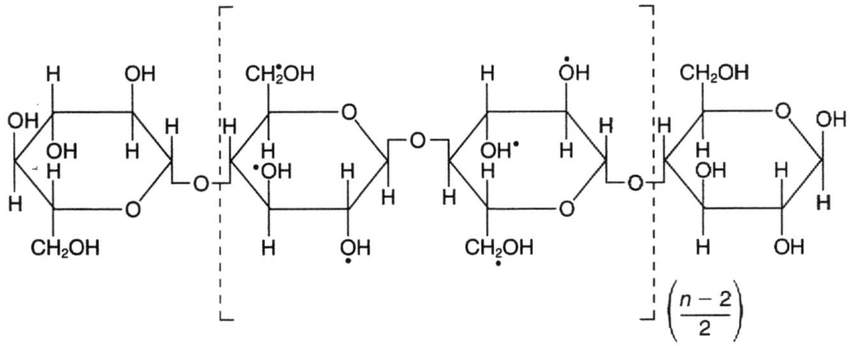

Figure 4.5 Idealized structure of cellulose, showing hydroxyl groups (*) available for substitution.

- methyl chloride, to yield methyl cellulose;
- ethylene oxide, to yield hydroxyethyl cellulose;
- propylene oxide, to yield hydroxypropyl cellulose;
- propylene oxide and methyl chloride, to yield hydroxypropyl methyl cellulose;
- monochloracetic acid, to yield SCMC.

The cellulose molecule consists of chains of anhydro-glucose rings (*β*-D-glucopyranosyl rings), with each ring having three hydroxy groups available for substitution (Fig. 4.5).

The degree of substitution (DS) of the molecule is the average number of groups which have been substituted, the maximum number being three. Different substituents have different polarities, so the DS varies for optimum solubility and performance. For the less polar groups, differing ranges of DSs are used, depending on the properties required from the resulting cellulose ethers. Typical levels of DSs for widely used industrial polymers are:

- methyl cellulose, 1.2–1.9 methoxy;
- hydroxyethyl cellulose, 0.9–1.1 hydroxyethyl;
- hydroxypropyl methyl cellulose, 1.2–1.7 methoxy and 0.2–0.3 hydroxypropyl.

With the more carboxymethyl groups (as in carboxymethyl cellulose, Fig. 4.6), a DS of 0.4–1.5 provides suitable water solubility. Commercially available grades are:

- medium, DS, 0.5–0.7;
- high DS, 0.7–1.2.

The number of anhydroglucose rings (or repeating units) in the cellulose chain is the degree of polymerization (DP). Commercial products have a DP in the range 50–2000, which corresponds to molecular weights of about 8000–325 000.

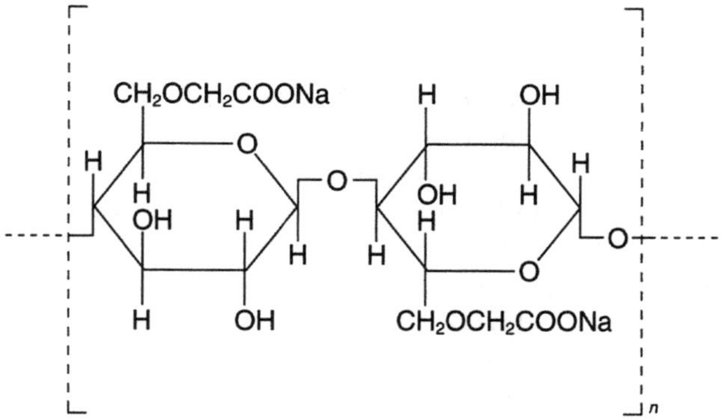

Figure 4.6 Idealized structure of sodium carboxymethyl cellulose with degree of substitution of one (DS = 1.0).

For high viscosity/high DP and high DP/temperature stable cellulose ethers, cotton-based cellulose is used as raw material, whereas for lower DP grades, lower-cost wood-pulp-based cellulose is preferred. Cellulose ethers are made from alkylene oxides by reaction with alkali cellulose in specially designed pressure vessels and are purified by washing the resulting reaction mixture with hot water. SCMC is manufactured in heavy-duty mixers, normally without pressure. Technical grades of SCMC can contain up to 50% residual sodium chloride. For food use, this salt is removed by washing with a solvent–water mixture, but a substantial proportion of SCMC is used as a technical grade for several important applications. The amount of active material is related to the application, as follows:

- 50%–90% active SCMC: detergents, mining;
- 90%–95% active SCMC: textile sizes;
- 95%–99% active SCMC: industrial uses (paint thickeners, etc.);
- ≥99.5% active SCMC: food, cosmetics, pharmaceutical, toothpastes.

SCMC is used in detergents to prevent soil redeposition after removal by synthetic detergents. This behaviour is believed to be due to the electrostatic repulsion forces between the negatively charged dirt particles and the similar charge induced by the SCMC attached to the fabrics. In oil-well drilling for exploration and production, SCMC is often employed for the suspension of the thixotropic clays ('bentonites') in the drilling muds, to lubricate the drilling bits and as a carrier to remove rock cuttings from the bore hole. In general, grades of SCMC used here have a high DS, so they are little affected by the presence of high salt concentration. SCMC also has good resistance to thermal degradation, which becomes significant in deep drilling (sometimes down to 7000 m, where bottom-hole temperatures are over 150°C).

SCMC has been used in the 'warp sizing' of textile fibres. In this important use of water-soluble polymers a dilute emulsion of a low-viscosity lubrication oil is suspended in water and stabilized by the polymer. The suspension is applied to the yarn as it passes at very high speed through the many metal and ceramic components of the spinning machine, to reduce abrasion and improve the quality of the yarn. The film-forming properties of SCMC are important in this warp-sizing application because it improves machine efficiencies (i.e. reduces the number of breaks in the yarn, which cause machine stoppages) and cost-effectiveness in use (a lower spinning size pick-up can be used so that the length of yarn on the beam for the weaving loom can be increased). SCMC also has a low biological oxygen demand, compared with starch, so that, when the fabric is washed after weaving, the problems of dealing with the resulting effluent are reduced. With some fibres, the use of SCMC has been replaced by that of other speciality polymers, notably, notably special grades of polyvinyl alcohol (Section 4.4.3), which are used on polyesters, and also speciality acrylate copolymers for other synthetic fibres.

In papermaking, SCMC is used, like other water-soluble polymers, for its film-forming and grease-resistant properties in pulp mixtures and in coating compounds. A high grade (with over 95% active material, and low salt content) is used to reduce the possibility of corrosion of the polished steel coating rollers of the papermachines, where accurate control of the viscosity and flow properties of the coating composition is essential for maintaining the quality of the printability of the paper surface.

There are many other applications of SCMC and related polymers which depend on the film-forming, viscosity-increasing, protective colloid-suspending and hydrophilic properties of these polymers.

4.4 Synthetic polymers

4.4.1 Hydrogel polymers

These are prepared by synthetic chemical reactions from a range of specialty vinyl and acrylic monomers. The simplest definition of a hydrogel is ‘a network which swells in water’. However, they are better considered as polymeric materials which are able to swell in water and retain a significant fraction (typically 30%) of water within their structure, but do not completely dissolve in water. Hence, hydrogels are characterized by a pronounced affinity of their chemical structure towards water, in which they merely swell, but do not dissolve. This is owing to the existence of cross-links which, at least in water, bind macromolecules or their segments either by permanent bonds – ‘fixed’ chemical links – or through more extensively organized regions, which can be considered as ‘effective’ cross-links, as formed from molecular associations, usually as hydrogen bonds. These effective cross-links are formed by bonds of sufficient strength so that changes in the physical structure occur only slowly, and usually under stress.

Within the above definitions, natural polymers [such as gelatin (a protein derived from animal collagen) and agar and alginates (both carbohydrates, derived from vegetable sources)] and synthetic polymers can be included. However, most of the recent scientific studies have been devoted to synthetic polymers. The largest group comprises the homopolymer and copolymer derivatives of acrylic and methacrylic acids:

$$\left[-CH_2-\underset{\displaystyle COOR''}{\underset{|}{C}R'}- \right]_n$$

where $R' = -H$; $-CH_3$; $R'' = -H$; $-CH_2CH_2OH$; $-CH_2.\underset{\displaystyle CH_3}{\underset{|}{C}}H.OH$; $-CH_2.\underset{\displaystyle OH}{\underset{|}{C}}H.CH_2OH$, etc.

The most frequently used monomer is hydroxyethyl methacrylate (HEMA):

$$CH_2{=}\overset{\displaystyle CH_3}{\overset{|}{C}}H.CO.O.CH_2.CH_2OH$$

and, colloquially, hydrogels prepared from this monomer are known as poly-HEMAs (usually, these are copolymers of controlled composition).

During preparation of the polymers, alteration of the ratio of hydrophobic and hydrophilic monomers allows the degree of hydrophilicity to be varied. The degree of hydrophilicity is also affected by the relative hydrophilicity of the chemical cross-linking agents employed. Although this possibility, in principle, allows wide variation in control of hydrophilic–hydrophobic behaviour, there are, in practice, certain constraints arising from the possible inhomogeneity of copolymer composition, especially if the copolymer reactivity ratios are significantly different. Suitable monomers are available for hydrogel polymers, including several long-chain organo-fluorine unsaturated monomers, to obtain a range of hydrophobic–hydrophilic properties. The resulting polymers have limited use in hydrogel polymer systems. In practice, the availability of monomers with convenient reactivity ratios imposes some limitation on the available properties.

Other synthetic hydrogels are based on derivatives of acrylamide polymers:

$$\left[-CH_2-\underset{\displaystyle CO\underset{\displaystyle R''}{\underset{|}{N}}-R'}{\underset{|}{C}R}- \right]_n$$

where $R = -H$; $-CH_3$; $R', R'' = -H$; $-CH_3$; $-C_2H_5$; $-CH_2.\underset{\displaystyle OH}{\underset{|}{C}}H.CH_3$; etc.

These polymers swell to a greater or lesser extent in water, depending on the hydrophilic power of the side group. With lower hydrophilicity, the polymer will swell in water, but with further increase in hydrophilicity, the polymer becomes water-soluble. In general, increased hydrophilicity is related to a greater proportion of hydroxyl groups in the polymer, since this increases the polarity of the molecule.

Other polymer molecules containing polar groups, such as polyvinyl pyridine and poly-*N*-vinylpyrrolidone are water-soluble and behave as hydrogels only if they are cross-linked chemically by a second reaction following that of polymerization of the starting monomer. The most commonly used cross-linking agents are derivatives of ethylene glycol dimethacrylate:

$$CH_2{=}\underset{\displaystyle R'}{C}.COO(CH_2.CH_2O)_x.CO.\underset{\displaystyle R''}{C}{=}CH_2$$

or of methylene-*bis*-acrylamide:

$$CH_2{=}CH.CO.NH.CH_2.NH.CO.CH{=}CH_2$$

Hydrolysed polyacrylonitrile can also be considered as a hydrogel system, when it is converted to a polyacrylamide–polyacrylate copolymer:

$$\left[CH_2{-}\underset{\displaystyle CN}{CH}\right]_n \longrightarrow \left[CH_2{-}\underset{\displaystyle CN}{CH}\right]_a\left[CH_2{-}\underset{\displaystyle CONH_2}{CH}\right]_b\left[CH_2{-}\underset{\displaystyle COOH}{CH}\right]_c$$

polyacrylonitrile

The unhydrolysed nitrile sequences are more extensively organized regions of the molecular chain which prevent the complete solution of the hydrolysis products even though the hydrophilicity of both the polyacrylamide and the polyacrylate sequences is high. The molecular structures and solution behaviour of certain types of polyvinyl alcohol are similar, in many features, to those of poly-HEMA-type hydrogels. Polyvinyl alcohol is, in terms of annual tonnage produced commercially, the most important water-soluble polymer (Section 4.4.3). It is a group of copolymers formed by the partial or complete hydrolysis of polyvinyl acetate:

$$\left[CH_2{-}\underset{\displaystyle O.CO.CH_3}{CH}\right]_n + nCH_3OH$$

polyvinyl acetate

$$\downarrow$$

$$\left[CH_2{-}\underset{\displaystyle OH}{CH}\right]_n + nCH_3O.CO.CH_3$$

polyvinyl alocohol   methyl acetate

Polyvinyl alcohol becomes water-soluble when more than approximately 80% of the acetate groups have been removed from the starting polyvinyl acetate but it can be insolubilized by further reaction with cross-linking agents such as formaldehyde (HCHO), glyoxal (OHC.CHO) or glutaraldehyde ($OHC.CH_2.CHO$), each of which will react, more or less controllably, to provide acetal cross-links between the linear polymer molecules, for example:

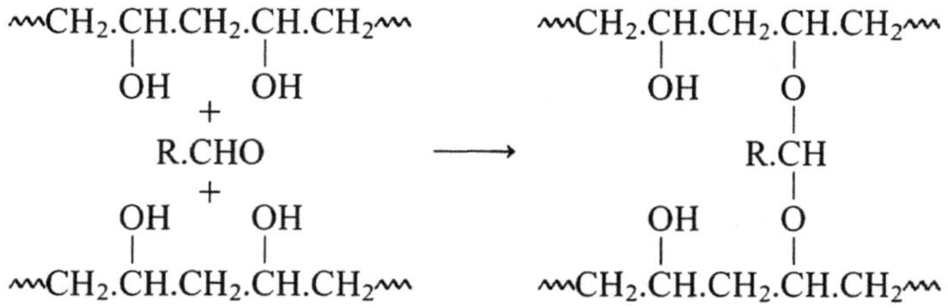

The affinity for water and the polarity of the hydrophilic groups are important in relation to the properties of the polymers. Some hydrogels are almost completely neutral, for example poly-HEMA, or sucrose methacrylate, whereas others are acidic, for example, polyacrylic and polymethacrylic acid and their copolymers, or basic, such as polyvinylpyridine and its derivatives. In addition, it is possible for mixtures of positively and negatively charged polymers to form water insoluble, but water-swellable, complexes which behave as hydrogels.

Other macromolecular species form hydrogels, including several cross-linked natural polymers, and physical blends can also be prepared, for example from poly-HEMA and collagen, which can have improved mechanical strength compared with the parent polymers. Because of their high degree of hydration as water-plasticized polymeric networks, hydrogels have rather low mechanical strength, and this strength decreases as the water content is increased. Hydrogels are prepared both by bulk and by solution polymerization, and suspension polymerization may also be employed to prepare macroporous hydrogels with a high degree of cross-linking and hence with a low degree of swelling in water.

Because of this low mechanical strength the fabrication of hydrogels into useful shapes presents practical difficulties. Thin layers of hydrogels can be cast from aqueous solution on to other films and fibres, and other forms can be prepared notably optically pure blocks and films, porous sponges and almost unswellable porous beads. Transparency is important, since hydrogels are used in soft contact lenses. These are prepared by several different methods: by a spin-casting technique, by standard methods of casting into moulds made to patterns of ophthalmic prescriptions or sometimes by controlled machining of individual frozen hydrogel blocks. In all these methods care has to be taken to achieve high surface polish. In contact lenses, an advantage to the use of hydrogels is that these polymers are permeable to oxygen, which is necessary for the cornea of the eye. Hydrogel lenses are for this reason fitted close to the cornea surface, unlike hard contact lenses, and remain practically static on the cornea, whereas other types move as the eye blinks. The corneal surface depends more on the oxygen which passes across the lens with hydrogel lenses, usually made from polymethyl methacrylate.

The next stage in the use of hydrogel polymers, and their many variations, in body implants involves their use, in association with protein-type polymers, in possible applications as semi-synthetic cartilage in joint replacements.

Further information on the specialty hydrogels used in superabsorbent polymers (SAPs) for the preparation of personal products for medical and other purposes is described in Section 4.5.

4.4.2 *Polyacrylamide hydrophilic polymers*

The polyacrylamide group of hydrophilic polymers, briefly mentioned above is less clearly definable than the other groups mentioned but is of considerable and increasing technical and industrial importance. These polymers are not employed in solid form, but their solutions and dispersions in water have many applications in effluent treatment, pulp and paper production, mineral ore processing and in some types of oil-well drilling.

Acrylamide monomer, $CH_2{=}CH.CO.NH_2$, is , when pure, a white crystalline powder of moderate toxicity (it can be absorbed through the skin, causing interference with nerve functions), which can be polymerized in aqueous solution to yield polymers of very high molecular weight. These polymers,

$$\left[\begin{array}{l} CH-CH_2 \\ | \\ CONH_2 \end{array}\right]_n$$

if suitably prepared and purified, are not toxic. Pure polyacrylamides are relatively inert and are little used in this form, but can be converted by hydrolysis of some of the amide groups into the acid form, giving aniomic polymers which have many applications. Alternatively, by special polymerization techniques, similar anionic polymers can be prepared by copolymerization of acrylamide monomer with unsaturated acid, and cationic polymers from acrylamide and cationic comonomers. The most common comonomers are:

- weak acids (acrylic and methacrylic acids);
- strong acids [vinyl sulphonic acid, 2-acrylamido-2-methyl-propane sulphonic acid (AMPS)];
- bases (dimethylaminoethyl methacrylate, dimethylaminoethyl acrylate, diallyl dimethylammonium chloride and their quaternary salts).

Proportions of comonomer vary with the application. Some 30% or even more of a particular 'polyacrylamide' may consist of other monomers (some of which, including those above, may be comparatively expensive) which will alter the charge density of the polymer side chains and hence the solution properties of the resulting polymer for specific applications.

The behaviour of acrylamide copolymers in solution is complex. The viscosity depends not only on the solution concentration, molecular weight and temperature but also on the rheology, including the shear rate at which the viscosity is measured and the previous history of the solution. The viscosity of acrylamide homopolymers is, broadly speaking, independent of pH, but that of their partially hydrolysed forms is sensitive to the presence of other ions, especially metallic cations such as aluminium. Solution viscosity is reduced by high shear and by oxidation – both of which can occur in certain oil-well-drilling applications.

High molecular weight polyacrylamides can form gels in water: these gels can also be cross-linked, either by copolymerization or by chemical post-treatment (Section 4.4.1). These gels are employed in analytical separation techniques, notably in gel electrophoresis.

Most applications of polyacrylamides (both homopolymers after hydrolysis and copolymers) depend on:

- high solubility in water;
- possible high molecular weight (for many uses, especially those involving flocculation, a large molecule is essential for satisfactory and cost-effective performance);
- the ability to produce specific polymers for particular end uses, by copolymerization – cationics are used mainly as papermaking additives; high molecular weight anionics are used in mining, water treatment and oil-recovery processes; nonionics are used mainly in mining and ore treatment.

In general, polyacrylamides are used in relatively small quantities (often as little as 1–10 parts per million), to improve processes, rather than in formulation and finished products. Partially hydrolysed polyacrylamides are used to clarify potable water. In municipal sewage treatment, an anionic polyacrylamide is added to improve flocculation of the suspended solids, and to reduce the oxygen demand. Cationic polyacrylamide may then be added to the resulting sludges to improve filtration rates, so that the dewatered solids can be converted into solid cake more readily. In mining operations, polyacrylamides are used to recover valuable solids from washings, slurries, and waste water by assisting rapid settling and dewatering of concentrates and tailings.

Typically, anionic polyacrylamides are used to:

- increase the filter capacity of metal sulphide ore concentrates;
- increase the solids content of magnetite (iron ore) products;
- aid coal washing, by recovery of tailings (fines) and reduction of waste water pollution;
- improve separation of 'red mud' (iron oxides) in the production of alumina.

Polyacrylamides are also used to improve the extraction of ores of copper potassium, titanium dioxide, uranium, zinc and some precious metals.

In papermaking, cationic polyacrylamides are added at the sheet-forming stage to reduce the loss of fibre and pigment fines through the formation wire. The selected polyacrylamides flocculate the fibres and pigment, improving drainage of water, so reducing the drying times of the formed paper, improving production speeds and economizing with heat energy.

Other applications of acrylamide copolymers include: soil modification in civil engineering projects, to consolidate load bearing, sealing and grouting of dams and ponds; increasing the viscosity and altering the flow properties of water-based latexes and adhesives; improving the handling of slurries.

4.4.3 *Polyvinyl alcohol*

This is mainly used as a component of warp sizes, especially for polyester fibres (Section 4.3). Other principal industrial applications include water-based adhesives, paper coatings, cements, binders for non-woven fabrics, and binders for the phosphorescent pigments of television picture tubes. Polyvinyl alcohol film, made by solution of extrusion casting, is a high-clarity film. The water solubility can be altered by controlled cross-linking. For some specialist purposes, water-soluble polyvinyl alcohol film is converted into bags, for example for fishery bait and also for laundry bags for collection of highly soiled laundry in hospitals, where the entire bag and its contents can be sealed and placed unopened into a washing machine.

Like many other hydrophilic polymers, polyvinyl alcohol is non-toxic and is, therefore, often used as a component of the adhesive used in food packaging.

4.4.4 *Polyvinyl pyrrolidone*

Polyvinyl pyrrolidone (PVP)

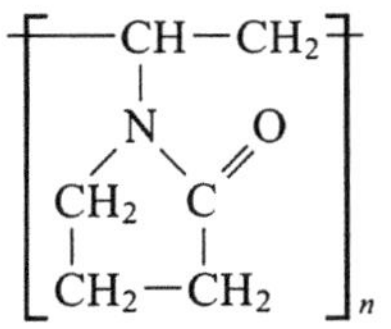

was developed in Germany in the 1930s as a by-product of the somewhat explosive (and at times heroic) chemicals from coal-based-technology of the period, using, as the basic monomer, acetylene. Acetylene can be obtained from the reaction of water on calcium carbide, which was manufactured from lime and coke heated in an electrical furnace at 2000°C. This batch process uses a large amount of electricity and is environmentally unattractive because of the large amount of calcium hydroxide slurry produced as a by-product. Acetylene can also be obtained from natural gas, but is more expensive than ethylene. The only production of PVP (in Germany and the USA) is derived from the original technology. The manufacturing route is by solution polymerization of the monomer *N*-vinyl-2-pyrrolidone (this should be, correctly, called *N*-vinyl-2-pyrrolidone), from acetylene.

Polyvinyl pyrrolidone is offered commercially in a range of molecular weights, designated by a *K*-value which derives from solution viscosity measurements of molecular weight. The *K*-values of the range of polymers available fall between 15 and 90: the numbers are essentially arbitrary but indicate the molecular size of the polymers, taking into account certain assumptions about the relationships between measured viscosity in aqueous solution and shape and size of the polymer molecules being measured.

Because of the complex manufacturing process PVP is relatively expensive but it has certain important properties which govern the main applications. These are that the polymer is:

- compatible with many biological systems, has low toxicity, and is non-irritant to the skin;
- soluble in water and many organic solvents, notably water-miscible solvents such as ethanol;
- able to form good films, with food adhesion to other substrates.

As a consequence of these properties, PVP is used in some cosmetics, such as hair and skin-care preparations (although in this field it has been partly replaced by various commercial ranges of copolymers with hydrophilic properties, notably those of maleic anhydride with methyl or isobutyl vinyl ether, or with butadiene, which are lower in cost). PVP has an unusual complexing ability, notably with iodine and iodine compounds: the PVP–iodine complex [which has some structural similarities to those of iodine with starch and with polyvinyl alcohol (Sections 4.2.1 and 4.4.3)], when in solution, is used as a simple and safe germicide.

Industrial applications of PVP are sparse, mainly because of cost. It is used in some non-gelatin-based specialist photographic systems and it has been proposed as a component of oil-well-drilling muds, where the thermal resistance of PVP could be useful, although it is unlikely to be commercially attractive. Cross-linked grades of PVP are available, with lower solubility in water, which are used in limited amounts in pharmaceutical applications as tablet binders.

4.5 Superabsorbant polymers

Water-swellable polymers and their applications have developed rapidly in the decade since the first edition of this book was written, as their importance in a variety of personal products (more than 90% of the market) has been recognized. They swell rapidly in water and, more especially, in salt-containing solutions, especially body fluids. Well-designed superabsorbent polymers (SAPs) will absorb, and hold under pressure, about 30 times their own volume of polyelectrolyte solution (notably urine), so that they are widely used as the absorbent interior of baby nappies and incontinence pads, replacing the much bulkier, and less effective, cellulose-based pads. They are generally terpolymers of acrylic acid ($CH_2{=}CHCOOH$) sodium acrylate [Fig. 4.7(a)], and a cross-linking agent (such as 1, 1, 1-trimethylolpropane triacrylate [Fig. 4.7(b)], methylene bisacrylamide or ethylene glycol diacrylate. A typical structure of an

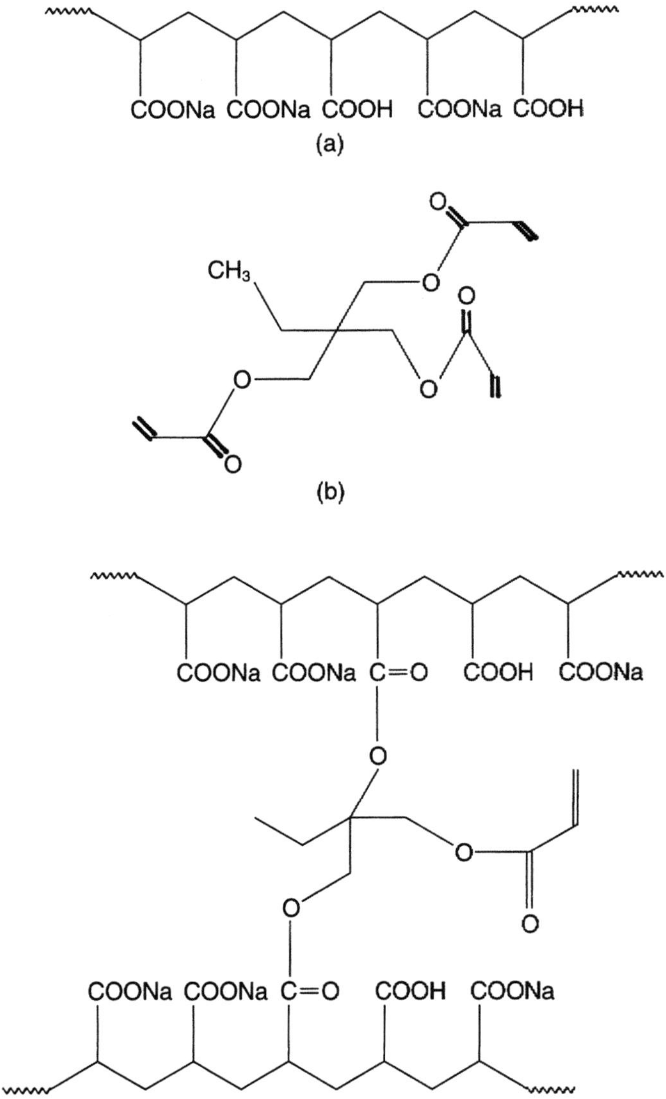

(c)

Figure 4.7 (a) Acrylic acid–sodium acrylate copolymer; (b) trimethylolpropane triacrylate; (c) typical structure of cross-linked superabsorbent polymer based on acrylic acid–sodium acrylate copolymer and trimethylolpropane triacrylate as cross-linking agent.

acrylic-based SAP is shown in Fig. 4.7(c). These SAPs are related to the hydrogel polymers (Section 4.4.1) but their composition is directed toward specific technical properties. In addition to modifying the swelling and mechanical properties, the cross-linker affects the amount of soluble polymer formed during the polymerization. Other commercially available polymers are graft copolymers with starch or polyvinyl alcohol as the grafting substrate.

The swelling and elasticity of these polymers depend on the structure and composition of the polymer network and, primarily, on the cross-linking density. Each of the several methods for preparing SAPs on a basis of acrylate polymers is aimed at adjusting the balance of properties of the superabsorbant by control of the network structure.

Basically, prepared by variations of well-established methods for polymer manufacture, including redox initiation and free radical initiated polymerization of acrylic acid and its salts, in aqueous solution. Chain transfer agents are included to reduce side reactions, including chain branching, and to control the degree of polymerization. Alternatively, 'inverse emulsion polymerization', in which monomers are polymerized whilst dissolved in hydrocarbon droplets dispersed in water, may be employed, although this method, although widely used for preparing very high molecular weight acrylamide copolymers for water treatment, is only employed to a limited extent in production.

The principal manufacturing techniques for SAP production include inverse suspension polymerizations and graft copolymerization. In inverse suspension polymerization, a well-established method for the production of other (hydrophobic) polymers, notably polyvinyl chloride, small droplets of aqueous monomer solution are dispersed into a second phase, usually a hydrocarbon mixture, before polymerization. The hydrocarbon mixture (usually kerosene) chosen depends on the solubility of the monomer(s) in this phase. The extent of neutralization is greater in suspension polymerization because of this solubility, which can lead to reaction and handling problems. Conventional free radical polymerization is employed. In spite of these difficulties, there are some advantages in the use of suspension polymerization, notably improved mixing of reagents during and after reaction (so improving process control), with easier removal of heat of polymerization. The significantly lower viscosity obtained assists stirring and heat pumping and improves the rate of heat removal from the reaction. Elongated polymers can also be prepared by alterations in the viscosity of the system.

In graft copolymerization water-soluble polymers such as starch (Section 4.2.1) and polyvinyl alcohol (Section 4.4.3) are grafted (by chemical reaction of polar vinyl monomers on to the main polymer chains) in order to modify their properties. Some processes are assisted by increased viscosity of the monomer solution, which is obtained with water-soluble graft polymer structures. Earlier types of SAPs, known as 'superslurpers' were made from starches grafted with acrylonitrile: the more recently developed acrylic acid types are derived from these. The choice of graft initiators is important – metal ions, notably Ce^{++++}, complex with the hydroxyl sites of the substrate polymer and act as a keying point for the free radicals formed.

Control of particle size and shape of the polymers formed is important, since suspensions of relatively large particles with a narrow particle size difficult to control in the early stages of polymerization. Particle size ranges are controlled by choice of suitable dispersing agents. Those employed include sorbitan long-chain esters, sucrose fatty acid esters and combination suspending agents based on clays such as bentonite and kaolin, modified by reaction with fatty amines when larger particle sizes are required. The particle shape is also important as it affects the rate at which liquid is absorbed and the methods of making the final product. The shape can be affected by the viscosity of the monomer phase: elongated particles can be produced by increasing the viscosity of the reaction mixture above 50 poise, by addition of water-soluble thickeners. In practice, granules of irregular shape are prepared: the usual shape is an irregular granule, formed by grinding lumps of dry solid polymer.

The choice of cross-linking agent is also important: relatively small amounts (1%–2% on monomer) play a major role in determining the properties of the final copolymers. The agents mentioned above are all used, depending on their reaction rates with acrylic acid and sodium acrylate. The choice of cross-linking agent will also depend on the method used to neutralize the carboxylic groups – a high-pH process may require a hydrolytically stable compound, such as tetraallyloxethane, $(CH_2{=}CHCH_2O)_2CHCH(OCH_2CH{=}CH_2)_2$, rather than a diacrylate. However, different cross-linking agents are used with inverse suspension polymerization methods, and the resulting performance of the SAPs obtained may be affected.

Superabsorbent fibres have been available for many years and have been used for medical purposes, as they can be formed into absorbent pads with a relatively firm structure. One simple type, which can be prepared in the laboratory, is based on a simple water-soluble polyacrylic acid, with minor amounts of methyl acrylate and OH-substituted comonomers (typically 2-HEMA) (Section 4.4.1) reacted in aqueous solution with free radical catalysts. The resulting polymer solution is shaped into a fibre (by pouring through an orifice) and fixed by salt precipitation and/or heating, then dried, forming ether crosslinks. Such fibres are fragile, and an external plasticizer such as polyethylene glycol may be incorporated to improve their flexibility. Mixtures of granular SAPs with short plastic fibres resemble fluff-balls and can be placed into absorbent cores so that they remain in position. They can be used in industrial filters to remove water from fuels or hydraulic oils.

The growth of the SAP industry in the past decade (over 500 000 tonnes of these polymers were produced in 1995, and this rate of production continues to rise, making them, in annual tonnage, the largest group of synthetic water-soluble polymers), with 90% of output directed towards the perceived need for absorbent polymer products, including baby nappies, adult incontinence products (more necessary with an ageing population) and other personal care products, has been directed towards providing thinner and more robust devices at lower cost. Other applications of SAPs include reduction of water absorption into buried communication cables, reduction of leakages in some food packages and, in horticulture control of the moisture content of soils. This has resulted in significant research effort directed towards meeting the changing needs, with considerable changes in the techniques of polymer manufacture and evaluation.

Acknowledgement

Dr P.M. Budd (Manchester University) is thanked for his assistance in the provision of Fig. 4.1.

Further reading

Davidson, R.L. (ed.) (1980) *Handbook of Water Soluble Gums and Resins*, McGraw-Hill, New York.

Finch, C.A. (ed.) (1983) *Chemistry and Technology of Water Soluble Polymers*, Plenum Press, New York.

Finch, C. A. (ed.) (1973) *Polyvinyl Alcohol: Properties and Applications*, John Wiley, Chichester and New York.

Finch, C.A. (ed.) (1992) *Polyvinyl Alcohol: Developments*, John Wiley, Chichester and New York.

Finch, C.A. (1996) *Industrial Water Soluble Polymers* Royal Society of Chemistry, London.

Kudela, V. (1987) Hydrogels, in *Encyclopedia of Polymer Science and Engineering* (eds H.F. Mark, N.M. Bikales, C.G. Overberger G. Menges, *et al.*) vol. 7, John Wiley, Chichester and New York, pp. 783–807.

Buchholz, F.L. and Peppas, N.A. (eds) (1994) *Superabsorbent Polymers: Science and Technology*, ACS Symposium Series 573, American Chemical Society, Washington, DC.

Molyneux, P. (1983) *Water Soluble Synthetic Polymers: Properties and Behaviour*, 2 vols, CRC Press, Boca Raton, Fl.

Ward, A.G. and Courts, A. (eds) (1977) *Science and Technology of Gelatin*, Academic Press, London.

Alleavitch, J., Turner, W.A. and Finch, C.A. (1989) Gelatin in *Ullmann's Encyclopedia of Industrial Chemistry*, vol A12, VCH Verlagsgesellschaft, Weinheim pp. 307–317.

Whistler, R.L. and BeMiller, J.N (eds) (1973) *Industrial Gums*, 2nd edn (1993, 3rd edn), Academic Press, New York.

5 Polymers with electrical and electronic properties

M.T. GOOSEY

5.1 Introduction

Since the original version of this chapter was written some ten years ago, the utilization of polymers with special electrical and electronic properties has continued to grow significantly and there have been many changes and developments which need to be reported. Consequently, this chapter has been rewritten with particular reference to those developments that have occurred over the past ten years.

There are now many types of polymers having special properties that make them of interest for electronics, optoelectronics and associated technology. These include materials possessing unique combinations of electrical, optical, thermal, chemical and physical properties which enable them to be utilized in applications that have not normally been associated with more conventional polymers. Examples include electrical conductivity and photoconductivity, non-linear optical effects and dielectric properties such as piezo-electricity and pyroelectricity. Many of these polymeric materials offer significant potential advantages over the traditional materials used for the same application and, in some cases, such as photoresists and non-linear materials, applications not possible by other means have been achieved.

Two of the most important developments over the past ten years have been the considerable progress in conductive polymers and in photoresist technology for submicron featured semiconductor device fabrication. Conductive polymers are now available commercially in large quantities and there are numerous examples of practical applications. Also, the fabrication of today's very fine featured semiconductor devices would, for example, be impossible without the major development work that has occurred during the past ten years to stretch the capability of photoresist chemistry to its current submicron limits.

Whilst there have been significant developments and growth in many areas it is also quite clear that other areas really failed to show significant progress and in fact this includes some of the original examples cited in the earlier version of this chapter. Although they may still be of some interest to a number of readers the emphasis of this chapter is now directed to those new developments of the past ten years that have shown the most promise both from a technological and from a potentially practical perspective.

This chapter therefore provides an introduction and review of special polymers that are currently being used or are being investigated for their novel properties. No in-depth knowledge of specific topics is assumed and the text is intended to provide an introduction and overview of each specific area.

Wherever possible, suitably updated references have been cited and the reader is encouraged to use these for further insights into specific topics.

5.2 Conducting polymers

5.2.1 Introduction

Polymers have traditionally been utilized in electrical and electronic applications as insulators where advantage is taken of their very high resistivities and excellent dielectric properties. Familiar uses include cable sheathings, dielectric layers and films as in capacitors, printed circuit substrates and encapsulants. However, it is some of their other useful properties, such as strength, flexibility, elasticity, stability, mouldability and ease of handling that have prompted the large amount of research and development on conductive polymers. The combination of these properties with the ability to conduct electricity is very attractive and there has been significant continuing activity in this area over the past ten years.

Polymers can be made to conduct electricity relatively easily by compounding them with high loadings of conductive metals such as copper, silver and gold, or, where cost considerations are important, by carbon. However, although these materials find wide application, conduction is by way of the filler particles and not the polymer itself. Whilst these materials find use in conductive coatings for electromagentic shielding of equipment and in die attach adhesives for semiconductor devices, some of the beneficial properties of the unfilled polymer are diminished by the incorporation of the large amounts of conductive species required to impart significant conductivity. These materials will not be discussed further here but useful insights into their formation and application are given elsewhere [1–3].

5.2.2 Conductive polymers via controlled pyrolysis

Early conductive polymers were based on refractory pyropolymers which largely resembled graphite and were virtually useless in terms of polymer processing requirements. An example of these types of polymers are the poly (acene-quinone radical) polymers (PAQRs) which have structures as follows:

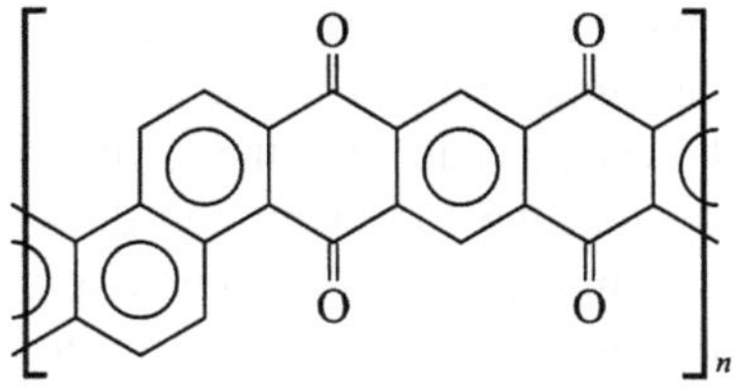

These have conductivities in the range 10^{-6}–10^{-7} S m^{-1} but are also intractable, infusible and have poor mechanical properties. Although largely surpassed with the discovery of conduction in doped conjugated polymer systems, they did receive some renewed interest with a view to practical application because they had the benefit of being environmentally stable unlike the early doped systems. For example, an easily processable conductive pyropolymer based upon a phthalonitrile resin prepared from 4,4′-bis (3, 4-dicyanophenoxy) biphenyl has been reported [4]. With conductivities as high as 2.7×10^{-4} S m^{-1} and stability after exposure to air for two years and water for 80 days, applications in electromagentic shielding and electrostatic dissipation materials have been proposed. However, the recent progress in producing stable and processable doped conducting polymers probably means that interest in these materials will again be limited.

Over the past twenty years or so a number of polymer systems have been identified which, when partially oxidized or reduced with dopants, become electrically conducting. These polymers are structurally similar, having a conjugated backbone along which, under appropriate conditions, conduction can be induced via their delocalized orbitals. Initially, one of the key problems with these materials was their poor stabilities, but with careful selection of the appropriate polymer, the development of new materials and optimization of processing, this limitation can be overcome. Numerous applications of these materials have been proposed and conductive polymers do now find commercial applications. The fact they can also exhibit reversibility of the doping–undoping process has made them of interest for energy storage and display applications. Some of the more important polymers of this type and their potential applications are now reviewed after a brief introduction to their conduction mechanisms.

5.2.3 Conduction mechanisms

The ability of materials to conduct electricity varies widely, allowing their classification into good conductors (metals), semiconductors and non-conductors (insulators) and can be explained simplistically by using band-gap models to describe the energy levels of a material's electrons. The outermost shell of electrons in a material contains the valence electrons and these can be placed in the valence band which defines their lowest energy states. In order for conduction to occur, an electron must obtain sufficient energy to promote it to the conduction band. The energy difference between the valence band and the conduction band is known as the forbidden gap and determines whether a material is a conductor, a semiconductor or an insulator. For a conductor, the highest energy level of the valence band and the lowest energy of the conduction band are similar, meaning that the forbidden gap does not exist and electrons pass easily into the conduction band. For non-conductors the separation between the two bands is large and promotion of an electron to the conduction band is not possible. With semiconductors, the forbidden gap is moderate and limited conduction occurs. This is shown schematically in Fig. 5.1.

Electrical conduction in organic materials may occur through the movement of either electrons or ions. In each case, the conductivity (σ) is equal to the

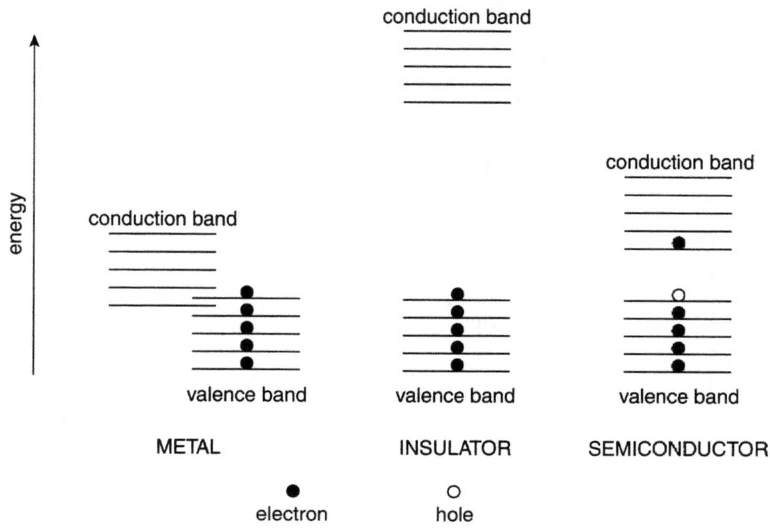

Figure 5.1 Representation of energy band for metal, insulator and semiconductor.

product of the carrier mobility, μ, its charge, q, and the number of carriers or the concentration, n, as in

$$\sigma = \mu q n.$$

Depending upon the specific mechanisms involved, all three of these variables may change with the environment and can be modified to some degree. Mechanisms which have been employed to explain the electrical conductivity of organic materials include [5]: ionic conduction, band-type conduction, hopping conduction, excitonic conduction, quantum mechanical tunnelling between metallic domains and, more recently, theories based on conformational structures and conformational defects such as solitons, polarons and bipolarons introduced during doping. It can be seen, therefore, that there are many possible mechanisms contributing to conduction in doped conjugated polymers and it is far from clear that all of the mechanisms have been fully elucidated. More detailed discussion of these mechanisms is beyond the scope of this text but for further insights the reader is referred to the published literature [6].

The range of material conductivities and those of conductive polymers are shown in Fig. 5.2.

5.2.4 Polyacetylene

Polyacteylene is now of historic significance since it was in 1977 that high levels of conductivity were first reported in the partially oxidized material. Polyacetylene also represents the simplest of the conjugated organic polymers and has therefore received much attention as a fundamental material. It has both

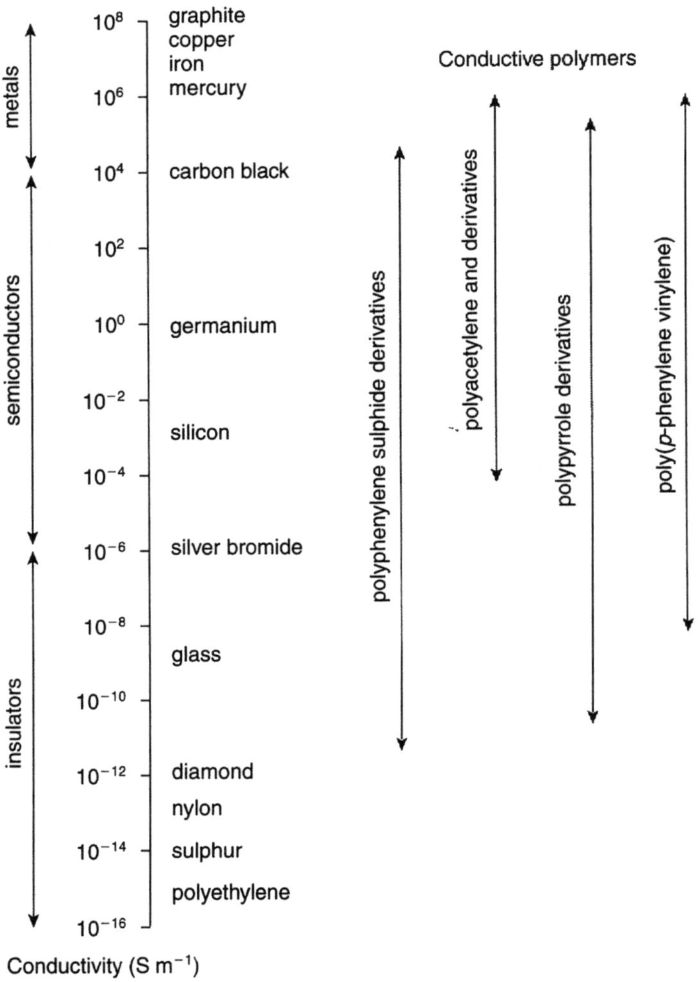

Figure 5.2 Conductivities of various elements, compounds and polymers.

cis and *trans* forms which can be prepared as silvery, flexible, polycrystalline semiconducting films:

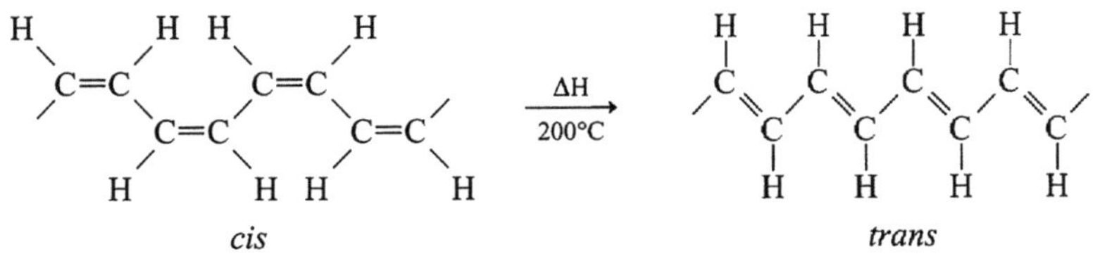

Polyacetylene films can be made relatively simply by wetting the inside walls of a glass reaction vessel with a solution of $(C_2H_5)_3Al$, and $(nC_4H_9O)_4Ti$ and admitting gas at pressures from a few millitorr to one atmosphere. As prepared, this co-called Shirakawa-type polymer is highly crystalline, completely insoluble and comprises fibrils with diameters of 5–20 nm. On exposure to air, both neutral and doped films become brittle after a few days, losing their characteristics silvery lustre. This slow oxidation is also accompanied by a decrease in electrical conductivity and the appearance of a carbonyl stretching band in their infra-red spectra.

As prepared, pure polyacetylene has a relatively low conductivity in the region of 10^{-3}–10^{-4} S m^{-1}. By chemically or electrochemically doping polyacetylene films, one can greatly enhance their electrical conductivities. Dopants may be either electron attracting (p-type dopants) or electron donating (n-type dopants). Doping is typically achieved by exposing *cis* or *trans* films to vapour of the dopant. The doping of *cis* films has been more extensively studied than has that of *trans* films because their doped conductivities are two or three times greater than those of their *trans* analogues. For both *cis* and *trans* polyacetylene the increase in conductivity with dopant concentration levels out at 1%–2%, and this is suggestive of a semiconductor-to-metal transition [7]. Typical p-type dopants are Br_2, I_2, AsF_5, $HClO_4$ and H_2SO_4, and n-type electron donating doping can be achieved by immersing films in solutions of sodium naphthalide in tetrahydrofuran (THF) or by the electrolysis of a solution of LiI in THF using the polyacetylene film as the cathode.

The poor physical and mechanical properties, intractability and instability of polyacetylenes have prevented their widespread application but have also prompted studies of copolymers and composites containing polyacetylene as the conductive component. Copolymers may be termed as random, block or graft depending upon their structure. Random copolymers of acetylene and methylacetylene with varying compositions have been synthesized, the methyl acetylene being selected because its polymer is soluble in organic solvents [8, 9]. The conductivities of I_2 and AsF_5 doped copolymers vary strongly with composition, conductivity decreasing over five or six orders of magnitude as the methyl acetylene content increases. Unfortunately, the decreased conductivities are accompanied by continued insolubility and enhanced air and heat sensitivities. With block and graft copolymers a number of techniques have been developed for copolymerizing polyacetylene with soluble carrier polymers [10, 11]. Other polyaryl acetylenes have been reported, although none has given high conductivities when doped [12].

The problems experienced with polyacetylene and its copolymers and blends have prompted extensive research into other conductive polymer systems. A popular method of producing conjugated electronically conducting polymers is via the polymerization of resonance-stabilized aromatic molecules. This approach received much attention following the discovery that oxidation of pyrrole at a platinum electrode in the presence of a supporting electrolyte produced free-standing electrically conductive films. Many other aromatic systems have since been studied and some of the more important ones are now reviewed.

5.2.5 Polyparaphenylenes

The polyparaphenylenes (PPP) form a series of polymers consisting of phenyl rings which can be doped to give moderate to high conductivities. Polyparaphenylene

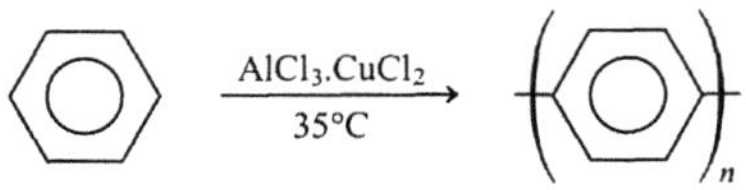

benzene poly(*p*-phenylene)

itself is prepared by the Friedel Crafts polymerization of benzene by using a mixture of aluminium chloride and cupric chloride. By oxidative coupling of the benzene rings at 35°C a black intractable powder is produced which can be purified by washing with hot hydrochloric acid followed by rinsing and drying to yield a red-brown product. Alternatively, an improved method for producing PPP by means of a soluble polymeric precursor has been reported which has an intermediate made by the bacterial oxidation of benzene and a resultant polymer with a high degree of polymerization [13]. PPP was the first example of a nonacetylenic hydrocarbon polymer that forms complexes having metal-like properties. As prepared, the polymer has an intrinsic conductivity of less than 10^{-12} S m^{-1} but, as with polyacetylene, PPP can be doped with AsF_5 to give conductivities of around 1 S m^{-1}. It may also be doped with alkali metals to provide highly conducting n-type materials which have a metallic gold appearance. When polyphenylene was exposed to a solution of potassium naphthalide in THF for about 100 h the conductivity increased from 10^{-10} S m^{-1} to 7.2×10^2 S m^{-1}. Similarly, increases in conductivity were also obtained using sodium naphthalide [14]. Conductive PPP can also be produced in a combined doping–polymerization reaction by exposing *p*-terphenyl to AsF_5 vapour. The colourless crystals darken and finally obtain a metallic blue lustre on exposure to the vapour. Potential applications for PPP are similar to those envisaged for polyacetylene, such as Schottky barriers in photocells. (A Schottky barrier is a metal semiconductor contact that has rectifying characteristics similar to a p–n junction.) The main advantage of poly-*p*-phenylene is that, because of its non-acetylenic composition, it has a much higher thermal stability (450°C in air and 550°C in inert atmosphere).

An interesting copolymer of poly-*p*-phenylene and polyacetylene has also been produced as a conducting polymer. Poly-*p*-phenylene vinylene (PPV) was made by a relatively new technique in which a soluble precursor polymer was prepared and processed prior to its conversion to the conjugated polymer [15]. Such a route allows variations to be made in the polymer morphology and hence the electrical conductivity. Before the development of this method it was initially prepared as low molecular weight oligomeric powders [16] owing to limitations in its solubility. When doped with AsF_5, PPV oligomers were reported to have conductivities of about 10^2 S m^{-1} [17] whilst films prepared through the precursor route attained conductivities an order or magnitude higher. Films prepared by the precursor method can be stretched to give a structure with a high degree of uniform molecular orientation. After doping, the conductivity in the direction parallel to the orientation has been reported to be as high as $\sim 2.5 \times 10^5$ S m^{-1} [18].

5.2.6 Polypyrrole

Polypyrrole can be prepared by the electropolymerization of pyrrole and it is then obtained as highly coloured dense conducting films. Early preparations of polypyrrole utilized sulphuric acid, but electrolytes comprising lithium chlorate, tetraethylammonium tetrafluoroborate or tetraethyl ammonium chlorate in acetonitrile have been used. By using electrodes made of platinum foils, gold or conducting glass, polypyrrole is obtained as a highly coloured dense conducting film.

Polypyrroles have prompted considerable research because they are a group of polymers that can be easily produced in the doped state. When a monomer solution is electrolysed in the presence of $Et_4N^+BF_4^-$ a blue-black conducting polymer is produced at the anode. The polymer can then be removed from the electrode and conductivities of up to 10^4 S m^{-1} have been reported [19]. One of the principal advantages of polypyrrole over other doped polymers is its excellent thermal stability in air. It is thermally stable up to 250°C, showing little degradation of its conducting properties below this temperate. Above 250°C the rate of weight loss increases, becoming very rapid over 600°C. Although chemical analysis of polypyrrole varies with preparative conditions, elemental analyses have indicated that the pyrrole rings remain intact with each unit being connected by its α carbons:

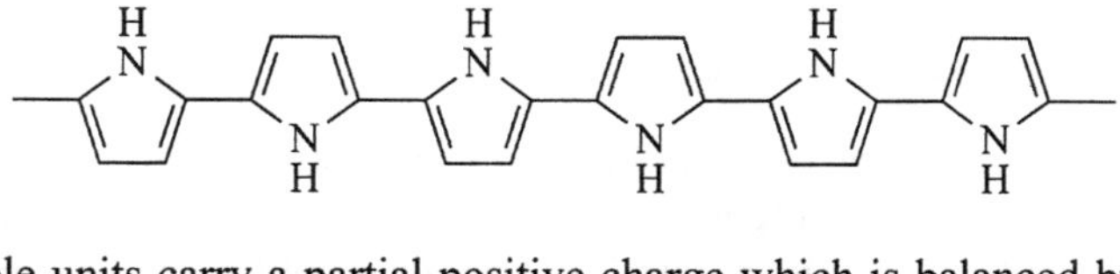

The pyrrole units carry a partial positive charge which is balanced by one BF_4^- ion to every four pyrrole rings [20].

Polypyrrole has also been produced with ClO_4^- and $CH_3C_6H_4SO_3^-$ as the counter ions in place of BF_4^-. With polypyrrole toluenesulphonate (PPTS) significant improvements in tensile strength and Young's modulus have been reported and this has prompted its thorough investigation [21].

As produced by anodic polymerization, polypyrrole is highly coloured in the conducting state. Murthy *et al.* [22] produced blue-black thin films in sulphuric acid with bulk conductivities of 6.5 S m^{-1}, yet it can also be cathodically reduced to give a green transparent insulating material. This type of electrochromic switching has led to the proposal of polypyrroles for use in display devices. Polypyrrole has also been produced on vitreous carbon electrodes which allows high voltages and currents to be used to grow thick films in relatively short times. Films so produced adhere well to the electrodes but can still be readily removed. They have excellent mechanical properties, being tough and flexible, and can be stretched up to 60%.

Derivatives and copolymers of polypyrrole have also been investigated. Films of poly-*N*-methylpyrrole have conductivities much lower than polypyrrole, whilst films produced from mixtures of pyrrole and *N*-methylpyrrole have conductivities which vary with composition over five orders of magnitude. The size of any substituent group plays a major part in determining the ultimate conductivity and can even inhibit film formation. For example, poly-*N*-phenylpyrrole and poly-*N*-butylpyrrole have reduced conductivities in the range of 0.1–1 $S\,m^{-1}$, whilst with bulky substituents such as tertiary butyl and cyclohexyl, brown-black powdery products are obtained instead of films. As with other materials, graft copolymers of polypyrrole have been made by the electrochemical oxidation of pyrrole in the presence of a chain that has been initially modified with a pendant pyrrole unit [23].

Other heterocyclic aromatic ring compounds also undergo simultaneous oxidation and polymerization. Thiophene, carbazole, indole, furan and thiophene, amongst others, all undergo electrochemical polymerization and oxidation to yield oxidized polymers of varying conductivities. In the case of 2,2′-bithiophene, the resultant oxidized film may be electrochemically reduced and in so doing undergoes a colour change from green to red.

5.2.7 Polyaniline

Polyaniline

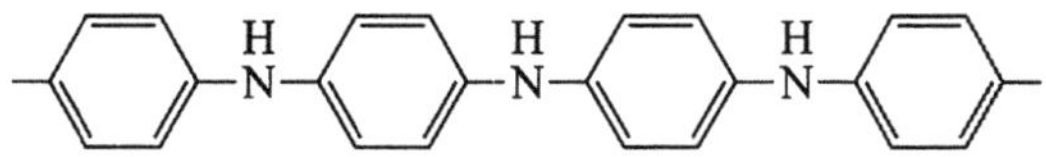

has received much recent attention because of potential applications such as in the fabrication of novel batteries. In 1990 two US companies, Lockheed and Hexcel Corporation, started a joint venture to manufacture polyaniline and blends with conventional polymers. Since then various other companies in the USA, Europe and Finland have also made major commitments to developing this material. Although polyaniline was first produced in the nineteenth century, interest in this polymer and its derivaties did not really begin to develop until the mid-1980s. Several chemical methods exist for the preparation of polyaniline although the preferred method utilizes electrochemical polymerization. This is because the resulting products are 'clean' and do not require extraction from the initial monomer/oxidation/solvent mixture [24].

Under appropriate conditions the so-called emaraldine base form of the polymer reacts with dilute acids to give corresponding emaraldine salts, with protonation of the nitrogen atoms. This reaction is accompanied by a 9–10 order of magnitude increase in conductivity to 1×10^2–5×10^2 S m^{-1} with chloride as the counter ion and even higher if the chloride is replaced with species such as camphorsulphonic acid [25].

Key attractions of polyaniline are the relatively low cost of the synthetic route from aniline and the fact that it has very important features that distinguish it from other conductive polymers. For example, its conductivity is not just determined by its degree of oxidation but also by the degree of protonation. Solution processing of polyaniline has been obtained via a number of routes and there has been recent activity on the use of large anions to confer solubility to the polymer.

5.2.8 Polyphenylene sulphide and its analogues

Polyphenylene sulphide (PPS)

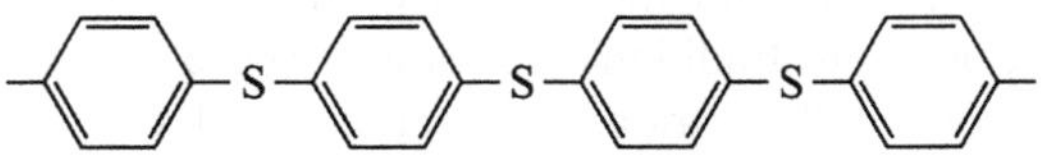

has been available commercially for many years. It is a white polymer with a melting point of 288°C and is intrinsically insulating having a conductivity of less than 10^{-14} S m^{-1}. Doping with AsF_5 leads to an increase in conductivity to around 10^2–10^3 S m^{-1}, although there is also a deterioration in mechanical properties and a colour change to dark blue. Solutions of AsF_5 doped PPS in AsF_3 have been produced and films with good mechanical properties and conductivities up to 2×10^5 S m^{-1} can be cast. Other dopants used with PPS include AsF_5, $NOPF_6$, FSO_3H and SbF_5. Direct ion implantation has also been used as a method of making thin PPS films conductive and these have been found to be relatively environmentally stable.

The selenium and tellurium analogues of PPS have also been synthesized and investigated [26]. As prepared, polyphenylene selenide (PPSe) and polyphenylene telluride (PPTe) are insulators with conductivities of less than 10^{-9} S m^{-1}. Exposure of PPSe to AsF_5 results in an insulator-to-conductor transformation with a maximum conductivity in the range 0.1–1 S m^{-1} being achieved. The thermal stability of PPSe is similar to that of PPS, whereas PPTe is decomposed by relatively mild conditions.

5.2.9 Other important conducting polymers

There is a large number of other conjugated polymers that have been exposed to dopants such as AsF_5 in an attempt to obtain conductivity. Worthy of note is the inorganic polymer polysulphur nitride. Polysulphur nitride is obtained as gold-coloured fibrous crystals from the solid state polymerization of S_4N_4 crystals. At room temperature its conductivity is in the region of 10^5 S m^{-1} and this can be increased an order of magnitude by chemical modification with bromine.

5.2.10 Applications for conducting polymers

When the original version of this chapter was written in the mid-1980s, conducting polymers had found only limited applications in some specialist areas and no major commercial uses had yet emerged. There were, however, many potential applications being proposed and investigated. Over the past ten years the situation has changed significantly, and conducting polymers are now seen in everyday commercial applications. It should be stated, however, that despite almost 20 years of intensive development activity it appears that many of the originally proposed applications were overoptimistic and have not yet become possible. Nevertheless, with growing experience of these materials, a number of more realistic applications has begun to emerge in which the novel properties of the conductive polymers are exploited rather than those that are readily obtainable in more traditional materials. These include applications in plastic batteries, solar cells, various semiconductor devices, sensors, electrochromic displays, EMI shielding and in the preparation of printed circuit boards. Some examples of these are now briefly discussed.

Rechargeable plastic batteries have been one of the most widely publicized potential applications for conductive polymers because of the perceived possibility of achieving much higher power-to-weight ratios. The approaches taken to the development of plastic batteries have included the use of conductive polymers both in the electrodes and in the electrolytes, although their most promising use has been as electronically conducting electrode materials. In polymers where conductivity can be introduced with both positive and negative doping procedures, the possibility exists for their use in both electrodes of a battery. Various battery demonstrators have been built and the Bridgestone Corporation of Japan released commercially available novel rechargeable button cell batteries based on polyaniline as early as 1987/88.

Conductive coatings have long been used in electromagnetic shielding applications, and the potential use of intrinsically conductive polymers in this application is an attractive one. With the commercial availability of polyaniline and related blends since the early 1990s has come the possibility of using these materials in shielding applications [27]. They are reported to show superior electromagnetic shielding properties compared with previously used materials [28]. The blends are easier to process than metal flake or fibre-filled polymers and are non-abrasive. They also exhibit higher and more uniform conductivities than are possible with carbon-filled coatings. It appears that polythiophene blends currently being developed may also be used for similar purposes [29].

One area where there has been commercial application of conductive polymer technology in reasonable volumes is in the so-called 'direct plate' processes used in the preparation of printed circuit boards. Having drilled holes in a printed circuit board it is necessary to provide metallization in the hole, and this is usually achieved by the combination of an initial electroless copper stage followed by an electroplated copper. By forming a conductive polymer film in the drilled hole it is possible to electroplate directly on to the film thereby avoiding the use of electroless copper. In one manifestation of this process, polymerization of pyrrole monomer is effected via residual manganese dioxide left in the hole from the previous chemical de-smear process. Although, this method is actually used in commercial volume production of printed circuit boards, concerns over both the volatility and the possible toxicity of the pyrrole monomer have prompted alternative conductive-polymer-based routes to be evaluated. For example, a less volatile and reportedly less toxic monomer, ethylene dioxy thiophene, has been developed by BASF [30] and Shipley [31] has patented the use of colloidal conducting polymers such as polypyrrole, thereby avoiding the use of the monomer for this application.

High-quality aluminium electrolytic capacitors have been made by using doped polypyrrole as the solid electrolyte [32]. These have excellent frequency characteristics and are very stable in the presence of both moisture and heat. Because the polymer film can be very thin it is possible to achieve high capacitance values.

5.3 Photoconducting polymers

The enhanced flow of current under the influence of an applied electric field that occurs when a semiconductor is exposed to visible light or other electromagnetic radiation is known as photoconduction. The simplest process of photoconduction occurs when a photon is absorbed and an electron is promoted from the valence band to the conduction band. This process generates an electron-hole pair, augmenting the concentration of intrinsic carriers. The requirement for this type of photoconduction is that the photon energy must exceed the band gap and consequently there is a threshold wavelength (i.e. energy) associated with the process. Carriers may also be generated from photon absorption by way of more complex routes such as in the generation of excitons. These are excited electronic states which are themselves unable to transport charge. However, if two excitons collide they can produce an electron–hole pair.

In many common polymers it is very difficult to observe photoconduction even when the samples are irradiated with photons having energies well in excess of their band gaps. This is largely because the carriers have very short lifetimes (≤ 1 ns) owing to rapid recombination and deep hole trapping.

The development of organic polymeric photoconductors was stimulated by the discovery that poly-*N*-vinyl carbazole (PVK) sensitized with certain pigments and dyes could exhibit sufficient photoconductivity to be usable in electrophotographic photoreceptors.

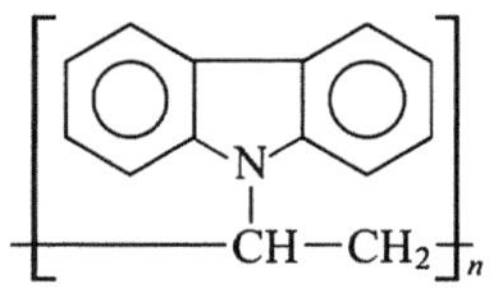

poly-*N*-vinyl carbazole

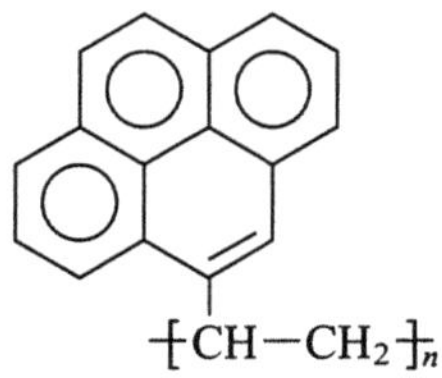

polyvinylpyrene

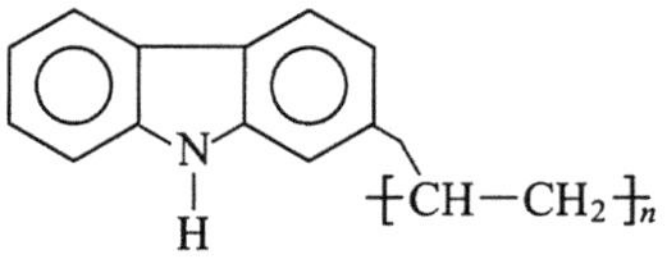

poly-2-vinyl carbazole

Subsequently, many hundreds of photoconductive polymeric compositions have been disclosed in the literature.

The excellent photoconductivities of PVK and related materials are generally thought to be a result of their ability to take up a helical conformation with successive aromatic side chains lying parallel to each other in a stack along which electron transfer is relatively easy. PVK absorbs ultraviolet light in the 360 nm region and forms an exciton state which ionizes in an electric field. The photogeneration efficiency of PVK can be greatly enhanced by the addition of an equimolar amount of the electron acceptor 2,4,7-trinitrofluorenone (TNF):

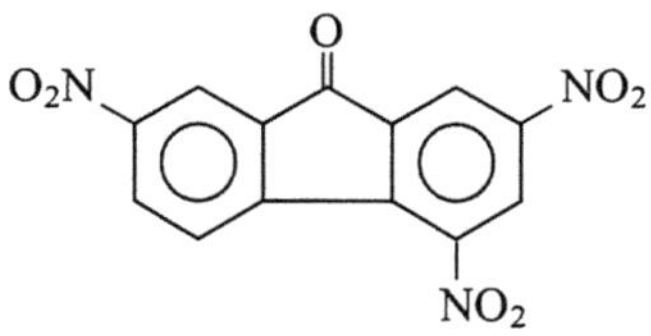

2,4,7-trinitrofluorenone

PVK is an insulator in visible light with a conductivity of around 10^{-14} S m^{-1} at 550 nm. The addition of TNF shifts the absorption of PVK into the visible by the formation of a charge transfer state, rendering it photoconductive at 550 nm [33]. PVK alone is a hole conductor, but the use of TNF creates electron carriers and in this case the conduction mechanism is actually electron-dominated.

Much work has been reported on the development and modification of poly-*N*-vinyl carbazole. Reported [34] carbazole derivatives include poly-*N*-epoxypropylcarbazole, poly-*N*-epoxy-3-bromocarbazole and poly-*N*-epoxy-propyl-1,3,6-tribromo-carbazole. All of these materials are sensitized well by pyryllium salts and they are recommended for the fabrication of organic electro-photographic layers. An important non-carbazole-derived photoconductive material based on a doped phosphene structure has also been reported [35].

Inorganic polymers also exhibit photoconductivity. For example; Poly[bis(2-naphthyloxy)phosphazene] is intrinsically an insulator with a very low photosensitivity towards both UV and visible radiation. However, when doped with TNF in a 1 : 1 molar ratio it becomes a strong photoconductor with a sensitivity similar to polyvinyl carbazole doped with TNF in the same ratio. Bis[catena-poly(methyl arsenic)] and poly[bis(*p*-tolylamino)phosphazene] also exhibit photoconductivity, as do the glow discharge polymerization products of tetramethylsilane, tetramethyltin and tetramethylgermanium. For more detailed insights into the current status of photoconductive polymers the reader is referred to the review by Stolka [36].

Photoconduction forms the basis of the electroreprographic industry and, although perhaps the most complex of the various conduction mechanisms in polymers, it is widely exhibited commercially. In the photocopying process, or xerography as it is sometimes known, a photoconductive material is coated onto a metal drum and uniformly charged in the dark by a corona discharge. The drum is then exposed to the bright image of the item to be copied and the illuminated areas of the photoconductive material become conductive and dissipate their charge to the earthed metal drum. The remaining photoconductor in the dark areas is still charged and able to attack a black positively charged resin-coated toner powder forming a latent image. The latent image is then transferred to a negatively charged piece of paper which is heated to fuse the resin, making the image permanent. Early photoconductive materials included selenium-based compounds such as As_2Se_3. This is difficult to handle because it has to be applied by vacuum sublimation and is rather brittle. Poly-*N*-vinyl carbazole based materials are now replacing the selenium based compounds. The major advantages of organic photoconductors include panchromaticity, flexibility, solution coating capability and low cost. Other reported applications include photothermoplastic imaging, holographic recording and optical switching devices.

5.4 Polymers in optoelectronics

With continued growth in the use of light in fields such as telecommunications and computing there has, for some time, been a need to identify, evaluate and use many new materials. Polymers have figured significantly in the development of optical technology both from a passive and from an active perspective.

In optical communications polymers find extensive use in passive applications, with examples including optical fibres, optical interconnection networks deposited over active electronic devices and graded refractive index materials.

Polymeric optical fibres are attractive because they possess a number of advantages over glass fibres. They are, for example, easier to handle and splice and are lighter and less expensive to produce. Although polymeric optical fibres are traditionally very 'lossy' compared with glass fibres, they have been improved significantly in recent years. The key to achieving lower losses lies in the selection of polymer, because in order to be highly transparent they must be of high purity and amorphous. Polymers used as cores in optical fibres include polymethylmethacrylate, polystyrene and certain highly amorphous polycarbonates.

In order successfully to utilise optoelectronics in information transmission and processing it is necessary to provide suitable interconnects between various components, devices and the transmission system. With the greatly increased circuit densities and data processing speeds encountered in state of the art technology there is a major opportunity to overcome some interconnect problems through the use of optical interconnect technology. It is in this area that there is perceived to be a great opportunity for polymeric integrated optics, and connectors, couplers, and polymeric waveguides have all been proposed.

Polymeric materials have also been used to produce so-called graded index materials in which the refractive index of the material varies continuously within the medium. These materials display very useful properties that are not attainable with conventional constant index materials, and many applications have been reported. They are normally produced by the copolymerization of two or more monomers with different refractive indices, and, by controlling the resultant polymer's composition in the axial, radial or spherical direction, controlled index polymers can be produced. The materials most commonly utilized in these applications are generally acrylates, acetates and styrene derivatives with various substituents added to modify their refractive indices.

In active applications there has been growing interest in materials that have so-called 'non-linear' optical properties which makes them of potential use in optical switching and waveguiding for telecommunications applications [37, 38]. Initially, work was largely focused on second-order processes occurring in inorganic dielectrics and semiconductors and nearly all of the non-linear optical materials commonly in use were inorganic solids. However, organic and polymeric solids offer exceptionally large non-linear second-order optical properties, a larger variety of asymmetric crystal structures and higher laser-beam-damage thresholds. Applications envisaged for organic and polymeric crystals include amplifiers, frequency doublers, waveguides, Q-switches (a method of building up power output in lasers) and filters.

Non-linear second-order optical properties include second harmonic generation and the linear electro-optic effect. Second harmonic generation is the ability of a material to double the frequency of light passing through it, whilst the linear electro-optic effect is a change in the refractive index of a material under the application of a low-frequency electric field. In order for these effects to occur, two conditions must be met. First, the material must not have a centre of symmetry, and, second, for maximum second harmonic generation a crystal should possess propagation directions where the crystal birefringence cancels the natural dispersion, leading to the condition of equal refractive indices at the fundamental and second harmonic frequencies. This is known as 'phase matching' and can lead to a high percentage of the incident light being converted to the second harmonic.

Polydiacetylene has been shown to exhibit large second-order and third-order non-linear optical properties when its monomer is polymerized to give highly conjugated liquid crystal polymers [39]. Such di-substituted diacetylene monomer units crystallize as linear arrays, where the substituent groups represent optically non-linear organic groups. Polymerization may be initiated by various methods, such as thermally or by UV and gamma irradiation, and the result is a crystalline polymer chain with long-range-three-dimensional order. To exhibit efficient second-order processes the polymer should have:

- optical transparency over the frequency range of the fundamental and second harmonic waves to avoid optical loss;
- long-range structural order;
- phase matching where the optical birefringence conceals the natural dispersion.

In addition to the π electron system of the substituent groups, the conjugated backbone of the polymer possesses a π electron band structure that in general exhibits optical transparency from approximately the middle of the visible region to the near infra-red, giving refractive indices from 1.6–2.0. The natural structural anisotropy of the ordered polymer chains leads to considerable optical birefringence important to phase matching.

Phase-matched second harmonic generation in single-crystal polymers was first observed in 2-methyl-4-nitroaniline substituted diacetylene polymers and has been clearly demonstrated to increase in the solid state as the degree of polymerization increases [40]. A number of other diacetylene structures have been synthesized recently and these exhibit phase-matched second harmonic generation orders of magnitude greater than that of lithium iodate.

As well as being used as optical quality single crystals, disubstituted diacetylene polymer thin films for waveguide applications can be obtained by evaporation, solidification or Langmuir–Blodgett deposition of the diacetylene monomers. This type of approach allows film thickness to be controlled at the molecular level. Another important feature of these diacetylene films is that they can be patterned by means of selective polymerization techniques such as developed by means of UV, X-ray and electron-beam lithography. For a more detailed review of this topic the reader is referred to the texts by Hornak [41] and Goodwin *et al.* [42].

5.5 Polymers with piezoelectric, pyroelectric and ferroelectric properties

5.5.1 Introduction

Electrical effects in materials have been observed over the centuries, with early examples including the discovery by ancient Greeks of the attractive power of 'electron' (amber) around 400 BC. The ability of tourmaline crystals to attract and repel hot ashes was also discovered at an early stage by the Indians and

Ceylonese and became known in Europe around 1700. Brewster observed the effect in a number of crystals and introduced the word 'pyroelectricity'. (Pyroelectricity may be defined as an inherent electrical polarization whose magnitude varies as a function of temperature). The Curie brothers discovered the piezoelectric effect in 1880 when they demonstrated that quartz and some other crystals showed opposite charges on portions of their surfaces when compressed in a particular direction. The charge was proportional to the pressure and disappeared when the pressure was released. (Piezoelectricity may be defined as electric polarization produced by mechanical strain, the polarization being proportional to the strain.) It was not until almost 40 years later that a practical application was proposed, and during the following decades a number of both natural and synthetic single crystals were also found to be piezoelectric. In 1947 it was discovered that a ceramic material, barium titanate, could be made piezoelectric by poling (a term used to describe methods of inducing polarization). A very strong effect was also found in lead zirconate titanate (PZT).

Piezoelectric properties of polymers have been investigated since the 1920s, and in 1941 a small but detectable piezoelectric effect was found in the natural protein, keratin. A major advance occurred in 1969 when Kawai discovered the strong piezoelectric effect in polyvinylidenefluoride (PDVF) [43] which is much greater than for other polymers. In 1971 the pyroelectric properties of PVDF were also first reported [44] and, as a consequence, considerable research and development has continued up to the present day.

5.5.2 Polyvinylidene fluoride

Polyvinylidene fluoride, $(CH_2CF_2)_n$, is a semicrystalline polymer (50%) consisting of a carbon backbone with alternate carbon atoms having either two hydrogen or two fluorine atoms attached. The polymer forms lamellar crystals typically 10 nm thick by 100 nm long dispersed in an amorphous phase which has supercooled liquid properties and a glass transition temperature of about 40°C. The molecule, which may extend through several crystalline and amorphous regions, can have three morphological forms known as α, β and γ PVDF. The preferred form for piezoelectric activity is the β phase, which has a large dipole moment, and the proportion of the phase present may be increased by special treatment during the extrusion process. PVDF films are often uniaxially or biaxially stretched or rolled, not only to convert one form to another but also to improve their mechanical and electrical properties. In order to impart piezoelectric properties to PVDF a number of processing steps have to be undertaken. The material is uniaxially stretched and annealed, and then poled either thermally or by a corona discharge procedure.

In the case of thermal poling, electrodes are depositied on the PVDF and it is subjected to a field of 0.3–1.4 Mv cm^{-1} at temperatures around 70°C–100°C for one hour. The effects of applied field and poling temperature upon the piezo coefficient are shown in Figs 5.3(a) and (b), respectively.

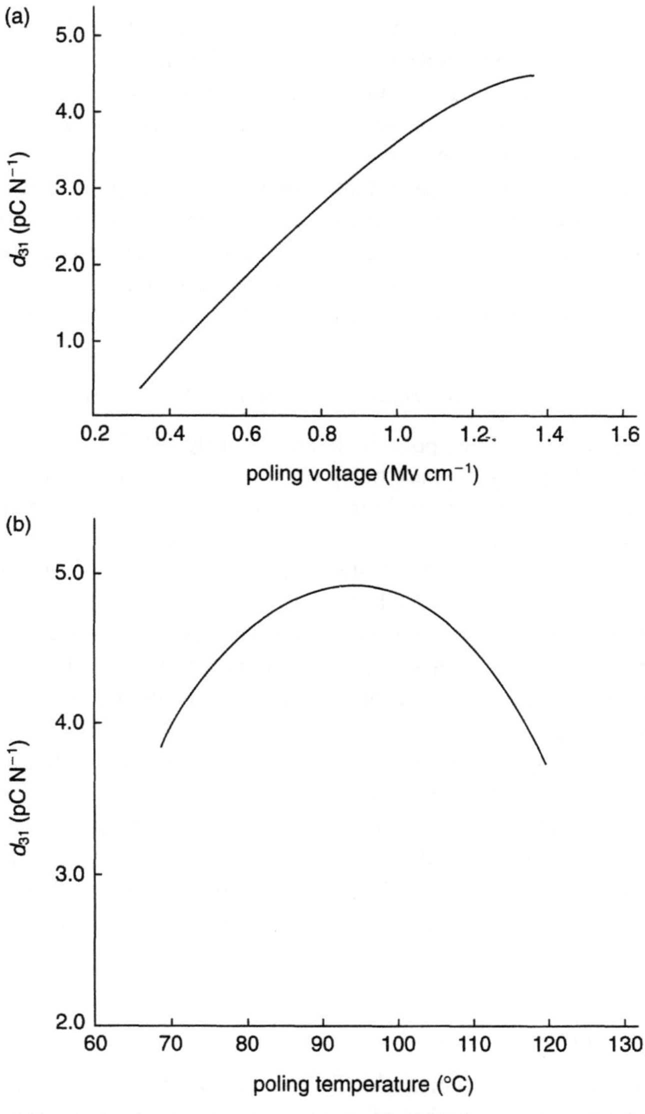

Figure 5.3 Variation of piezo coefficient, d_{31} with (a) applied field (poling temperature 90°C); and (b) poling temperature (field 1.2 Mv cm^{-1}). Source: Jacobs, J.E. and Renerly, S.A. *Infra-red Physics*, **19**, 1979.

For corona poling, the non-metallized or half-metallized sample is subjected to a corona discharge from a needle electrode a few centimetres away. A charge builds up on the film, generating a field in the sample which is sufficient to align the polar chains even at room temperature. With either method the dipolar alignment occurs by reorientation of the randomly orientated crystalline regions in the direction of the field. The orientation is caused by chain rotation about the

C—C chain axis rather than by reorientation of the crystallites which are interconnected by the long molecular chains and thus hindered in their rotation. The parallel orientation of these dipoles leads to a residual polarization in the film and this is the basis for several proposed mechanisms explaining its piezoelectric and pyroelectric behaviour.

A ferroelectric has the property that a single crystal is polar and its polarity can be reversed by applying a sufficiently high electric field. Most ferroelectrics show a phase transition at a temperature known as the Curie point, above which they become paraelectric, and although various observations support the suggestion that PVDF is ferroelectric its Curie point has not been detected.

5.5.3 Polyvinylidene fluoride–trifluoroethylene copolymers

As produced from the melt, polyvinylidene fluoride (PVDF) film is largely in the non-polar α form, the β phase being obtained only after subsequent processing operations. If, however, trifluoroethylene (TrFE) is copolymerized with vinylidene (VDF) fluoride in concentrations as little as 7% by weight, a copolymer is formed with crystallites completely in the β form [45]. This obviates the requirement for stretching after synthesis, although by doing so the activity may be enhanced somewhat. Similar solution cast films are also obtained in the β phase. Consequently, the copolymer is much more versatile than is PVDF because it can be processed by conventional methods such as injection moulding.

Unlike PVDF, copolymers of this type have been shown to demonstrate the ferroelectric-to-paraelectric transition [46] and it is thus possible to examine ferroelectric as well as piezoelectric and pyroelectric properties in undrawn samples. For a copolymer with a composition of 56% vinylidene fluoride and 44% trifluoroethylene, differential scanning calorimetry has indicated a phase transition near 70°C, whereas with 90% VDF the transition shifted to 130°C [47].

5.5.4 Other materials

Whilst PVDF and its copolymer with TrFE are the best known and most studied polymers of this type, a number of other materials exhibit piezoelectricity and pyroelectricity and are worthy of mention. If a polymer has a large dipole moment in its molecule that can be aligned to form a polar crystal there is every likelihood it will exhibit strong piezoelectricity and pyroelectricity. Polar polymers such as polyvinyl chloride, polyacrylonitrile and nylon 11 have relatively large dipole moments but they exhibit weaker piezoelectric activities than does PVDF because the dipoles cannot be aligned so well by an electric field. Polyvinylidene cyanide also has a large dipole moment but it is unfortunately unstable, undergoing degradation in contact with atmospheric moisture. In order to circumvent this problem, vinylcyanide has been polymerized with vinyl acetate to impart some degree of thermal and mechanical stabilization. The copolymer possesses a large dipole moment owing to the C—CNs of the vinyl cyanide but the C—CNs are also prevented from interacting by the isolating spacer effect of the vinyl acetate. This combination allows easy rotation of the dipoles, and strong piezoelectricity may be induced by poling. Vinylidene cyanide–vinyl acetate copolymers are reported to exhibit almost the same piezoelectric activity as PVDF over the temperature range 20°C–200°C [48]. This type of copolymer shows better acoustic impedance matching with body tissues in medical applications and a greater thermal stability than does PVDF because of its low density and high glass transition temperature [49]. It is thus a promising material for medical ultrasonic transducers.

Although trifluorethylene is a component of the ferroelectric (VDF–TrFE) copolymer, polytrifluorethylene (PTrFE) had not been recognized to carry a spontaneous polarization until its pyroelectricity was reported by Oka and Koiziumi [50]. The temperature dependence of the pyroelectric coefficient shows a rather anomalous behaviour when it is compared with other poling-induced pyroelectric polymers. With PVDF the pyroelectric coefficient increases with increasing temperature whereas that of PTrFE changes sign twice over the temperature range 100K–350 K, suggesting a more complicated dependence of the polarization upon temperature. Also, the pyroelectric activity is strongly enhanced in samples having low initial crystallinity, whereas higher degrees of crystallinity produce higher activity in PVDF.

5.5.5 *Applications*

Many applications have been proposed for these types of polymers, the principal material of interest being PVDF and its copolymer with TrFE. Polymers offer the advantages of mechanical flexibility, high shock resistance, low cost, availability in large sizes, low density and better chemical stability than conventional ceramic materials. The types of application can be broadly broken down into several convenient categories, namely: sound transducers in water and air, biomedical devices, pyroelectric sensors and electromechanical devices. Some brief examples are now given.

PVDF, with its low acoustic impedance and high piezoelectric coefficient, is ideally suited as a transducer for broad-band underwater receivers in hydrophones [51]. Compared with conventional ceramic materials, PVDF is only one-quarter the density, making it useful for lightweight hydrophone applications. Its softness and flexibility give it a compliance 30 times greater than ceramic. Various device configurations have been proposed to utilize PVDF in a hydrophone structure, including compliant tubes, rolled cylinders, discs and planar stacks of laminated material [52]. Marconi developed a 100-element 360° scanning sonar transducer utilizing PVDF, comprising 100 elements equally spaced around the circumference of a 300 mm ring. In operation a burst of sound is sent out and the array is sequentially scanned to provide a 360° view of the close acoustic impedance match between it and body tissues. Monolithic silicon–PVDF devices have been produced where a sheet of PVDF is bonded to a silicon wafer containing an array of metal oxide semiconductor field effect transistor (MOSFET) amplifiers [53]. The device is so arranged that when an acoustic wave is detected, the electrical signal resulting from the piezoelectric action in the PVDF appears directly on the gate of an MOS transistor. Because of its piezoelectric–oxide–semiconductor field effect transistor structure such a device is known as a POSFET. PVDF is generally an ideal material for transducers operating at frequencies above 0.5 MHz, and applications include hydrophones and pulse echo probes for medical and non-destructive testing. Kimura *et al.* [54] reported a 64 element linear array transducer operating at 5 MHz which offered a wide bandwidth pulse response, sharp ultrasonic field distribution and a high-energy-conversion efficiency.

While PVDF has piezoelectric properties similar to those of its ceramic analogues, the pyroelectric coefficient is too low to be useful except for specialized applications. Where the main requirements are speed and resolution, and where sensitivity is a minor consideration, PVDF has been used to form an array of 50 detectors for the energy-profile determination of various wavelength laser beams [55]. PVDF was also employed in a pyroelectric vidicon to give a high-resolution picture, because its low thermal conductivity reduces thermal spreading [56]. It is easier to fabricate than the transitional target material triglycine sulphate.

A considerable amount of interest has been expressed in the use of PVDF as a replacement for elements in telephone handsets. For a while, Plessey Telecommunications produced PVDF bimorph microphone units. PVDF has also been employed in high-fidelity tweeters and headphone transducers for audio applications [57]. More recently, a range of sports practice equipment has been developed in the US which incorporates PVDF sensors. In one example sensors are able to measure the reaction time and blocking power of a baseball pitcher's throw.

Photoresists for semiconductor fabrication

5.6.1 Introduction

Photoresists are radiation-sensitive materials that undergo a reaction upon irradiation such that differences in the exposed and unexposed areas can be exploited to allow selective processing and masking of an underlying substrate. They are typically coated onto a surface and exposed with ultraviolet light through a pattern or mask. Upon exposure, a solubility change occurs in the layer, rendering it more or less soluble than the unexposed regions. Photoresists are used in many fields for the production of semiconductor devices, printed circuit boards, other electronic components, engravings, print illustrations and small mechanical parts. The image formed by development of the resist after exposure produces a mask against etchants such as acids, alkalis and various plasmas.

There are many types of photoresist and they can function by a variety of mechanisms. Some typical mechanisms are:

- cross-linking of a linear polymer backbone by the photo-induced decomposition of a photosensitizer to generate active species;
- cross-linking of a polymer containing photosensitive groups within its own structure;
- polymerization of a monomeric material to yield a reduced-solubility higher molecular weight polymer;
- enhancement of solubility by a photoinduced molecular rearrangement;
- enhancement of solubility via a reduction in molecular weight by bond scission.

The origins of photoresist technology go back as far as the 1820s when Neipce is credited with the earliest recording of detail of such a system. He discovered that bitumen layers derived from India asphalt became insoluble to a mixture of mineral spirit and lavender oil after several hours exposure to sunlight. The bitumen contains unsaturated compounds that are destroyed by exposure to short wavelengths, causing a degree of cross-linking. Neipce also found that if he coated a metal plate it could be etched with acid where unprotected by the resist and used for printing. Since Neipce's discovery of photoresists they have found application in a wide range of industries. They have been used for many years in the electrical and electronics industries to produce printed circuit boards, but it is with the fabrication of silicon integrated circuits that photoresist technology has been taken to the limits of development. In the following sections I use the applications of photoresists in the electronics industry to illustrate some of the various types of polymeric photoresists available and to discuss the mechanisms of their operation.

The processing of semiconductor devices is made possible only by the use of polymeric coatings and their subsequent degradation or cross-linking by light. The manufacture of a functioning semiconductor device, such as the familiar microprocessors and memory circuits found in personal computers, is achieved by treating a substrate of silicon (or in some special cases gallium arsenide or germanium) in a number of etching, doping, passivation and interconnection steps. A typical integrated circuit (Fig. 5.4) consists of various patterned thin films of metals, dielectrics and semiconductors on the substrate, and these patterns, which are now routinely less than 1 μm wide, are formed using polymeric photoresists. The resist is used in a photolithographic process in which radiation of a suitable wavelength changes a polymer's molecular weight or solubility such that selective dissolution of either exposed or unexposed areas can be facilitated.

In microlithography a film of photoresist is spun on to the substrate at several thousand revolutions per minute, giving a dry film thickness in the range

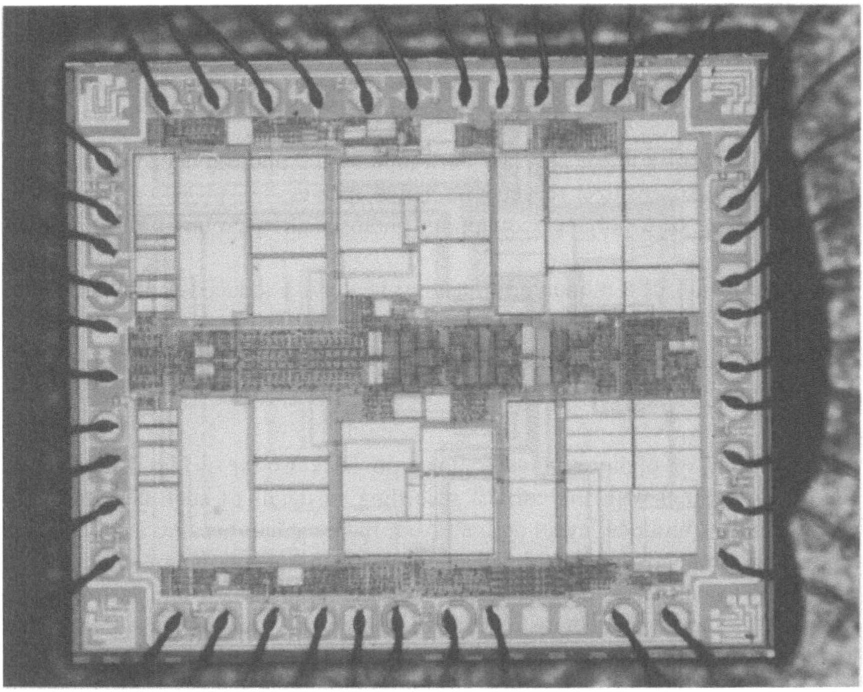

Figure 5.4 Typical integrated circuit surface: an SL6619 FSK data receiver made by GEC-Plessey Semiconductors (photograph courtesy of GEC-Plessey Semiconductors).

0.5–2.0 μm. The film is then exposed through a mask in a process analogous to exposure of photographic paper through a negative. A photoresist may be classified as either negative or positive working, depending upon the type of response induced by irradiation. A resist is negative acting if exposure causes cross-linking of the polymer and thus reduced solubility, or positive acting if exposure causes chain scission, degradation or a molecular rearrangement that leads to enhanced solubility. After exposure, the more soluble areas are removed with a suitable solvent in a process known as development, whilst the insoluble areas remain to protect selectively the substrate surface from subsequent etching, doping and sputtering operations. To be successful, therefore, a photoresist for semiconductor microlithography must not only have good solubility–insolubility characteristics but also good adhesion and resistance to the semiconductor processing conditions, etchants and dopants. A schematic representation of the lithographic process is shown in Fig. 5.5.

The majority of current microlithography for semiconductor processing is carried out using radiation in the near-ultraviolet wavelength range (350–450 nm). The principal wavelengths in use being the g and i line emissions of a mercury lamp at 405 nm and 365 nm, respectively. However, the increasing sophistication and degree of integration required in semiconductor fabrication

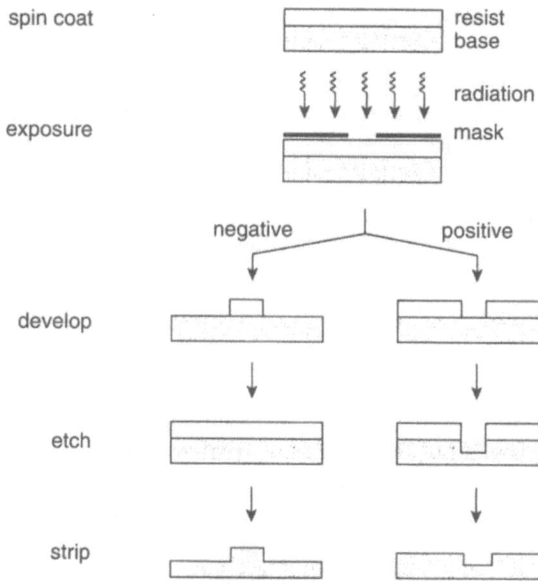

Figure 5.5 Lithographic processes.

now calls for features of 0.3 μm width or less to be produced. The ability to resolve such small dimensions is related to the wavelength of the radiation used and it is generally agreed that current near-UV lithography is becoming resolution-limited by its relatively long wavelength. By going to shorter wavelengths, sub-0.3 μm resolution is more readily achievable and there is currently much interest in using not only deep UV, for example 248 nm but also X-rays, ion beams and electron beams for exposing resists. The chemistry of the components used in the various types of resists reported above will now be introduced. For further information on the application of photoresists in semiconductor fabrication see the review by Clements [58].

5.6.2 Negative photoresists

Although there are numerous commercial photoresists available, they all tend to be derived from the same basic chemistry. A typical negative working photoresist is based on a cyclized or partially cyclized rubber. Natural rubber was initially used as the major component of these resists but has now been replaced with synthetic rubber prepared by the Ziegler–Natta polymerization of isoprene, which gives *cis* 1–4 polyisoprene structure (*cis* 1–4 polyisoprene is the major constituent of natural rubber). Poly (*cis* isoprene) has a low glass transition temperature and is too soft for use in lithography, so partial cyclization is performed to yield a polymer with good adhesion and film-forming properties. The azide-type photosensitizers used with this type of resist have good thermal stability and decompose efficiently upon irradiation to give highly reactive intermediates known as nitrenes. Mercury lamps are used to provide the ultraviolet exposure, and the azides can be tailored molecularly to have absorbance maxima at or near the mercury lamp emission lines. The nitrenes generated by the ultraviolet exposure undergo insertion reactions with carbon–hydrogen bonds to yield secondary amines, and with carbon–carbon double bonds to produce aziridines. The overall effect of the nitrene formation is to produce a cross-linked, reduced-solubility matrix.

Because of their very high reactivities, bisazide compounds have also been used with polymers other than the cyclized rubbers such as poly (parahydroxystyrene).

5.6.3 Positive photoresists

Positive photoresists have found wider use than negative resists because of their better resolution capability and higher thermal stability. As with the negative resists, there are various proprietary formulations available which are all variations on a basic theme. At the typical UV wavelengths used it is not possible to enhance solubility by photo-induced bond scission because the light is not energetic enough and consequently an alternative mechanism is utilized. Positive resists usually comprise a large photoactive molecule that is insoluble in basic solvents and also sufficiently bulky to inhibit dissolution of a base soluble polymer. During exposure the photoactive compound breaks down to yield a base soluble product, and the exposed region can be developed out using a basic solvent. The base soluble polymeric component is typically a novolac resin of relatively low molecular weight:

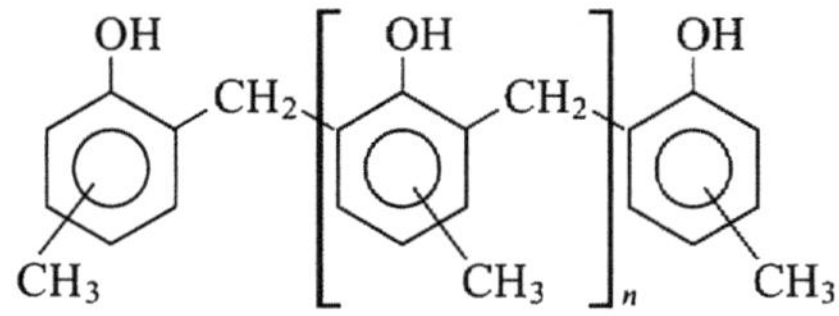

$n = 2 - 12$

Novolac resins are the condensation products of phenols and formaldehyde and are inherently soluble in alkaline solutions. In resists their dissolution is inhibited by the presence of the photoactive compound which is normally a 1–2 naphthoquinone diazide derivative. On exposure to UV this compound rearranges by nitrogen loss to give a ketene and finally an indene carboxylic acid:

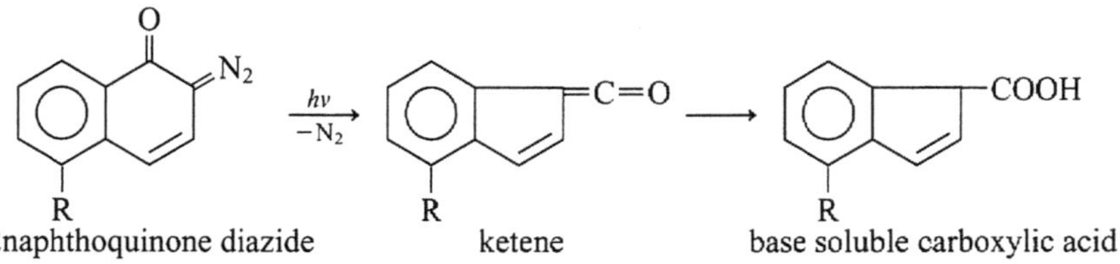

The carboxylic acid then readily dissolves in solutions containing metallic or quaternary ammonium hydroxides. In the unexposed areas, dissolution of the novolacs is prevented by the hydrophilic nature of the naphthoquinone diazide which often takes the form of a large polyfunctional derivative:

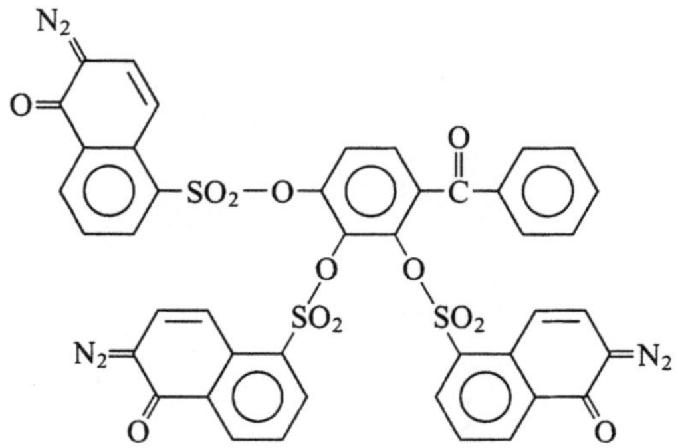

For further information on the chemistry of positive working resists see Hacker [59].

5.6.4 New developments in resist technology

With the demand for greater device integration and more memory the number of devices to be accommodated on a piece of silicon has increased dramatically, yet the device area has not been afforded a concomitant increase. Consequently, the size of features on a chip has continued to fall, and routine production is now in the 0.5 μm range for devices such as the 16 Mb DRAM. The next generation of DRAM will be 64 Mb and be based on the use of 0.35 μm lithography, whereas the 1 Gb DRAM scheduled for production early in the twenty-first century will need to have 0.15 μm features. One of the main factors influencing the resolution capability of photoresists is the wavelength of radiation used to expose them. At one time it was predicted that i line lithography would not be adequate for 0.5 μm lithography and therefore new shorter wavelength exposure sources and resists would be required to meet the needs of future device technology. Although this is definitely true, the capabilities of conventional i line resists have been extended to well below 0.5 μm and the ultimate limit now appears to be in the 0.25–0.3 μm region (Fig. 5.6). However, below this limit, smaller geometries will require new lithographic techniques and new types of resist. In this section I introduce briefly the chemistry of resists for use with the shorter wavelength deep UV, X-ray and electron beam exposure equipment.

Conventional i line lithography utilizes the mercury emission at 365 nm and, in order to give enhanced resolution, it is necessary to move to shorter wavelength exposure. One solution that has already been thoroughly investigated is the use of deep UV exposure at 248 nm from a KrF excimer laser source. At this

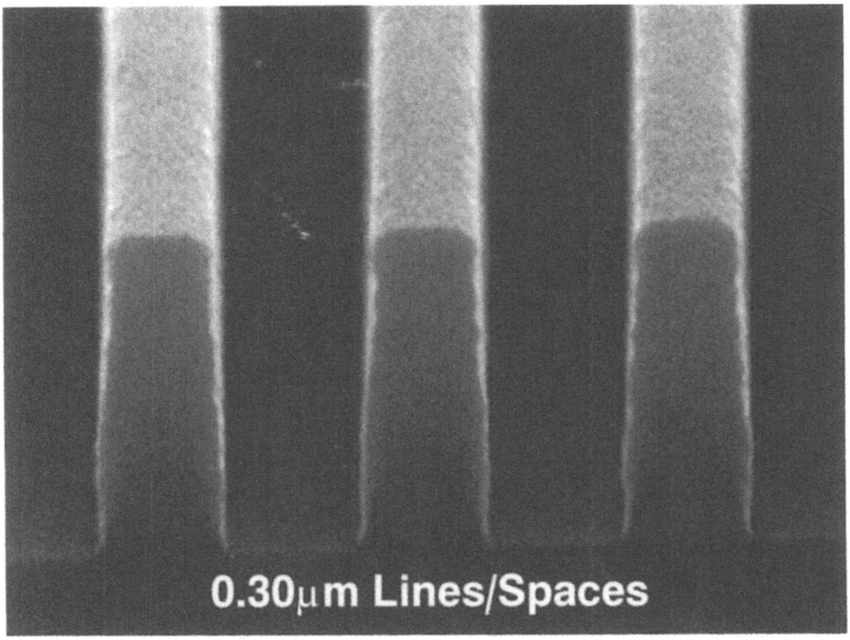

Figure 5.6 ~0.30 μm lines and spaces produced with Shipley high-performance i line resist.

wavelength the conventional positive photoresists described above cannot be used, and alternative 'chemically amplified' resists have been developed. In these formulations a photoacid generator is decomposed by exposure and a thermal, acid catalysed reaction gives the desired chemical transformations. An example of a positive type of resist utilises poly(*p*-vinyl)phenol that has had its solubility reduced by the introduction of appropriate additional protective chemical groups. Upon exposure, a photoacid generator gives a strong Bronsted acid which leads to the de-protection of the polymers and to enhanced solubility. A good introduction to the chemistry and operating principles of chemically amplified resists is given by Lamola *et al.* [60] and an example of 0.25 μm features produced with such a resist is shown in Fig. 5.7.

Below 248 nm the next preferred exposure wavelength is the 193 nm line from an ArF laser, and again resist systems specific to this wavelength have to be used. An example of a positive chemically amplified resist for this wavelength utilizes an alkyl sulphonium salt as the photoacid generator and a methacrylate terpolymer containing a tricyclodecanyl group [61]. This formulation possesses good transparency and thermal stability coupled with high acid generation efficiency. A formulation reported by Takahashi *et al.* [62] was based upon the use of adamantyl methacrylate. This had a high sensitivity and was able to produce 0.17 μm lines and spaces. Several articles give a good overview of the status of UV lithography [63].

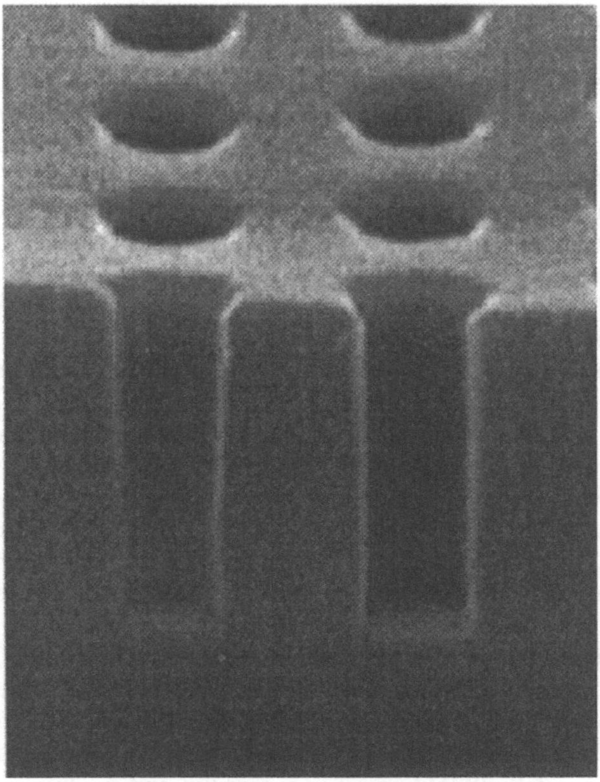

Figure 5.7 0.25 μm contact hole imaging produced with Shipley UVIIHS™ DUV. Positive photoresist.

The relentless move to smaller and smaller features has presented a dilemma for the electronics industry since it has been faced with the choice of continuing to develop shorter wavelength optical lithography or switching to alternative routes. The main alternative from a viable production perspective is the use of X-ray lithography. This technique was first investigated over 20 years ago but progress was impeded by the lack of adequate X-ray sources. In X-ray lithography a beam of soft X-rays with wavelengths of at most a few nanometers is used to expose a resist via a silicon or silicon carbide mask: there are no lenses as in conventional optical lithography. Typical X-ray wavelengths are the 0.43 nm and 0.46 nm lines emitted by Pd and Rh sources and these are strongly absorbed by bromine. Consequently brominated resists have been used successfully as X-ray resists [64]. In addition to absorbing strongly, X-ray resists must also be reactive. Thus ferrocene, although being a strong X-ray absorber, has found little use as a resist component because of its low reactivity.

The basic materials used for the formulation of X-ray resists are based upon those in other lithographies and include polymethyl methacrylate (PMMA) and related polymers. Much of the work to develop resists for X-ray lithography has been focused upon making these materials absorb more strongly. For example, the halogens have relatively high X-ray cross-sections and so materials such as poly-dimethyltetrafluoro propyl methacrylate, poly-2-fluoroethyl methacrylate and poly-2-chloroethyl methacrylate have been evaluated. It is worth noting that 0.5 μm CMOS circuits have been produced entirely via the use of synchrotron radiation and conventional positive and negative i line resists. More recently, chemically amplified resists have also been investigated and these appear to be the only materials capable of giving the required speed, dry etch stability and resolution. Further insights into resist materials and processes for X-ray lithography are given by Seeger and by Tolfree [65, 66].

An alternative approach to the use of X-rays utilizes an electron beam to irradiate the resist and is known as electron beam lithography. Electron beam lithography is also of much shorter wavelength than standard microlithography and therefore offers greater resolution capabilities. Exposure of the resist can be performed with a beam of electrons whose position is controlled by a computer driven beam deflector, obviating the need for a mask. As with conventional lithography, positive and negative acting resists have been formulated. The shorter wavelength and higher energy of the electrons means that covalent bond scission can now be accomplished and thus polymer degradation by exposure to the beam forms the basis for a positive photoresist system. PMMA and its analogues such as poly(hexafluorobutylmethacrylate) have been widely studied as positive resists. Upon exposure to an electron beam PMMA undergoes extensive main chain scission, although depolymerization back to the monomer does not occur. Unfortunately, PMMA is not very sensitive and so other materials have been more successfully utilized. Polyolefin sulphones were investigated by Bowden [67] and found to be extremely sensitive positive resists because the electron beam caused chain scission back to the olefin and SO_2.

Polymers used for negative working electron beam resists usually contain reactive cross-linking groups such as epoxy or allyl groups, and a typical example is a copolymer of glycidyl methacrylate and ethyl acrylate. Polystyrene and its derivatives such as poly(chloromethyl styrene) have also been employed as electron resists and although polystyrene itself is not sensitive enough, the incorporation of chlorine has a marked effect in improving sensitivity. Thus a copolymer of styrene (80%) and 4-chlorostyrene (20%) has a factor of 10 times greater sensitivity than polystyrene alone. It is now commercially available. Coploymers of glycidyl methacrylate and 3-chlorostyrene have also been investigated [68] and shown to have high sensitivity, good adhesion and plasma resistance. Again these materials are commercially available.

The main disadvantages of electron-beam lithography are that it is very slow and very expensive and it is thus unlikely to find extensive use in volume device production. It has, however, been used in the commercial fabrication of application specific integrated circuits (ASICs) as well as for the fabrication of masks for both optical and X-ray lithography.

It can be seen from the above that there is no longer a single way forward for semiconductor lithography and various possibilities are available for future device manufacture. These include deep UV, X-ray and, for a while at least, extensions of existing i line technology. Which, if any, becomes the preferred route remains to be seen. For further overview of the lithographic possibilities for manufacturing future semiconductor devices see Broers [69].

5.7 Summary and conclusions

In this chapter I have given a brief introduction to polymers with special electrical and electronic properties, including intrinsically conducting polymers and polymers with ferroelectric properties. I have also included details of polymers and polymer formulations that exhibit special properties that make them of potential interest in electronic and optoelectronic applications. The intention has been to give initial insights into just a few of the many special polymer materials associated with current electrical and electronic devices. There is no doubt that today's sophisticated electronics could not exist without the use of these special polymers. The relentless progress in semiconductor fabrication will, for example, ensure that many new polymer materials and processes are needed for the fabrication of future microprocessors, memory devices and all kinds of optoelectronics. The future for polymers in this area is guaranteed to be an interesting and exciting one.

References

1. David, R.F.S. (1975) *Solid State Technol.*, **40** (September).
2. Marshall, C. (1978) *Electrocomponent Sci. Technol.*, 5, 171.
3. Estes, R.H. and Kulesza, F.W. (1982) *Int. J. Hybrid Microelectron*, **5**(2), 336.
4. Walton, T.R. (1987) *J. Appl. Polym. Sci.*, **33**, 971.
5. Kaiser, A.B. (1991) *Synth. Met.*, **45**, 183.
6. Anderson, R. and Roth, S. (1994) *Brazilian Journal of Physics*, **24**, 3.
7. Chiang, C.K. (1978) *J. Chem. Phys*, **65**, 9058.
8. Bargon, J., Mohmand, S. and Waltman, R.J. (1983) *Mol. Cryst. Liq. Cryst.*, **93**, 279.
9. Chien, J.C.W., Wnek, G.E., Karasz, F.E. and Hirsch, J.A. (1981) *Macromol.*, **14**, 479.
10. Baker, G.L. and Bates, F.S. (1984) *Macromolecules*, **17**, 2626.
11. Galvin, M.E. and Wnek, G.E. (1985) *Polym. Bull.*, **13**, 109.
12. Simisnesu, C., Dimitrescu, S. and Percee, V. (1978) *J. Polym. Sci. Polym. Symp.*, **64**, 209.
13. Ballard, D.G.H., Curtis, A., Shirley, I.M. and Taylor, S.C. (1983) J. *Chem. Soc., Chem. Comm.*, **17**.
14. Ivory, D.M., Miller, G.G., Sowa J.M., *et al.* (1979) *J. Chem. Phys*, **73**, 3 August, 1506.
15. Wessling, R.A. and Zimmerman, R.G. (1968) *US Part.*, **3401**, 152.
16. McDonald, R.N. and Campbell, T.W. (1960) *J. Am Chem. Soc.*, **82**, 4669.
17. Wnek, G.E., Chien, J.C.W., Karasz, F.E. and Lellya, C.P. (1979) *Polym. Commun.*, **20**, 1446.
18. Murase, I., Ohnishi, T., Noguchi, T. and Hirooka, M. (1984) *Polym. Commun.*, **25**, 327.
19. Diaz, A.F. Kanazawa, K., Castillo, J.I. and Logan, J.A. (1981) In *Conductive Polymers* (ed. R. B. Seymour), Plenum Press, New York, p. 149.
20. Young, R.J. (1984) *Plastic Rubber Internal*, **9** (1 February), 29.
21. Wynne, K.J. and Street, G.B. (1985) *Makromolecules*, **18**, 2361.

22. Murthy, A.S.N., Shri, Pal. and Reddy, K.S. (1984) J. *Mat Sci. Letters*, **3**, 745.
23. Nazzal, A.I. and Street, G.B. (1985) *J. Chem. Soc., Chem. Commun.*, 375.
24. Genies, E.M. *et al.* (1990) *Synth. Met.*, **13**, 193.
25. Cao, Y. and Heeger, A.J. (1992) *Synthetic Materials*, **52**, 193.
26. Acompora, L.A., Daggar, D.L., Emma, T., *et al.* (1984) *Polymers in Electronics* (ed. T. Davidson), ACS Symposium Services, **242**(36), 462.
27. Vikiparta, K. (1996) *Compliance Engineering* (*European Edition*), **XIII**(2), 47.
28. Plastic Technology (1991) *Synth. Met.*, 15.
29. *European Chemical News*, (1991) 22 April 38.
30. Jonas, F. and Wolf, G. (1995) *US Patent* 5403467, 4 April.
31. Allardyce, G., Bass, K., Graves, J. and Shelnut, J. (1995) US Patent 5415762, 16 May.
32. Kodoh, Y., Tsuchiya, S. Kojima, T., *et al.* (1991) *Synth. Met.*, **41–43**, 1133.
33. Hoegel, H., Barchietto, G. and Tar, D. (1972) *Photochem. and Photobiol.*, **16**(4), 35.
34. Undzenas, A.J., Perelman, L.A., Ruskys, D.D. and Popov, A.F. (1984) *Sov. Phys Collect.* (*USA*), **24**(1), 73.
35. Di Marco, P.G., Giro, G., Gleria, M. and Lora, S. (1986) *Thin Solid Films* **135** (2 January), 157.
36. Stolka, M. (1995) In *Special Polymers for Electronics and Optoelectronics* (eds J. Chilton and M. Goosey), Chapman & Hall, London, Ch. 8. ISBN 0412 58400X.
37. Williams, D.J. (ed.) (19..) *Nonlinear Optical Properties of Organic and Polymeric Materials*, ACS Symp. Series 233, ACS, Washington, DC.
38. Zyss, J. (1985) *J. Mol. Electron.* **1**, 25.
39. Bloor, D. and Chame, R.R. (eds) (19..) *Polydiacetylenes*, Martinus Nijhoff, Dordrecht.
40. Garito, A.F., Singer, K.D., Hayer, K. *et al.* (1980) *J. Opt. Soc.*, **70**, 1399.
41. Hornale, L. (ed.) (1992) *Polymers for Lightwave and Integrated Optics: Technology and Applications*, Marcel Decker, New York.
42. Goodwin, M., Bloor, D. and Mann, S. (1995) In *Special Polymers for Electronics and Optoelectronics* (eds J. Chilton and M. Goosey), Chapman & Hall, London, Ch. 4.
43. Kawai, H. (1984) *Jap. J. Appl. Phys*, 8, 975.
44. Bergman, J.G., McFee, J.H. and Crane, G.R. (1971) *Appl. Phys Lett.*, **18**, 203.
45. Doll, W. and Lands, J. (1970) *J. Marcomol. Sci.* B 4, 897.
46. Lovinger, A.J., Davis, G.T., Furakawa, T. and Broadhurst, M.G. (1982) *Macromolecules*, **15**, 323, 329.
47. Fuakawa, T., Johnson, G.E., Blair, H.E., *et al.* (1981) *Ferroelectrics*, **32**, 61.
48. Miyata, S.A. Yoshikawa, M., Tasaka, S. and Ko, M. (1980) *Polymer J.* **12**, 857.
49. Tasaka, S., Miyasato, K., Yoshikawa, M., *et al.* (1984) *Ferroelectrics*, **57**, 267.
50. Oka, Y. and Koizumi, N. (1983) *Jap. J. Appl. Phys*, **22**, 281.
51. Woodward, B. (1977) *Acustica*, **38**(4), 264.
52. Shotton, K.C., Bacon, D.R. and Quillian, R.M. (1980) *Ultrasonics*, **123** (May).
53. Swartz, R.G. and Plummer, J.D. (1979) *Trans. Instr. Elect. Electron. Devices*, **26** November.
54. Kimura, K. Hashimoto, N. and Ohigashi, H. (1983) *Trans. Instr. Elect., Sonics and Ultrasonics*, **SU 32** (July), 4.
55. Gallantree, H. and Quillian, R.M. (1976) *Marconi Reva*, **39**(4), 189.
56. Jacobs, J.E. and Remely, S.A. (1979) *Infra-red Phys*, **19**, 1.
57. Tamura, M., Yamaguchi, T., Oyaba, T. and Yoshimi, T. (1975) *J. Audio Engng. Soc.*, **23**, 21.
58. Clements, S. (1985) In *Plastics for Electronics* (ed. M.T. Goosey), Elsevier Applied Science, London.
59. Hacker, N. (1995) *Processes Photoreact. Polym.* (eds V. Krongautz and A. Trifanuc), Chapman & Hall, New York, pp. 368–403.
60. Lamola, A.A., Szmanda, R.C. and Thackeray, J.W. (1991) *Solid State Technology*, p. 53. (August).
61. Nakano, K., Maeda, K., Iwasi, S., *et al.* (1995) *Proc. SPIE-Int. Soc. Opt. Eng.*, 2438 (Advances in Resist Technology and Processing XII), 433–44.
62. Takahishi, M., Takechi, S., Kaimoto, Y., *et al.* (1995) *Proc. SPIE-Int. Soc. Opt. Eng.*, 2438 (Advances in Resist Technology and Processing XII), 422–32.
63. *European Semiconductor* (1996) **8**(4), 61–5.
64. Yamsaka, T., Tsunoda, T. and Goto, Y. (1979) *Photog. Sci. Engng.*, **23**, 196.
65. Seeger, D. (1993) *IBM J. Res. Develop*, 435, **37**, 3.

66. Tolfree, D.W.L. (1996) *Materials World*, **4**, 189.
67. Bowden, M.J. (1981) *Solid State Technol.*, **73** (June).
68. Thompson, L.F. and Doerries, E.M. (1979) *J. Electrochem. Soc., Electrochem. Sci. and Technol.*, **126** (10), 1699.
69. Broers, A, (1992) *New Scientist* 18 April, 23.

6 Ionic polymers

B.N. HENDY

6.1 Introduction

Ionic polymers contain ions which are chemically bound within their structure. The term is not meant to apply to polymers that have been made by ionic methods, nor to those that, fortuitously, have ionic end groups. We will discuss mainly the types which fall within the scope of conventional organic polymer chemistry. Ionic polymers are specialist materials, some of which have limited, though important, commercial applications. Many are modifications of conventional polymers but a few are uniquely designed structures. Those containing few ions are thermoplastics called **ionomers**; those containing many ions are water-soluble and called **polyelectrolytes**.

6.2 Classification

Ionic polymers may be classified according to the nature of their bound ions, specifically the type of ion, its position within the structure and the amount present along a given length of polymer chain. They may be further classified according to the nature of the counter ion or by the nature of the supporting polymeric backbone.

6.2.1 Type of bound ion

The bound ion may be either an anion or a cation (Table 6.1). Most commonly it is the former but both types can occur together and the polymer is then said to be **ampholytic**. In each case the polyion may be strong or weak according to the degree to which it ionizes.

Table 6.1 Examples of bound ions

	Nature of polyion		
Type of polyion	Strong	Intermediate	Weak
Anion	Sulphonate $-SO_3^-$	Phosphonate $-PO_3H^-$	Carboxylate $-CO_2^-$
Cation	Quaternary $-NR_3^+$		Amine $-NH_3^+$

6.2.2 *Position of the bound ion*

The bound ion can be **pendant** to the polymer's covalent backbone or it can be **integral** or enchained. Pendant ions, such as in a polysulphonate, are the most common. Ionenes have integral ions:

polysulphone ionene

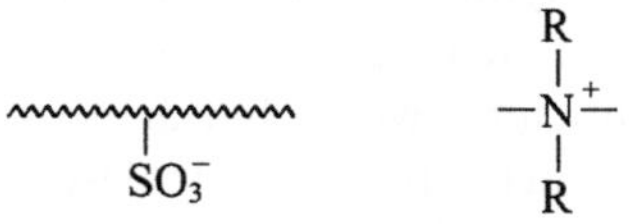

pendant integral

6.2.3 *Amount of bound ion*

The polymer may contain a few bound ions per unit mass or it may contain many. The ion content, or change density, may be expressed by relating the number of ionic bonds to that of the covalent network bonds thus leading to a scale of values ranging from zero for compounds such as polyethylene to 100 for ionic crystals such as sodium chloride, but it is more informative to quote the equivalent weight of the parent polyacid or polybase. It is particularly useful to express the ionic content as being the number of equivalents in each kilogram of the polymer. A copolymer consisting of an ionic monomer unit with a non-ionic is frequently characterized by the amount of the ionic unit it contains as a percentage. Conventional organic polymers can be modified with ions to cover a large range of ion contents, as shown in Table 6.2.

6.2.4 *Type of counter ion*

The counter ions neutralize the charge on the bound ions and may be grouped into three types: (a) univalent, (b) divalent or trivalent and (c) polymeric. The properties of an ionic polymer are profoundly affected by which type of counter ion it contains. (Section 6.4.2) Polymers with polymeric counter ions are often

Table 6.2 Ionic contents of organic ionic polymers

Designation	Chemical type	Formula	Mol %	Weight %	Equivalents (kg^{-1})
Ionomer	copolymer	$-(CH_2CH_2)_{49}C(CH_3)(COO^-)-$	2	6	0.7
Polyelectrolyte	homopolymer	$-CH_2CH(COO^-)-$	100	100	14

called **polyelectrolyte complexes, polysalts, polyion complexes, simplexes**, or **coacervates**.

6.2.5 The backbone

Almost any conventional polymer can be modified with ions to form an ionic polymer of the pendant type. Ionic polymers of the integral type, such as the ionenes, are specialized structures whose backbones can exist only in the ionic form. In either case the polymer may be thermoplastic or elastomeric, and it may have other features such as cross-linking or branching.

6.3 Synthesis

Specific syntheses will be described later but certain generalities will be discussed here. Ionic polymers can be synthesized by any of the methods applicable to conventional polymers but the choice of method will be influenced as much by the ionic function as by the type of backbone. The methods can be divided mainly into three types:

- direct synthesis;
- post-functionalization of a standard preformed polymer;
- post-functionalization of a special preformed polymer.

Direct synthesis is often used for polyelectrolytes. Typically, an aqueous solution of an ionic vinyl monomer (e.g. acrylic acid) is homopolymerized in aqueous solution. However, direct synthesis of less ionic polymers, which involves copolymerizing ionic with non-ionic monomers, is often not so straightforward because the two types of monomer are so different in their physical and copolymerization characteristics. Copolymerizations in aqueous emulsion or suspension can, for instance, be spoiled because the ionic monomer separates from its non-ionic comonomer by dissolving in the aqueous phase. Examples of direct copolymerization include ethylene with small amounts of methacrylic acid in the gaseous phase and butadiene with small amounts of methacrylic acid in aqueous emulsion at low pH to suppress the solubility of the acid monomer.

Post-functionalization of standard preformed polymers is a frequently used synthetic method, provided that the polymer is reactive. Examples are the sulphonation of polystyrene and the grafting on to polybutadiene of thioglycollic acid using free radicals. In some post-functionalization reactions there can be problems in achieving a uniform fully reacted product. This problem arises because the covalent polymeric substrate, the reactant and the ionic polymeric product have very different solubilities so that care is required to achieve a homogeneous reproducible reaction.

Many polymers are insufficiently reactive to be post-functionalized, or are damaged by the reaction conditions, and by-products may be formed which cannot be removed from the product. Under these circumstances it is common practice to incorporate reactive units into the polymer by copolymerization. Examples of such precursor copolymers which are described later (Section 6.9) are the sulphonatable ethylene–propylene rubber and poly (arylene-ether sulphone) (e.g. Victrex™). By using special post-functionalizable copolymers such as this, one can also sometimes avoid the difficulties mentioned above of copolymerizing ionic and non-ionic monomers. To this end there are many examples where carboxy polymers are formed by hydrolysing copolymers containing acrylate esters, acrylonitrile or maleic anhydride. As described later (Section 6.8), tetrafluoroethylene copolymerized with a sulphonyl fluoride is hydrolysed to form a sulphonic acid ionomer (Nafion™).

Carboxy polymers are probably the most common ionic polymers of all. Those which are used commercially are often made by the direct method and obtained in the acid form which is afterwards neutralized. Another common type are the polysulphonates, but they are generally made by sulphonating preformed polymers under mild conditions to avoid degradation and sulphone formation. Examples of sulphonating agents are: the complex of sulphur trioxide and triethyl phosphate, mixtures of concentrated sulphuric acid and acetic anhydride (acetyl sulphate), and concentrated sulphuric acid (which sometimes can also serve as the solvent). Alkyl sulphonate groups can be introduced by using sodium sulphite as in the case of the polyether from epichlorodryin:

$$-\mathrm{OCH_2}\underset{\substack{|\\ \mathrm{CH_2Cl}}}{\mathrm{CH}}- \xrightarrow{\mathrm{Na_2SO_3}} -\mathrm{OCH_2}\underset{\substack{|\\ \mathrm{CH_2SO_3Na}}}{\mathrm{CH}}- + \mathrm{NaCl}$$

The direct synthesis of polysulphonates is unusual, but free radical copolymerizations of various vinyl monomers with ethylene sulphonic acid and styrene sulphonic acid in aqueous and organic solvents have been studied. The direct synthesis of phenol-formaldehyde resins containing pendant methylene sulphonate is an old practice which apparently is now outmoded.

Cationic polymers are less common, though equally important. Pendant cations are usually of the quaternary ammonium type and made by a multi-step post-functionalization wherein a precursor containing labile chloride is reacted with a tertiary amine:

$$\sim\sim\underset{\substack{|\\ \mathrm{CH_2Cl}}}{\sim}\sim\sim \xrightarrow{\mathrm{N(CH_3)_3}} \sim\sim\underset{\substack{|\\ \mathrm{CH_3\overset{+}{N}(CH_3)_3}}}{\sim}\sim\sim + \mathrm{Cl^-}$$

Sometimes the inverse of this is done when a preformed polyamine is functionalized with an alkyl halide:

$$\sim\sim\underset{\substack{|\\ \mathrm{SO_2.NH(CH_2)_3.N(CH_3)_2}}}{\sim}\sim\sim \xrightarrow{\mathrm{CH_3I}} \sim\sim\underset{\substack{|\\ \mathrm{SO_2.NH(CH_2)_3.\overset{+}{N}(CH_3)_3}}}{\sim}\sim\sim + \mathrm{I^-}$$

6.4 Physical properties and applications

Synthetic ionic polymers span a vast range of physical properties but they have three distinctive features which dictate their usage: (a) ionic cross-linking; (b) ion-exchange capability; and (c) hydrophilicity.

6.4.1 Ionic cross-linking

Ionic cross-linking expresses itself as a kind of phase separation on the molecular level. This arises because the ions are held by chemical bonds within a low-dielectric medium with which they are incompatible, and they force the polymer backbone into conformations which allow them to associate together. This situation is similar to the behaviour of simple amphipathic molecules which, when placed on a liquid surface as in a Langmuir trough, form structured layers, or, when dispersed in the liquid, form micelles. Another similar situation arises with block copolymers where blocks of one type associate together to the exclusion of the other. What distinguishes ionic copolymers is that the force driving the separation is so great that separation happens even though the ions are dispersed uniformly along the backbone and not in structural blocks. Because these ionic associations involve ions from different chains they behave as cross-links; however, because they are thermally labile they can be broken on melting. Ionomers therefore behave as cross-linked, melt-processable thermoplastics, or, if the backbone is elastomeric, as thermoplastic rubbers. This is similar to the behaviour of some commercial block copolymers.

Two types of ionic aggregates are found in ionomers, called **multiples** and **clusters**. These two coexist in equilibrium within the matrix of covalent backbone material, which for an ionomer is the major component. The multiplets are very small and very numerous and are visualized as associations of ion pairs up to eight in number which are purely ionic and devoid of trapped segments of the covalent backbone. The clusters are large associations of these multiplets between which are trapped segments of the backbone. Many types of physical measurement provide evidence for these structures, the important techniques being small-angle X-ray scattering (SAXS), and infra-red and Raman spectroscopy.

Figure 6.1(a) is the X-ray scattering pattern of low-density polyethylene homopolymer, the peak at 20° (wide angle) is the result of polyethylene being 50% crystalline. Figure 6.1(b), which is that of polyethylene copolymerized with sodium methacrylate, is similar but has an additional peak at the relatively small angle of 4° arising from some repetitive structural feature of the ions with a periodicity of about 20 Å. That this peak is caused specifically by the ions and not by the comonomer is confirmed by its absence when the unionized acid form (—COOH) of the copolymer is tested [Fig. 6.1(c)].

The infra-red and Raman spectra of a number of ionomer types supports the theory of microphase separation into multiples and clusters. For example, the

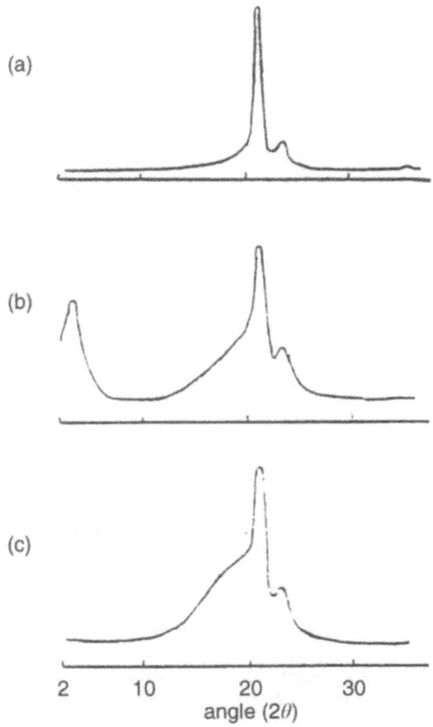

Figure 6.1 Polyethylene and its ionomer: X-ray diffraction of (a) low-density polyethylene; (b) poly(ethylene-co-sodium methacrylate) 5.8 mol%; and (c) poly(ethylene-co-methacrylic acid) 5.8 mol%. Source: Wilson, F.C., Longworth, R. and Vaughan, D.J., *Am. Chem. Soc. Divn. Poly. Chem., Polym. Preprents*, **9**, 505–14, 1968.

Raman spectra of a series of polystyrene ionomers containing 0–13.6 mol% of sodium *p*-styrene carboxylate shows distinctive bands which are ascribed separately to multiplets and clusters and which are absent from the acid forms of the copolymers, or from the homopolymer itself. From the relative intensities of these bands, the relative concentrations of ion pairs in the two species have been determined (Fig. 6.2). As the ion content builds up from zero, multiplets form but very quickly reach a saturation limit, thereafter growth of clusters is observed but there is no further increase in multiplets. The tendency to form clusters, and the multiplet solubility limit, varies according to the types of backbone and ions involved; backbones of low polarity such as poly(ethylene-co-sodium methacrylate) have a limit of 1 mol% whereas those of high polarity, such as sodium polysilicate, have a limit of about 20 mol%.

The phenomenon of ion aggregation affects the behaviour of all ionic polymers but it is with the slightly ionic types, the ionomers, where the effect is exploited to advantage to produce commercial, melt-processable, specialist thermoplastic materials. With highly ionic polymers, the polyelectrolytes, the

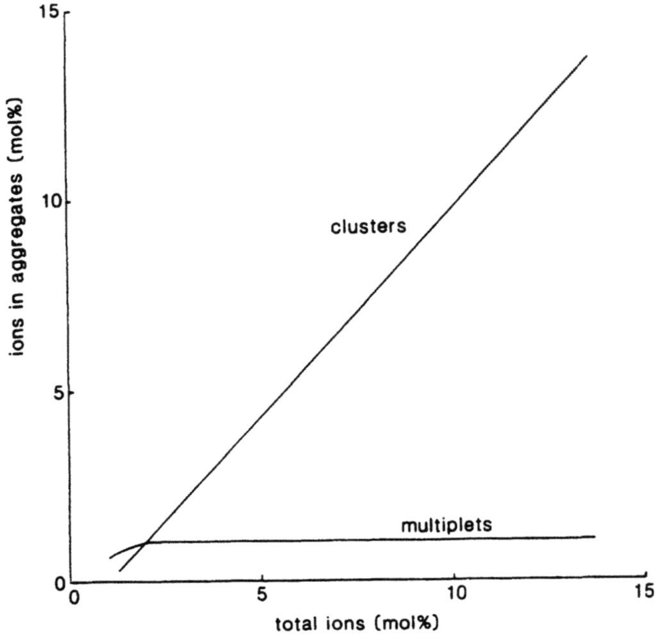

Figure 6.2 Distribution of ionic aggregates in poly(styrene-co-sodium-*p*-styrene carboxylate) by Raman spectroscopy. Source: Neppel, A., Butler, I.S., Brockman, N. and Eisenberg, A., *J. Macromol. Sci.-Phys.*, **B19**(1), 61–73, 1981.

ionic cross-linking is so extreme that the polymers decompose on melting or are too viscous for use as thermoplastics. Ionic aggregation profoundly affects most physical properties and some aspects of this will be discussed later but as an example and because of the ubiquitous nature of the glass transition temperature (T_g) the effect of ionic aggregation on (T_g) will be discussed.

(T_g) [1, 2] is a kind of melting transition which occurs when amorphous matter is heated, and is important because most polymers are partly, if not wholly, amorphous. What happens at the (T_g) is that the segments of the tangled polymer chain become suddenly free to rotate, so transforming the amorphous solid from a brittle, glassy, frozen liquid to a rubbery, highly-viscous liquid. In general with polymers it is observed that structural features that raise the (T_g) are those that interfere with segmental mobility, such as bulky side-groups and cross-linking; not surprisingly, ions also raise the (T_g), partly because of their bulk but mainly because they aggregate. The magnitude of the effect depends on the concentration of the ions and their nature, and is in the range 2°C–10°C per mol% of ionic comonomer. When the (T_g) is plotted against ion content, the curve often shows a discontinuity which is ascribed to the multiplet–cluster transition (Fig. 6.3). Studies of a number of fully ionized polyanions have led to the relationship (T_g) $\propto cq/a$, where c is the concentration of the ions, q is the

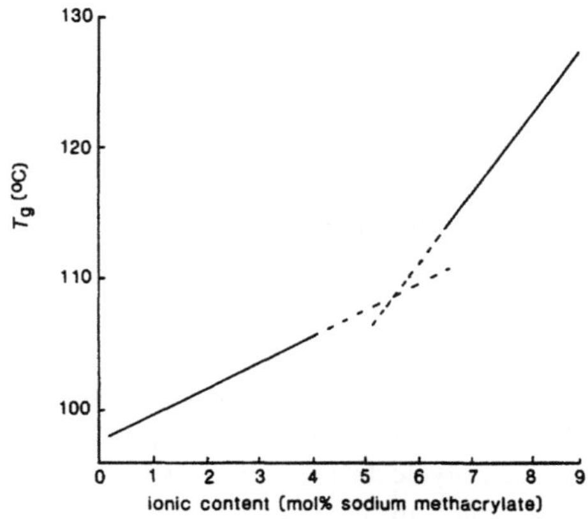

Figure 6.3 Glass transitions of poly(styrene-co-sodium methacrylate) Ionomers. Source: Eisenberg, A. and Trepman, E., *J. Polym. Sci.: Polym. Phy. Edn.*, **16**, 1381–7, 1978.

charge on the cation and a is the sum of the ionic radii. Straight-line relationships have been obtained for a number of polymer types. Partly neutralized carboxylates do not fit this relationship well, however, because accompanying the strong ionic cross-link is the weaker hydrogen-bonded cross-link of the undissociated acid groups which also raises the (T_g). Sulphonate ions generally raise the (T_g) more than carboxylate ions do because they form stronger ionic associations.

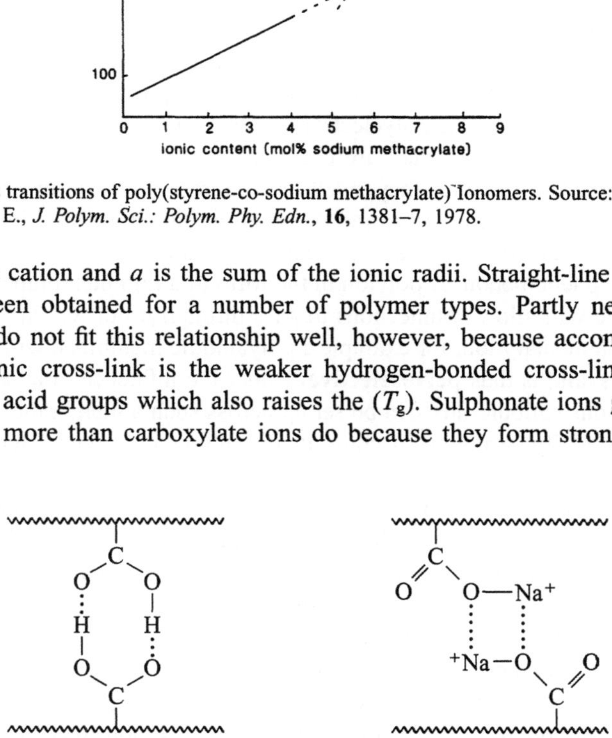

6.4.2 Ion exchange

The property of ion exchange has important consequences, and three different types of application depend on it. They are (a) ion-exchange resins; (b) ion exchange membranes; and (c) heterogeneous catalysis.

Ionic polymers contain two types of ion, those which are bound within the structure, and the counter ions which are free. A polyion immersed in a medium in which it is insoluble will exchange its counter ions for similar ions from the surrounding medium and an equilibrium will occur. The kinetics of the process will depend on the physical form of the insoluble polyion, its porosity and surface area, for example, but the fundamental fact is that the counter ions have free movement into and out of the polymer whereas ions of the same charge type as the bound ion do not. This barrier to a complementary ion is known as **Donnan exclusion**.

Ion-exchange resins are insoluble beads of an ionic polymer. They are best known for their use in the conversion of tap water to 'distilled' water by the process of **de-ionization** or **demineralization**. This works by contacting the water with two types of resin, one a polyacid and the other a polybase, so that the net effect is to exchange M^+ and X^- from the tap water for H^+ and OH^- (i.e. H_2O) from the beads.

To be worthy of their role, ion-exchange resins have to be highly ionic, therefore they are polyelectrolytes and at the opposite end of the ion-content scale to the ionomers. As polyelectrolytes, they are inherently water-soluble, but to make them insoluble they are chemically cross-linked during manufacture. When immersed in water they swell to a degreee which is inversely related to the density of the cross-links, and the swelling facilities rapid ion exchange.

Ion-exchange membranes are relatively modern application of the ion exchange phenomenon. A polyion in the form of a thin membrane exposed to an electrolyte will allow counter ions to pass through it but will act as a barrier to the complementary ion, for example a polyanionic membrane will allow passage of cations, and is thus **permselective**. Under the influence of an electric current a continuous cationic flux can be set up across such a membrane. This principle is commercially exploited in chlor-alkali cells as described later (Section 6.8) and in electrodialysis, and is being used in the development of permselective separators for batteries and fuel cells. A related application is an ion-selective electrode where an ion-exchange membrane encasing an electrode, when immersed in an ionic medium, will develop a potential proportional to the activity of the ion to which it is permselective.

The ion content of an ion-exchange membrane is often intermediate between that of an ionomer and an ion-exchange resin. A high ion content is not necessary for a membrane because its function is not governed by its exchange capacity. Although a high ion content can be advantageous by decreasing the electrical resistance it can be counterproductive if the membrane is so highly swollen by water that its permselectivity is impaired. Chemical cross-linking is used with ion-exchange resins to limit this swelling, or to prevent the resin from dissolving, but generally this is undesirable for membranes because the cross-links interfere with the processes of membrane fabrication.

In an example of electrodialysis (Fig. 6.4) saline water is fed to a dialyser, and drinkable water is collected continuously. The dialyser is an electrical cell divided into a series of subcells which are separated by alternate polyanion and polycation membranes. The ions move under the influence of the applied electric current, and the complementary permselectivities of the membranes ensure that salt is removed from alternate subcells and concentrated in the others. Practical dialysers have up to 100 such subcells. Other applications of dialysis include the concentration of brine and the desalting of cheese whey.

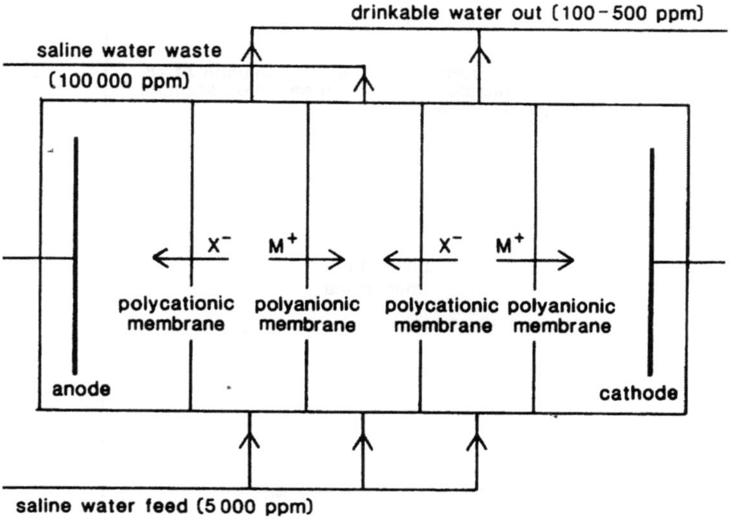

Figure 6.4 Desalination of water by electrolysis.

It is important not to confuse electrodialysis with ordinary dialysis which is not an ion-exchange process but is a separation based on differences in diffusivity between large and small molecules through a membrane which, generally speaking, is not ionic. **Donnan dialysis** (or **ion-exchange dialysis** is another diffusion-controlled separation process, but as it involves ion exchange it does need an ionic membrane (though without an applied current). In the illustration (Fig. 6.5), the cations diffuse each way across the membrane but the anions cannot, thus although the anions swap over, the amount of electrolyte on each side of the membrane is constant. Using on the right-hand side a high concentration of an expendable cation, M_2^+, a desirable cation M_1^+ which is in dilute solution the left-hand side can be exchanged across the membrane and made to accumulate on the right. This type of dialysis, using continuous cells, is under development as a means of stripping and concentrating radioactive ions from dilute radioactive wastes.

Piezodialysis is another ion-exchange separation driven by diffusion rather than an electric current. The membrane, known as a **charge mosaic** membrane is made up of minute patches of polyanion and polycation arranged side by side such that labile cations and anions can diffuse through it along the interfaces between the patches (Fig. 6.6). Membranes adequate for commercialization have not yet been developed.

Heterogeneous catalysis is another way in which the ion-exchange capability of ionic polymers is exploited. The ionic polymer has to be insoluble, but swollen by, the reactant solution. As will be described later (Section 6.8), dispersing a polysulphonic acid in an aqueous mixture can catalyse certain reactions after which the polyacid can be filtered off, cleaned and recycled.

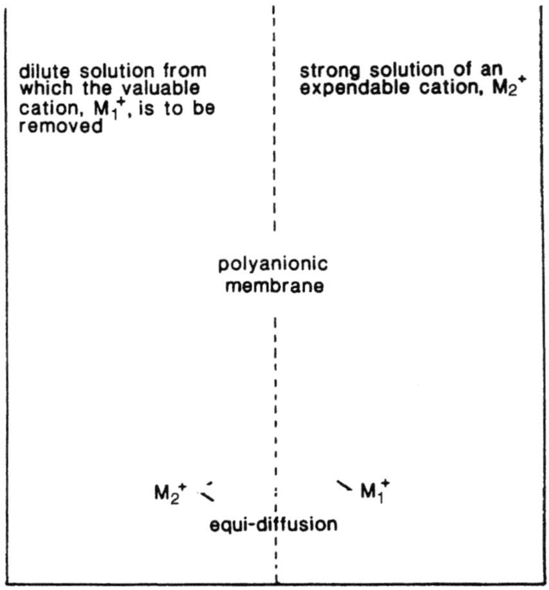

Figure 6.5 Ion-exchange dialysis.

6.4.3 Hydrophilicity

Ionic polymers are hydrophilic. Those of moderate ion content swell on contact with water; those of high ion content dissolve, unless they are cross-linked. Illustrative of these effects is the water uptake, on immersion for one day, of an aromatic polymer containing pendant sodium sulphonate groups (Table 6.3).

The consequences of an ionic polymer's hydrophilicity from the point of view of its properties and usefulness depend upon the ion content. In the case of ionomers, because they are used as thermoplastics, the absorption of moisture is

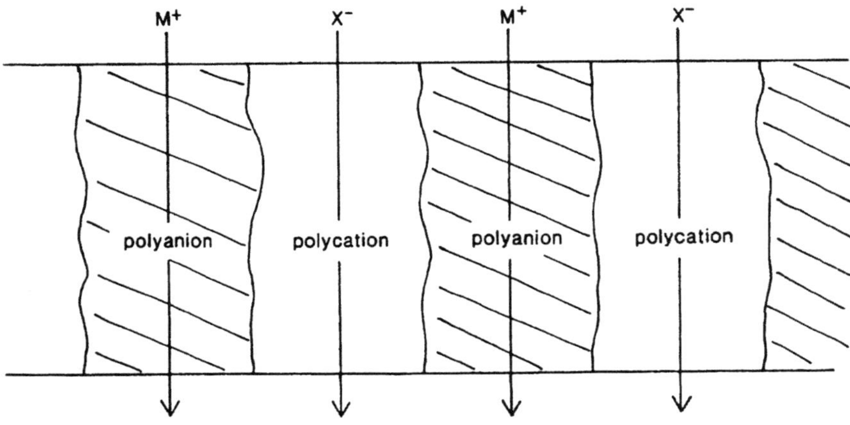

Figure 6.6 A charge mosaic membrane for piezodialysis.

Table 6.3 Water uptake of an aromatic sodium polysulphonate (a sulphonated aromatic polyethersulphone)

Ion content (equivalents kg^{-1})	Water uptake (% weight per weight)
0	2.5
0.1	3.3
0.5	6.1
1.0	16.4
1.5	25.0
2.0	40.0
2.5	$\propto$

undesirable. However, it is not a problem at very low ion contents provided the materials are dried before being fabricated. Generally, it is found that low levels of moisture in ionomers enhance the formation of multiplets and clusters, but saturation has the opposite effect; this has been studied by SAXS. When at low moisture contents the small-angle peak becomes intensified but as saturation it disappears. With moderately ionic polymers which swell without dissolving, the hydrophilicity has the advantage of making the structure more permeable for ion exchange. The hydrophilicity of moderately ionic polymers leads to another type of membrane application, that of **reverse osmosis.**

A common application of reverse osmosis is the purification of water. In this process a saline, brackish water or seawater is forced at very high pressure through a hydrophillic membrane, and water of drinkable quality emerges (Fig. 6.7); the process is often termed **desalination** (cf. the production of high-purity water by **de-ionization** using ion-exchange resins). The term reverse osmosis arises because the natural osmotic pressure tends to force the water in the opposite direction (i.e. from the region of low ionic concentration to that of high ionic concentration) and to achieve the desired flow the applied pressure has to exceed the osmotic pressure. Although the membrane is ionic and ion exchange occurs, there is no overall transmission of salt because of Donnam exclusion (unlike electrodialysis no electric current is applied). An essential requirement

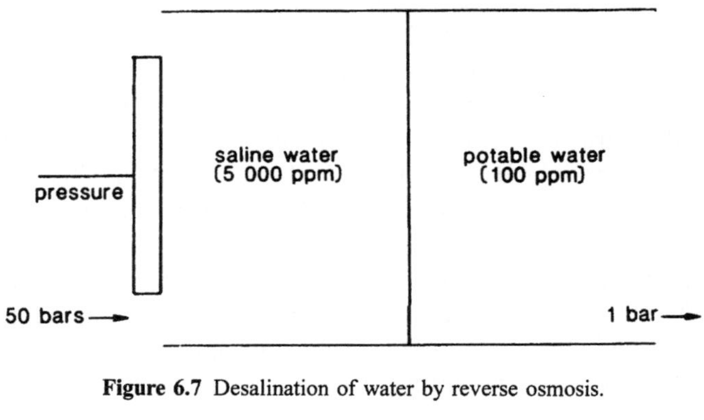

Figure 6.7 Desalination of water by reverse osmosis.

of a membrane for reverse osmosis is hydrophilicity, although it need be ionic; in fact, the traditional membrane material is the non-ionic cellulose acetate. Interest has developed recently in the use of certain ionic polymers, such as the sulphonated polyaromatics, because of their greater chemical stability and resistance to biological degradation.

To achieve high flux in reverse osmosis, a membrane of a given surface area has to be very thin (0.02–1.0 μm) but to withstand the high applied pressure it also has to be very strong. These two requirements, which are in conflict, are met by having the permselective membrane as a skin mounted on a support. This support is another membrane which is very porous like a filter paper, non-selective, and very thick (>100 μm). In some such bilayers each layer is of the same polymer so the membrane is called **integral** or **asymmetric** but in others the layers are of different polymers and the membrane is called a **composite**. Composite membranes are harder to make than are integral membranes but have the advantage of each layer being individually optimized – the skin for its permeability and the support for its strength. (Membranes used in other processes such as ultrafiltration and microfiltration are not made with ionic polymers. These separations are pressure-driven filtrations in which the membrane is like an ultrafine filter paper.)

While moderately ionic polymers only swell in water, highly ionic ones – polyelectrolytes – dissolve, except those which are covalently cross-linked. The polyelectrolytes have applications which depend upon their solubility and which are quite different from the thermoplastic applications of the slightly ionic ionomers or the membrane applications of the medium-ionic polymers. Basically, these applications depend upon the polyelectrolyte altering the fluid properties of an aqueous medium or modifying the behaviour of particles in aqueous slurries or colloidal suspensions.

Polyelectrolytes raise the viscosity of aqueous solutions and so act as **thickeners,** and the magnitude of the effect increases with the polymer's molecular weight. Traditionally, naturally occurring gums and acidic polysaccharides have been used as thickening agents in foodstuffs and pharmaceutical products. More recently, synthetic polyelectrolytes have been used in these roles. Synthetic polyelectrolytes are also much used to modify the flow characteristics of latex paints and similar proprietary fluids.

Polyelectrolytes will also stabilize particles in aqueous suspension thus acting as **dispersants**. In this role their hydrophobic backbones are adsorbed onto the surface of the particle by van der Waals' attraction while their ions form a hydrophilic surface which interacts with the water. A familiar example in polymer chemistry is the suspension of granular polymerization process when a hydrophobic vinyl monomer is mixed with water and agitated to form droplets which are stabilized by a polyelectrolyte. The formed polymer then mimics the shape of the droplet. A different example is in water conditioning when the microscopic nuclei of calcium carbonate, as it precipitates in hard water, absorb polyions which then prevents their growth into larger crystals.

Polyelectrolytes can also function as **flocculating agents**. In that role, depending on their ionic charge, they can interact with colloidal particles and neutralize the stabilizing hydrophilic charges. they have been used in this way to coagulate slurries and industrial wastes.

There are other applications of polyelectrolytes which stem from their behaviour in water in various ways. For example, they are used as sizes in the textile industry and in paper manufacture and as additives to drilling muds and to soil for conditioning purposes. A totally different kind of application is their use as dental cements, where a divalent cation is added to an aqueous solution of a polyanion to form a highly cross-linked precipitate of great strength.

6.5 Ionomers based on polyethylene

These are mainly copolymers with pendant carboxylate groups in which the polyethylene backbone is the major component. They are thermoplastic and can be processed in the usual ways but have certain properties superior to unmodified polyethylene, as will be illustrated below. Generally, they contain 3–10 weight% of copolymerized methacrylic acid of which 20%–80% is neutralized by either sodium or zinc. They were commercialized in 1964 by Du Pont under the trade name Surlyn™. As a consequence, further research was stimulated and the generic name 'ionomer' was coined.

They are manufactured by direct copolymerization of ethylene with methacrylic acid. This is a continuous process developed from the high-pressure method used to make low-density polyethylene by free radical initiation. The comonomers are mixed in the appropriate proportions, allowing for the greater reactivity of the methacrylic acid, and introduced with a peroxide initiator to a reactor at a pressure of about 2000 bars and a temperature of 250°C–280°C. At 15%–20% conversion to polymer, the residual monomer is removed then recycled with fresh monomer. It is important that the methacrylate units, and hence the ionic cross-links formed later, are randomly distributed and uniform in composition. The low conversion per pass ensures uniformity because the methacrylic acid reacts disproportionately rapidly and has to be replenished frequently. Other carboxylic acids have been copolymerized in this way, such as acrylic, itaconic and maleic acids; carboxylic precursors have also been copolymerized and afterwards hydrolysed, such as methyl esters and maleic anhydride.

Strictly speaking, these acid copolymers are not in themselves ionomers because they are only slightly ionized. In fact they have no outstanding properties and behave merely as typical copolymers of polyethylene, but when they are neutralized, and the carboxylate ion formed, the distinctive properties of the ionomers become manifest. The neutralization is a kind of post-treatment. It is conveniently done by melting the polyacid in a mill at 150°C and adding a base or basic salt in powder form or in solution. The base can be sodium methoxide

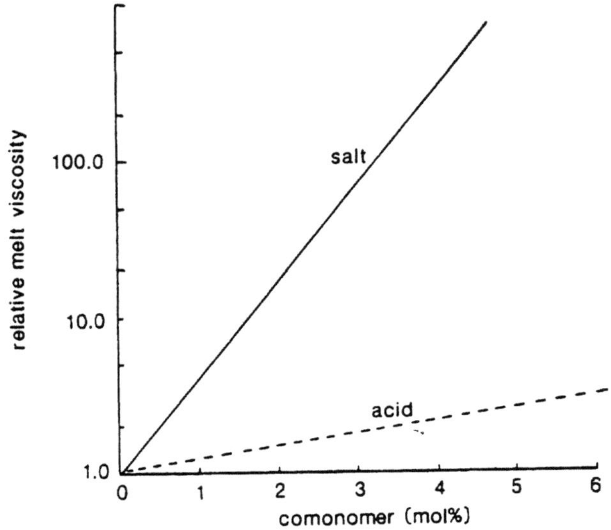

Figure 6.8 Melt viscosity of poly(ethylene-co-methacrylic acid) and its sodium salt relative to polyethylene. Source: Longworth, R., Chapter 2, in *Ionic Polymers* (ed. L. Holliday), page 96, published by Applied Science Publishers, London, 1975.

dissolved in methanol, which is then vaporized off, or it can be magnesium acetate dissolved in water, the water and acetic acid by-product being vaporized off, or it can be zinc oxide mixed with acetic acid. In these neutralizations the melt starts as a soft, fluid, opaque mass of polyacid and then becomes stiff, rubber and transparent as the ionomer forms.

The loss in fluidity, or gain in melt strength, is one of the significant properties of ionomers and a factor which determines their usefulness. The effect is a function of the amount of methacrylate ion and is caused by the formation of strong ionic cross-links (Section 6.4.1); that it is not a consequence of copolymerization is shown by comparing the melt viscosity of an ionomer with its acid precursor (Fig. 6.8). The melt viscosity copolymer containing, for example, 2 mol% of acid is increased, compared with an otherwise equivalent polyethylene homopolymer, by only a small amount (50%), but when neutralized, and therefore ionized, it is increased twentyfold. The slight increase with the acid is due to the weak hydrogen-bonded cross-links. The melt-strength of an ionomer is such that its molten film can be drawn down over the sharp edges of a nail without its being torn or punctured.

Another important characteristic of the melt behaviour is that the viscosity is very shear sensitive (Fig. 6.9). As already explained, at low shear the viscosity of an ionomer is very much greater than its acid precursor, but the situation is different at high shear because the two are more nearly the same. This is because of the physical binding of the ionic cross-links breaking down and is a useful feature in a number of melt-fabrication processes. Thus the high melt strength at

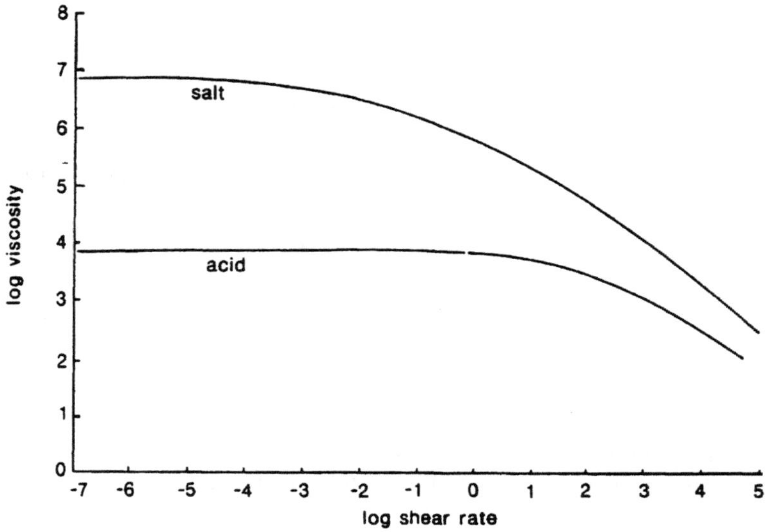

Figure 6.9 Melt viscosities (160°C) at various shear rates of poly(ethylene-co-methacrylic acid) (3.5 mol%) and its sodium salt. Source: Longworth, R., Chapter 2, in *Ionic Polymers* (ed. L. Holliday), page 102, published by Applied Science Publishers, London, 1975.

low shear makes ionomers useful for blow-moulding or extrusion and for any process where the melt is partially supported.

A striking property of the ethylene ionomers is that they are transparent, unlike polyethylene itself and the acid copolymers, which are not. The haze in polyethylene is a result of it being partly crystalline, the minute crystallites within it agglomerating into spherulites which are of a size to scatter light. However, in the ionomers the crystallites, although present, are unable to agglomerate because of the high viscosity of their surroundings. Crystallite formation is in fact helped by the ions, the domains of which serve to nucleate them. Thus, microcrystallinity is enhanced by the ions but macrocrystallinity, which causes haze, is inhibited. This clarity is useful in packaging applications.

An ionomer differs from polyethylene and the acid copolymer in many other ways. Generally, it is tougher and, as shown in Table 6.4, relative to the acid copolymer its tensile strength is increased by 40% and its stiffness more than doubled. These effects are much the same for sodium and zinc ionomers.

Surlyn™ polymers are supplied as small, free-flowing pellets. There are six grades of the sodium type and nine of the zinc type, some of which are approved for food contact by the Food and Drug Administration of the USA. They are described as flexible and tough with good impact toughness at low temperature and with good resistance to grease and solvents, to stress cracking and to abrasion. for outdoor use they are compounded with 3%–5% of carbon black and 0.1%–0.5% of an antioxidant. They are injection moulded into various objects. Their resistance to grease and solvents has made them useful in meat

Table 6.4 Physical properties of polyethylene (PE) ionomers.[a] Source: Rees, R.W., US Parent, 3264 272, 1966

Property	PE	Copolymer ($-CO_2H$)	Ionomer Na^+	Ionomer ZN^{2-}
Appearance	hazy	hazy	transparent	transparent
Melt index (g per 10 min)[b]		5.8	0.03	0.09
Yield strength (MN m^{-2})	8.3	6.14	13.2	13.3
Elongation (%)	600	553	330	313
Ultimate tensile strength (MN m^{-2}	12.4	23.4	35.9	29.8
Stiffness (relative)		1.0	2.8	3.0

[a] Base copolymer in poly(ethylene-co-methacrylic acid), 3.5 mol%.
[b] ASTM-D-1238-57T.

packaging while their high melt strength has made them useful for blow moulding into films, sheets, bottles and blister packs. They are very effective as external surface coatings on glass bottles to contain breakages.

6.6 Elastomeric ionomers

A number of ionic polymers exist in which the covalent backbone is a recognized elastomer. Their ionic content is small, therefore they are ionomers. A necessary feature of a practical elastomer is that it is lightly cross-linked and this is provided, at least in part, by the ions. Elastomeric ionomers which are cross-linked exclusively by their ions have the useful feature of being thermoplastic.

The carboxylated elastomers are close relatives of the polyethylene ionomers described above. Their backbone is typically polybutadiene containing some acrylonitrile and styrene to adjust its flexibility and toughness, and, additionally, up to 6% by weight of acrylic or methacrylic acid. Polycarboxylic acids of this type were first commercialized in 1949 by B.F. Goodrich under the trade name Hycar™. In the early days, the role of the carboxyl group was to make the polymer polar rather than ionic and the materials were vulcanized conventionally with sulphur; later, ionic cross-linking as a form of vulcanization was developed.

Butadiene ionomers are usually made by direct copolymerization with the acid monomer by means of emulsion methods. Typically, the monomers are slurried in water with sodium dodecylbenzene sulphonate as emulsifier and potassium persulphate as the free radical initiator. The product is a latex and usually marketed as such. The tendency of the carboxylate monomer to dissolve in the aqueous phase instead of remaining in the droplets of polymerizing butadiene is suppressed by making the aqueous phase acidic, so maintaining the monomer in the non-ionized form.

Butadiene ionomers are also occasionally made by hydrolytic post-treatment. Controlled hydrolysis of copolymers of acrylate esters, acrylonitrile, acrylamide or maleic anhydride with butadiene is one technique which has been used to generate simple pendant carboxylate. Pendant alkyl carboxylate groups, however, have been generated by grafting reactions [3] using free radical initiators to activate double bonds in the polybutadiene backbone. Thus structure has been produced by treating a solution of polybutadiene in benzene at 50°C with thioglycollic acid to give

$$-CH_2\underset{\displaystyle SCH_2CO_2H}{\underset{|}{C}}HCH_2CH_2-$$

and by milling it with β-mercaptopropionic acid, to give

$$-CH_2\underset{\displaystyle SCH_2CH_2CO_2H}{\underset{|}{C}}HCH_2CH_2-$$

The carboxylated polybutadienes when neutralized become vulcanized [4]. This can be done, for example, by treating with aqueous sodium hydroxide then heating, or by heating directly with zinc oxide. The ionic cross-link formed with the sodium ion is of modest strength at room temperature and dissociates at 100°C whereas that with zinc is much stronger, although the polymer is capable of substantial flow at higher temperatures. Some properties of sodium and zinc vulcanizates are compared with those of the acid precursor in Table 6.5.

High tensile strength is a characteristic of ionic vulcanizates. A standard butadiene elastomer vulcanized with sulphur gives a strength of 1.9–5.8 MN m^{-2} whereas an equivalent copolymer containing 1.5% of methacrylic acid, and vulcanized with magnesium oxide, gives 29.0 MN m^{-2}. The two types of vulcanizate respond differently to fillers such as carbon black. The one with sulphur is reinforced, that with ions is weakened.

Carboxylated polybutadienes as latices are well-developed items of commerce. They are used mainly in dipping and coating processes, applications which often do not involve ionized carboxylate, at least not directly. The applications centre on paper coating, adhesives, carpet-backing, binding non-woven fabrics, glove-dipping and also shrink-proofing woollen garments, where the carboxyl groups are believed to react with pendant amino groups in the proteins of the wool. One manufacturer supplies a range of twelve carboxylated latices which are copolymers containing 40%–66% of butadiene (Revinex™, Doverstrand); some of these are styrene–butadiene rubbers (SBRs) and other acrylonitrile–butadiene rubbers (NBRs). Another supplier of carboxylated latices makes as many as 21 grades of SBR and eight of NBR (Polysar).

Table 6.5 Vulcanization of carboxylate polybutadiene. The polymer is poly(butadiene-co-methyacrylic acid) of ionic content 1.2 equivalents kg^{-1} (6.7 mol% methacrylic acid). Source: Brown, H.P., *Rubber Chem. and Technol*, **36**, 931–62, 1963

Form	Tensile strength (MN m^{-2})	Elongation (%)
Acid	0.7	1600
Sodium salt	11.7	900
Zinc salt	41.4	400

Table 6.6 Comparative properties of a commercial carboxylated acrylonitrile–butadiene rubber vulcanized in various ways. Sources: B.F. Goodrich Chemical (UK) Ltd, Hycar 102, Manual HM-11 (Halato–Telechetic polybutadiene); Reed, S.F., *J. Polym. Sci., A-1*, **9**, 2147–53, 1971; Broze, G., Jerome, R. and Teysine, R., *J. Polym. Sci., Polym Lett.*, **19**, 415–18, 1981

Property	Peroxide only	Zinc oxide only	Peroxide + zinc oxide	Sulphur + zinc oxide
Carbon black (reinforcement) (%)	40	None	40	40
Ultimate tensile strength ($MN\ m^{-2}$):				
cold	19.8	39.6	31.8	26.8
hot	8.8	1.3	18.9	15.6
Tear resistance (relative):				
cold	1.0	1.6	1.8	2.1
hot	0.5	0.3	0.8	0.9
Compression set, 70 h at 121°C (%)	25	95	35	20
Hardness (durometer A)	67	73	87	85
Modulus at 100% elongation ($MN\ m^{-2}$)	2.5	2.2	13	13

Carboxylated polybutadienes have been much studied as solid, thermoplastic elastomers but have not been used much as such, principally because they have poor compresssion set, high stress relaxation and poor performance at higher temperature. Table 6.6 compares the properties of a solid commercial NBR vulcanized in various ways. Mixed vulcanizations are used for these rubbers to combine the advantages of the ionic method with the conventional methods. The mixed vulcanizates have notable resistance to abrasion, oil and fuel, and high tensile strength; they are used in applications such as industrial rollers and wheels, and shoe soles. There are fewer grades of dry NBRs available than of latices; examples are Krynac™ (Doverstrand), in three grades, and Hycar™ (B.F. Goodrich), also in three grades.

Elastomeric ionomers have also been developed from ethylene–propylene copolymers. These are saturated structures of ethylene (50%–75% by weight) and propylene, but generally some diene (4%–5%) is included to provide residual unsaturation and allow vulcanization by conventional processes; the diene is commonly ethylidenenorbornene (ENB). Du Pont made carboxylated ionomers by free radical grafting of maleic anhydride (0.5%–5%) on to the diene moiety of the polymer and neutralized the product with rosin salt [5]. The ENB can also be sulphonated, [6]:

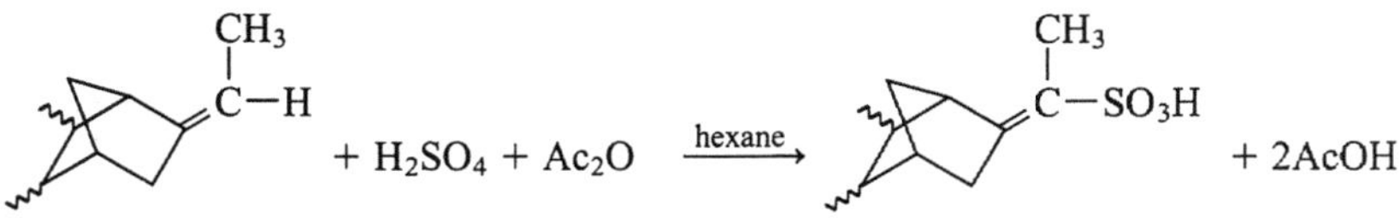

The acid sulphonates are not very thermally stable but their salts are. They can be neutralized by ion exchange with sodium hydroxide or acetate solutions, or more directly by titrating a solution of the polymer in toluene/methanol mixtures, or they can be milled with an acetate or stearate salt. Ionomer properties are said only to develop when the acids are neutralized, as with the carboxylates, although sulphonic acids would be expected to be ionized even in the acid form, unlike the carboxylates. The best properties are given by zinc salts, in particular when they are plasticized by zinc stearate. In the early 1980s polymers of this type (sulphonated EPDMs) called 'ionic elastomer' and 'thionic polymer' were being actively developed in the laboratory; a type of 'sulpho-EPDM' is about to be marketed in the USA for use as a roofing felt.

The halato-telechelic polybutadienes are another type of ionic elestomer which has been studied but not apparently marketed. These have backbones with integral ions and are made by linking up dicarboxy-terminated polybutadiene oligomers with divalent cations such as zinc, thus:

$$\text{+}\,{}^{-}O_2C\text{—(butadiene)}_x\text{—}CO_2\text{—}Zn^{+}\,\text{+}$$

The precursors can be made by oligomerizing butadiene anionically but are made more simply by free radical methods in solution using a carboxy-substituted azo-initiator. To ensure that the product is difunctional and of low molecular weight, the initiator is at relatively high concentration and a solvent such as *t*-butanol which does not chain transfer is used.

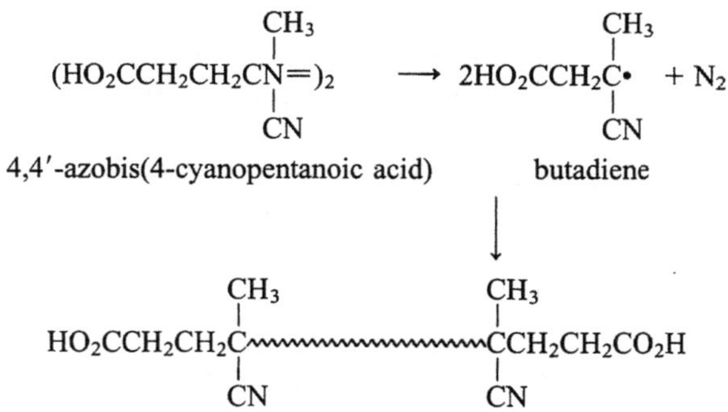

Some carboxy-terminated liquid oligomers of this type are marketed which have various speciality, but mainly non-ionic, applications, such as modifying epoxy resins. They are oligomers of butadiene or butadiene-co-acrylonitrile with carboxyl contents of 0.45–0.67 equivalents kg^{-1} and degrees of polymerization of 50–70.

Chlorosulphonated polyethylene [7] is a commercial elastomer which is not ionic as supplied but becomes a sulphonate ionomer when vulcanized. It is made by a random substitution reaction on polyethylene, typically:

$$\text{—}CH_2(CH_2)_3CH_2\text{—} + 2Cl_2 + SO_2 \rightarrow \text{—}\underset{\displaystyle Cl}{\underset{|}{C}}H(CH_2)_3\underset{\displaystyle SO_2Cl}{\underset{|}{C}}H\text{—} + 2HCl$$

This material is known as Hypalon™ (Du Pont) and is supplied in nine grades. The composition corresponds to about one carbon in 90 bearing a sulphonyl chloride and one is seven a chloride. Vulcanization is usually done with a metal oxide, notably lead or magnesium. Like the carboxy polybutadienes the ionic vulcanizates have poor compression sets but can be improved by mixed vulcanizates. This is done by adding conventional sulphur vulcanizing agents which react with ethylene groups formed by decomposition of some sulphonyl chloride groups at the processing temperature. Mixed vulcanization is less commonly done by adding a diamine which condenses with sulphonyl chloride to form sulphonamide cross-links.

6.7 Ionomoers based on polystyrene

Polystyrenes containing small amounts of ions are not marketed but have been well studied and have contributed greatly to the general understanding of ionomers. They may be divided into two types: those in which the ions are attached to the phenyl ring, which are usually made by post-functionalization, and those containing an ionic comonomer, which are usually made by direct copolymerization. The former have enabled comparisons to be made between different types of ion on the same backbone, the latter are the styrene analogues of the ethylene and butadiene ionomers described above.

Polystyrene when dissolved in concentrated sulphuric acid becomes highly sulphonated but it can be lightly sulphonated by treating its solution in 1,2-dichloroethane with small amounts of acetyl sulphate formed by mixing acetic anhydride and sulphuric acid (Table 6.7).

Polystyrene can be carboxylated by lithiation [8] followed by treatment with carbon dioxide to form the lithium carboxylate which can then be ion exchanged to the acid or other salt forms:

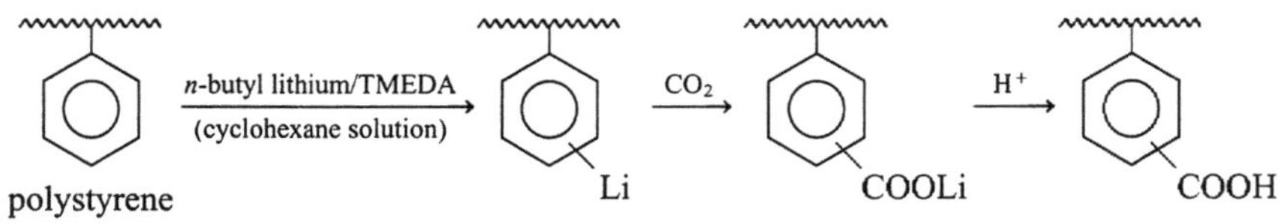

(TMEDA = N,N,N',N'-tetramethylethylenediamine). A preferred method giving exclusively *para* substitution is to lithiate poly (p-bromostyrene); random copolymers of *p*-bromostyrene and styrene are readily made by vinyl polymerization:

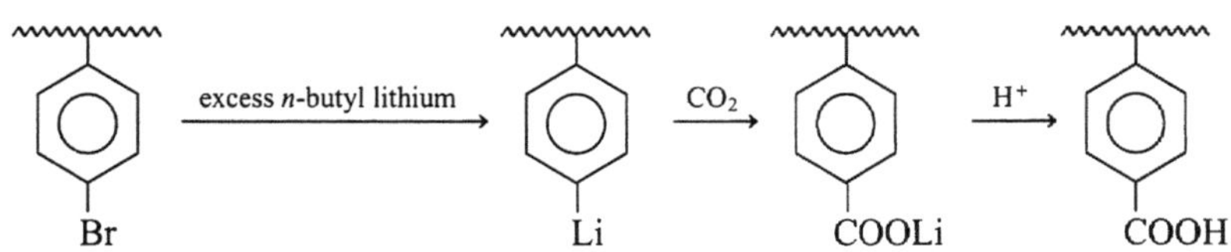

poly(styrene-co-*p*-bromostyrene)

The ionomers of polystyrene show the usual characteristics of ionic association leading to multiplets and clusters. A number of physical properties when

Table 6.7 Lightly sulphonated polystyrene [2 molar (with respect to styrene monomer) solution of polystyrene in 1,2-dichloroethane, at 50°C for 1 h. Source: Makowski, H.S., Lundberg, G.H. and Singhal, G.H., US Patent 3870 841, 1975

Sulphonating agent (mol%)[a,b]	Sulphonated product (mol%)[a,c]
1	0.32
3	2.53
5	4.37

[a] Mol% with respect to styrene monomer.
[b] CH_3COOSO_3H.

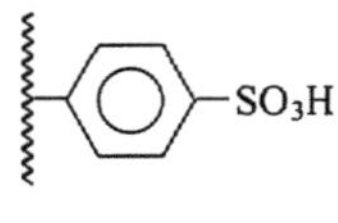

plotted against increasing ionic content show discontinuities arising from the formation of, first, multiplets and, then, clusters. From the graph of (T_g) (Fig. 6.3) and from Raman spectra (Fig. 6.2) two types of ionic structures are inferred.

Comparisons have been made between equimolar sulphonates (S-PS) and carboxylates (C-PS) in backbones of identical molecular weight. C-PS has ionomer properties only when neutralized, which for a weakly ionized acid is to be expected; S-PS to some extent is the same, which, as mentioned before, is surprising for an acid which is strongly ionized. Thus, as S-PS is progressively neutralized there is a gradual increase in the melt viscosity (Fig. 6.10); when 90% neutralized there is an inexplicable step increase.

The main difference between S-PS and C-PS is in the strengths of their ionic associations and the temperature at which these dissociate. This is shown by their melt viscosities; that of S-PS is the greater by a factor of 10^3 at 220°C (Fig. 6.11). Also, thermal-mechanical analysis shows that the associations in C-PS and S-PS persist for some 50°C and 120°C, respectively, higher than the T_g of the PS backbone.

Plasticization studies with S-PS has underlined the dichotomy between the non-polar backbone phase and the polar ionic phase. By suitable choice of plasticizer either one or the other may be selectively plasticized leading to distinctly different effects. Di-octyl phthalate, a non-polar plasticizer, affects the backbone but not the ionic cross-links, so is more effective in lowering the T_g than in affecting the melt viscosity; on the other hand, glycerol, a polar plasticizer, affects the ionic cross-links but not the backbone, so is more effective in lowering the melt viscosity than the T_g (Fig. 6.12).

Selective plasticization of S-PS has made possible a polystyrene with properties like plasticized PVC. Ordinarily, when polystryene, which is wholly amorphous material, is plasticized it forms a worthless, viscous liquid, but PVC, which differs by being partially crystalline, forms a tough, useful plastic. In the case of PVC, the crystals, which are unaffected by the plasticizer, hold the plastized amorphous parts together like cross-links. The ionic cross-links of

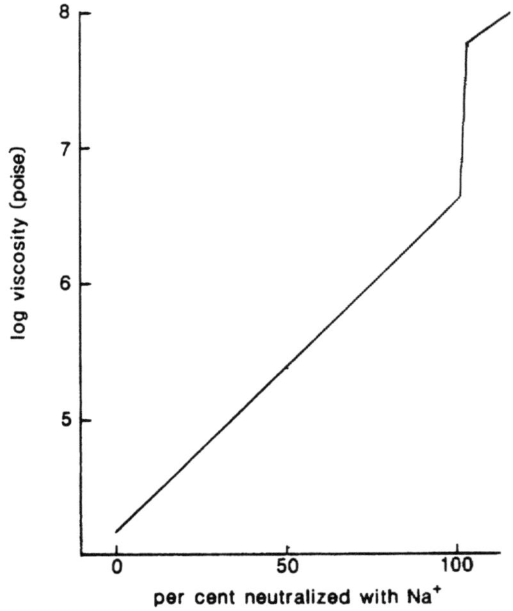

Figure 6.10 Melt viscosities at 250°C of partly neutralized poly(styrene-co-styrene-*p*-sulphonic acid) (2.5 mol%). Source: Lundberg, R.D., Makowski, H.S. and Westerman, L., *Adv. Chem. Ser.*, **187**, 67–76, 1980.

S-PS can behave like the crystals of PVC so that by careful choice of ionic content and plasticizer, polystyrene can be made to mimic PVC. These materials have been patented.

6.8 Ionomers based on polytetrafluoroethylene [9–11]

Polymers with a polytetrafluoroethylene (PTFE) backbone and pendant perfluorosulphonate or perfluorocarboxylate groups have become important commercial materials. The sulphonates were introduced in the early 1970s by Du Pont as Nafion™ and the carboxylates in 1978 by Asahi Glass as Flemion™. They are made by free radical copolymerization of tetrafluoroethylene (TFE) with vinyl monomers, giving non-ionic precursor copolymers which can be post-functionalized by hydrolysis:

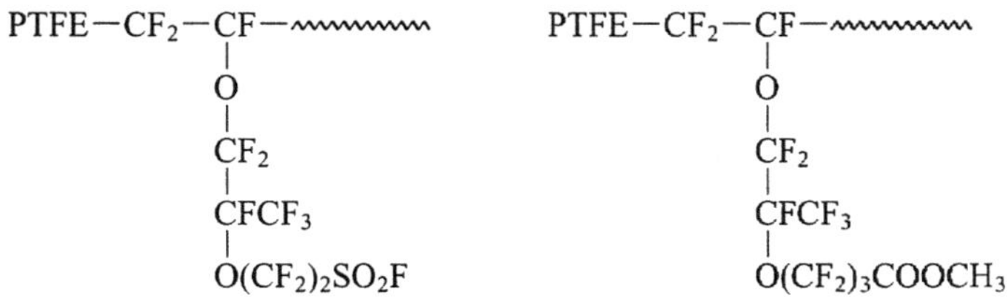

Nation™ precursor   Flemion™ precursor

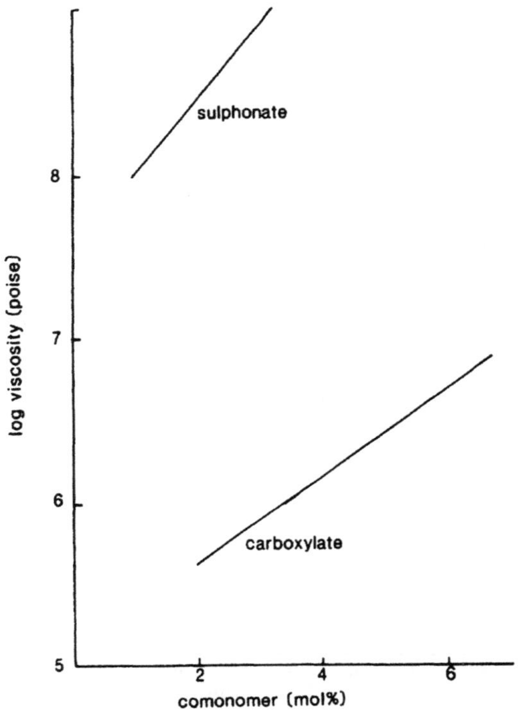

Figure 6.11 Melt viscosities at 220°C of poly(styrene-co-sodium *p*-styrene sulphonate) and co-sodium *p*-styrene carboxylate. Source: Lundberg, R.D. and Makowski, H.S., *Adv. Chem. Ser.*, **187**, 21–36, 1980.

The comonomers are themselves perfluorovinyl compounds and made from TFE by multi-step syntheses using hexafluoropropylene oxide.

Nafion™ was originally developed for use as a membrane in fuel cells but it has been more useful in various other electrolytic separation processes. These applications exploit its cation-exchange properties and its ability to survive in extremely aggressive chemical environments. An example of this is its use as a membrane in the chlor-alkali cell where it has been an outstanding success and is gradually replacing the traditional diaphragm and mercury cells: the only factor inhibiting its use is its high cost. In the chlor-alkali process a cell is partitioned by the membrane into an anode compartment, to which brine is added, and a cathode compartment, to which water is added (Fig. 6.13). On electrolysis, Na^+ passes from the brine around the anode and through the membrane to the water in the cathode where it forms sodium hydroxide. At the anode, chlorine is released from the brine. The process works because the membrane allows transfer of Na^+ ions by ion exchange, maintains the brine and water separate and acts as a barrier to Cl^- and OH^-.

From a structural point of view, the Nafions™ show the classic characteristics of ion-phase separation and contain multiplets and clusters. In common with

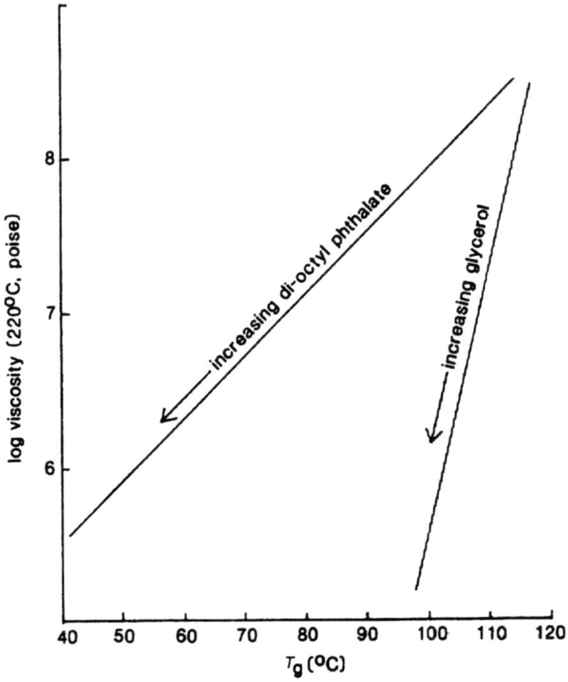

Figure 6.12 Selective plasticity of poly(styrene-co-sodium *p*-styrene sulphonate) (178 mol%). Source: Lundberg, R.D., Makowski, H.S. and Westerman, L., *Adv. Chem. Ser.*, **187**, 67–76, 1980.

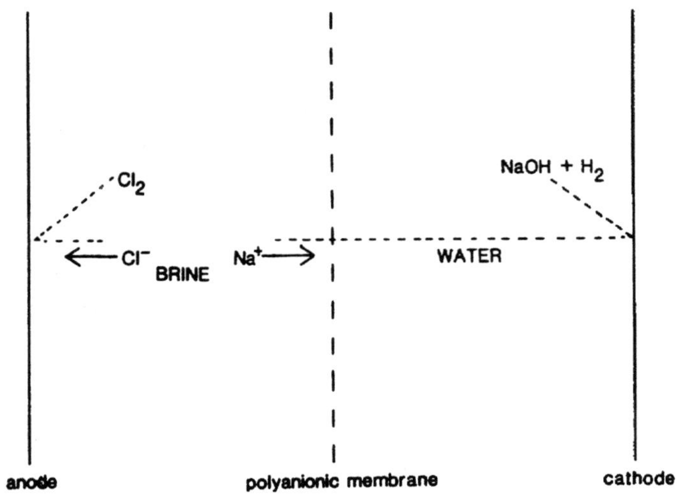

Figure 6.13 Chlor-alkali cell.

polyethylene ionomers, SAXS gives a small-angle scatter arising from the ionic phase which is absent from the non-ionic precursor copolymer, and a wide-angle scatter arising from the PTFE-type crystallinity in the backbone.

The Nafion™ membranes have ionic contents of 0.66–0.91 equivalents kg^{-1}. Four different types are made, one is a bilayer of these two extremes of composition, another is a bilayer with a perfluorocarboxylate polymer. They are made by melt processing the precursor copolymer which then, in membrane form, is hydrolysed.

The requirement for balancing the ionic content of an ion-exchange membrane with its permselectivity and hydrophilicity has been mentioned already (Section 6.4.2). It was to help with this that the perfluorocarboxylates, the Flemions™, were introduced, the idea being that a carboxylate at a given ion content would be less hydrophilic than a sulphonate. The bilayered Nafions™ arose from similar thinking and an attempt to combine certain advantages of the two types.

The performance of membranes in the chlor-alkali process is assessed by the efficiency with which power is consumed and the highest concentration of uncontaminated sodium hydroxide which can be produced. Rapid progress has been made in recent years, particularly since the introduction of carboxylate membranes, such that the efficiency is 95% and the concentration 33%.

Bilayer carboxylate membranes can be produced by surface modification of the sulphonte. Asahi Chemical reduce a Nafion™-type surface to sulphinic and sulphenic acids then oxidize it to a carboxylate layer of 2–10 μm thickness. Tokoyama Soda make Neosepta F™ membranes by treating a Nafion™-type precursor with an alcohol then oxidizing the surface in air:

$$\sim OCF_2CF_2SO_2Cl \xrightarrow{\text{reduction}} \sim OCF_2CF_2SOOH \xrightarrow{-SO_2 \text{ and oxidation}} \sim OCF_2COOH$$

Nafions™ are also supplied as powders which in their acid forms have been studied as catalysts. Being insoluble they function heterogeneously and can be recovered from the products and recycled. Also because they are *superacidic*, that is stronger than sulphuric acid, their catalytic power is high. Many types of reaction have been studied, such as Friedel–Craft reactions at low temperatures and esterification, but a recent example is the von Pechmann condensation of resorcinol with ethyl acetoacetate to form hydroxycoumarin [12]. The reactants are stirred with Nafion™ powder in the H^+ form which is afterwards filtered off, washed, then dried at 105°C; the Nafion™ loses no activity even when recycled five times.

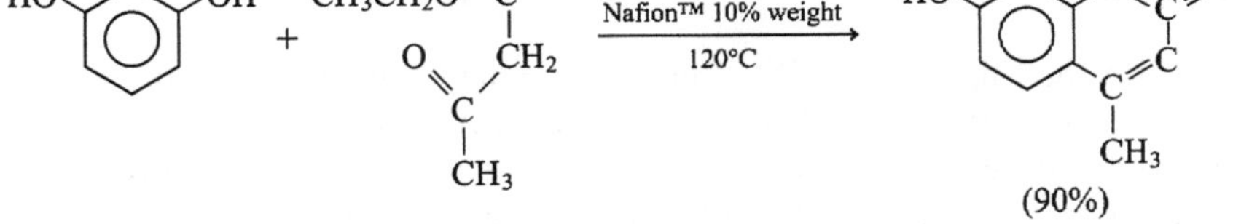

(90%)

6.9 Ionomers with polyaromatic backbones

Aromatic polyethersulphones are engineering-type thermoplastics with a phenylene backbone. They differ from polystyrene which has an aliphatic backbone, but they share with it an aromatic functionality which can be sulphonated. Currently, ionomers of polyethersulphones, principally sulphonates, are being developed as membranes, particularly for purifying water by reverse osmosis. They distinguish themselves in this application by being resistant to oxidation by the chlorine used in water treatment, to biological fouling, to harsh chemical cleaning operations and to compaction under the high operation pressures. The traditional polymer, cellulose acetate, is less robust, while Nafion™, though chemically very inert, is too expensive and difficult to fabricate in a form suitable for reverse osmosis. The sulphones with their completely amorphous backbones and solubility are easily cast into membranes from solvents. As mentioned previously, the optimum ion content is a balance of properties, in this case between a high flux of the permeant, water, and a low leakage of the rejected species, the dissolved salts. The favoured ion content is about 1 equivalent kg^{-1}.

The sulphonates are made by post-functionalization. The precursor polymers are commercial-type plastics made by an ether-forming condensation of bisphenol and a derivative of diphenyl sulphone. One of the two main types, Udel™ (Union Carbide)

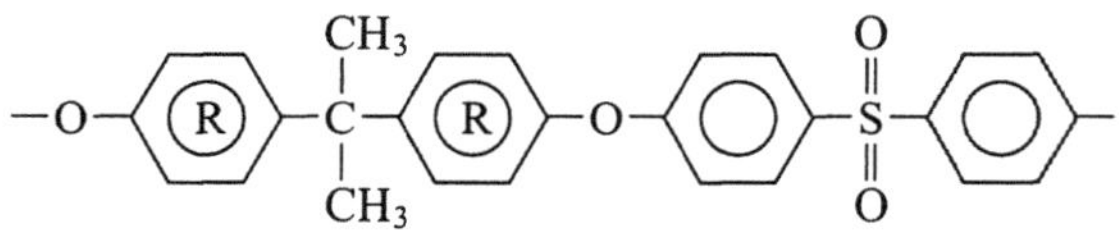

when dissolved in dichloroethane and treated with a complex of sulphur trioxide and tri-ethyl phosphate, becomes monosulphonated on the rings marked 'R' (the rings connected to the sulphone group being deactivated by the sulphone group) [13]. As only partial sulphonation is required, however, less than the stoichiometric amount of sulphonating agent has to be used. The other main type is Victrex™ (IC) which is a random copolymer of two units

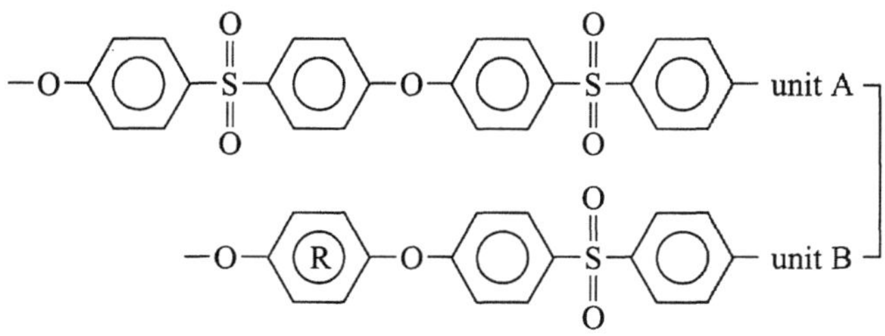

in which the ring marked 'R' of unit B becomes monosulphonated, the other aromatic rings, and the whole of unit A, being inert. Victrex™ is sulphonated by the simple process of dissolving it in sulphuric acid and the extent of reaction is predetemined by the proportion of unit B in the copolymer.

Poly (dimethylphenylene oxide) (PPO made by GEC) was sulphonated in the early 1970s and evaluated in membrane applications though not developed [14–15]. Its phenylene rings were monosulphonated in a process like that used for Udel™. Another polyphenylene oxide containing terphenyl units with pendant carboxyl groups, made as the ester in a multistage synthesis from partly aliphatic materials, is being evaluated by some Canadian workers.

The general physical properties of the sulphonated polysulphones show many of the characteristics of the simpler aliphatic ionomers. SAXS of sulphonated Udel™, however, is said not to show the small-angle scatter of the ethylene and PTFE ionomers. The strength of the ionic associations in sulphonated Victrex™ is indicated by the increase in T_g which occurs despite the already high value of 220°C for the parent polymer (Fig. 6.14).

6.10 Polyelectrolytes for ion exchange

The process of ion exchange was described earlier (Section 6.4.2). The resins used are highly ionic, covalently cross-linked, insoluble polyelectrolytes and are supplied as water-swellable beads which have either a dense internal structure (gel-type) or a porous, multichannelled one (macroporous-type or macrorecticular). Polystyrene (PS) cross-linked with about 8 mol% divinyl benzene (DVB) is the matrix on which most modern resins are based; the bound ions are pendant to the phenyl ring in the *para* position. Very many grades are made and used in a variety of specialist applications. The gel-type of PS resin was the first to be

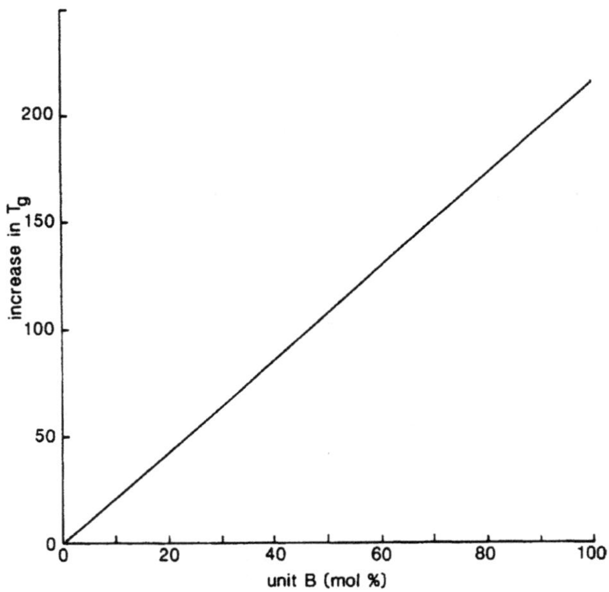

Figure 6.14 Glass transitions of Victrex™ sodium sulphonate relative to the percursor copolymers. The data are midpoints of the T_g process determined by differential scanning calorimetry.

introduced (1947) and the macroporous type came later (1959); the latter, with its effective high surface area, gave access to the exchange sites for larger ions and enabled the manufacturer to make stronger beads by increasing the amount of cross-linking.

The form of the beads is established by polymerizing the styrene mixture as droplets dispersed in water using a conventional free-radical process. The beads, which are of diameter 0.25–1 mm, are then post-functionalized to give the finished resin, during which process they retain their discrete-bead identity because of the cross-links. Porous beads can be made by adding homopolystyrene, which is soluble in the monomer mixture, and leaching it out later with for instance, toluene.

The most common resins are the cation-exchange type with a strong-acid (e.g. sulphonic) functionality. To make them, the precursor beads are dispersed in about 10 times their weight of concentrated sulphuric acid and heated slowly to 150°C. The reaction, although very simple in principle, involves delicate operations in order to achieve beads of suitable structure and durability. When fully monosulphonated, polystyrene has a theoretical ion content of 5.1 equivalents kg^{-1} (dry), but many practical resins have values of about 4.4–5.2 equivalents kg^{-1}; an example is Amberlite IR-120™ (Rohm and Haas). Weak-acid, cation-exchange resins generally have a matrix of polyacrylic or polymethacrylic acid cross-linked with DVB and are also made by aqueous granular polymerization but sometimes by using an ester which is afterwards hydrolysed. They have ionic contents of 9–10 equivalents kg^{-1}. For example Amberlite IRC-84™.

Strong-base anion-exchange resins are also very common. They are made by chloromethylating the PS–DVB presursor then aminating the product with a tertiary amine to form a quaternary amine group such as $N^+(CH_3)_3$ or $N^+(CH_3)_2(CH_2OH)$; they have ion contents of 3–4 equivalents kg^{-1}, for example Dowex SBR™ (Dow Chemical). Weak-base, anion-exchange resins are made similarly but are aminated with a primary or secondary amine to form secondary or tertiary amine groups, for example Amberlite IRA-93™ (Rohm and Haas).

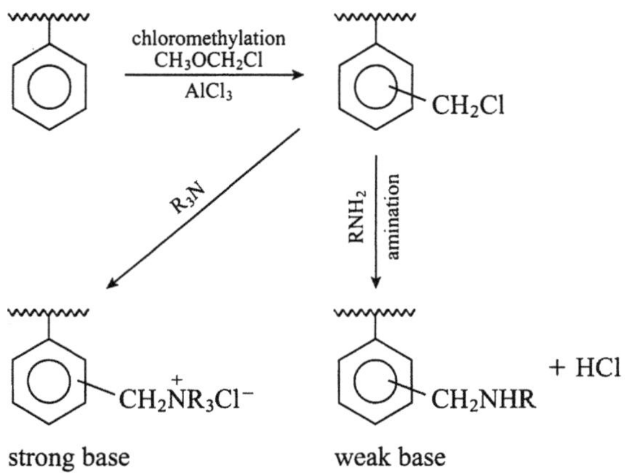

Resins used for demineralizing water are the strong-acid and strong-base types. The water has to make contact with both of them and so is passed through a bed of each type or through a mixed bed of both. A recent development is an ampholytic resin containing a weak acid and a weak base (Sirotherm™, ICI) which when cold absorbs ions but when hot releases them, thus when it is fully exchanged or exhausted it can be rinsed in hot water and re-used. Re-use of the strong-acid or strong-base types requires treatment with concentrated acid and alkali (to reverse the equilibrium), which is a more expensive process. The 'weak' resins do not totally demineralize the water like their 'strong' counterparts. However, the water they produce, containing 100–200 ppm of dissolved solids, is adequate for many purposes.

Some PS resins are specially prepared with chelating properties, making them highly selective towards certain ions. Thus a sodium amino-phosphonate resin, for example Duolite ES 467™ (Rohm and Haas) when added to a strong brine solution (25%) contaminated with 10 mg $^{-1}$ of calcium ion will selectively absorb the calcium ion until only 0.02 mg l^{-1} remains. This process is particularly useful for purifying the brine used in the chlor-alkali cell described earlier:

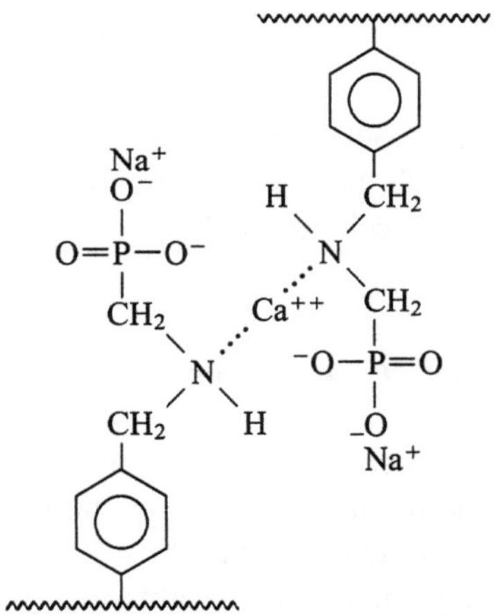

calcium amino-phosphonate complex

Some PS resins are made for use in non-aqueous media, for example Amberlyst A26™ (Rohm and Haas) which has a strong-base functionality.

Prior to the development of PS resins, phenol-formaldehyde (P-F) condensates were used as matrices but they have now been replaced. A few weak-base types still exist (e.g. Duolite ES 562™, Rohm and Haas) which are made by adding an amine during the polycondensation or after it is complete. These P-F condensates are used for enzyme fixation.

An expanding area of research is the heterogenization of homogeneous catalysts. Many of the organic polymer supports are based upon PS–DVB resins, thus besides their use in ion exchange they are of interest as catalysts. Some strong acid resins are sold for catalytic use, for example Amberlyst 15™ (Rohm and Haas) [16], and are used to manufacture oil-epoxide plasticizers and invert sugar from sucrose. Examples of other reactions for which these resins are very effective are: the alkylation and acylation of aromatic hydrocarbon, acylation of olefins, dehydration of *t*-butanol and the hydrolysis of esters. The effectiveness of a resin as an acid catalyst can be measured by the rate at which it decomposes *t*-butyl acetate to form acetic acid. The dry resin (0.5 g) is agitated in *t*-butyl acetate (10 ml) at 20°C for 2 h, samples are taken and the acetic acid produced determined by titration.

PS–DVB is not as stable in aggressive chemical environments at high temperature as is Nafion™, the catalytic use of which has been already mentioned. However, it can be improved if the sulphonate group is attached to an alkyl side chain [17]. The resin illustrated loses only 15% of its acidity when heated to 200°C for 50 h in water, instead of 70% lost by the ring-substituted analogue:

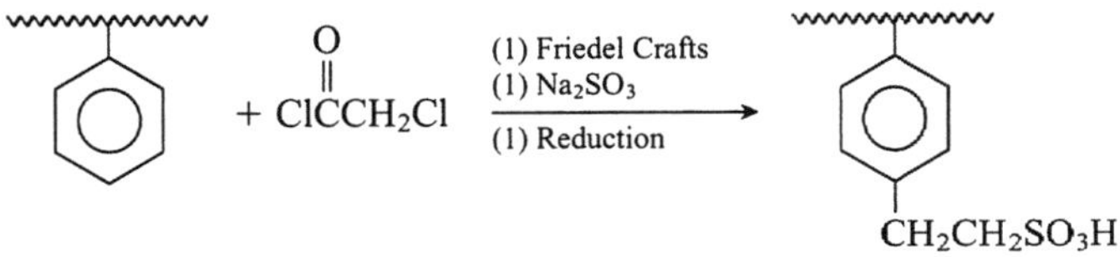

Cellulose-based ion-exchange polymers have been investigated. They are made by reacting ionic reagents with the hydroxyl groups present in the cellulose, for example to form sulpho-ethyl cellulose, $CH_2CH_2SO_3H$.

6.11 Polyelectrolytes based on carboxylates

Polyacrylate-type homopolymers are the most ionic of the organic polymers. They are water soluble and produce aqueous solutions with unusual and useful physical properties. They are generally made by free radical polymerization in aqueous solution. Very high molecular weights (e.g. 4×10^6) can be achieved which give very viscous solutions. Polyacrylic, polymethacrylic and polyitaconic acids are the three main types:

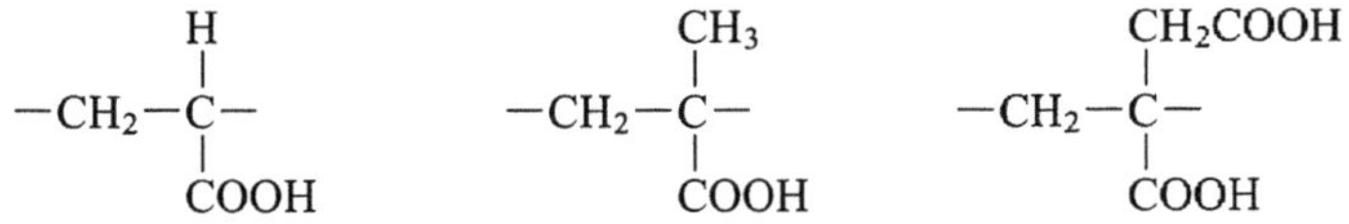

13.9 equivalent kg^{-1} acrylic | 11.6 equivalent kg^{-1} methacrylic | 15.4 equivalent kg^{-1} itaconic

Aqueous solution or dry powders of these materials are commercially available. Versicols™ (Allied Colloids) which are homopolymers of acrylic or methacrylic acids or their copolymers with acrylamide are sold in either form.

Texigels™ (Scott Bader) are aqueous solutions of similar materials and Texipols™ are their dispersions in an organic medium which can be pumped and compounded without the gelatin and viscosity problems sometimes encountered when these kinds of polymer are manipulated in water. Carbopols™ (B.F. Goodrich) are six grades of polyacrylic acid of varied molecular weight. Carbosets™ (B.F. Goodrich) are acrylic copolymers and have a similar carboxl content but generally can be dissolved in alkaline solutions. There are used mainly in coating applications; one such uses an aqueous solution of the ammonium or morpholine salt which on heating loses ammonia or morpholine and dries to form a water-repellent layer. They are designed to be covalently cross-linked with epoxides and so on but some applications use ionic cross-links made with zinc ions.

A different type of carboxylic polyelectrolyte is the structure below. It is a hydrolysed alternating copolymer of methyl vinyl ether and maleic anhydride. Gantrez An™ (GAF) and Sokalan CP2™ (BASF) are of the type

$$\begin{array}{ccccccc} & & OCH_3 & & & & \\ & & | & & & & \\ -CH_2- & & CH & - & CH & - & CH- \\ & & & & | & & | \\ & & & & CO_2H & & CO_2H \end{array}$$

A modern application of polyacrylic acid is for cements in dentistry. These are made by mixing an aqueous solution of the polymer with zinc oxide when the zinc salt precipitates out as highly cross-linked gel which rapidly sets very hard under oral conditions. A development of this is the replacement of the zinc oxide with a tooth-coloured glass powder which releases Al^{3+} and Ca^{2+} ions. These cements, called 'glass-ionomer' or ASPA (aluminosilicate polyacrylic acid), set very rapidly, bond well to tooth enamel and are compatible with living tissue [18].

6.12 Polymers with integral ions

6.12.1 Halato-telechelic polymers [19]

Halato-telechelic polymers (HTPs) are curious materials which are not at present of commercial value. Those based on dicarboxy polybutadiene have already been discussed in Section 6.6. Metal dicarboxylates like calcium sebacate

$$\cdots[\cdots{}^{-}OOC(CH_2)_8COO\cdots Ca^{+}\cdots]\cdots$$

are simpler examples which can behave either as crystalline monomeric salts or as amorphous polymers with a degree of polymerization of 25.

Fusion of the crystalline salt gives a viscoelastic melt which on solidification becomes an amorphous mass like a highly cross-linked thermoplastic. The halatopolymeric transformation occurs because crystallization of the solidifying melt is impeded by the high viscosity and the entanglements of the long alkyl

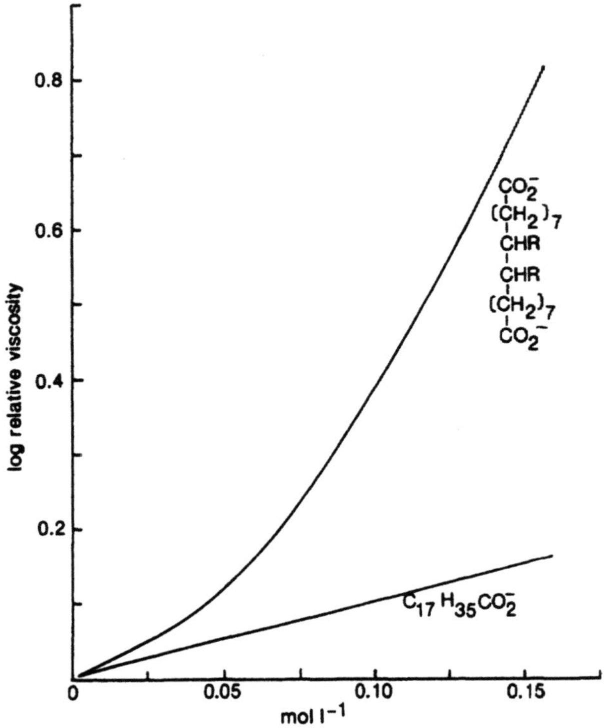

Figure 6.15 Relative viscosities of zinc carboxylate in pyridine. Source: Cowan, J.H. and Teeter, H.M., *Ind. and Engng. Chem.*, **36**, 148–52, 1944.

chains. Impurities and the absorption of water may, however, subsequently promote slow recrystallization.

Many metal dicarboxylates can be polymerized in this way but the requirement that they melt before decomposing limits their number. Aromatic acids and aliphatics with less than eight carbon atoms yield non-melting salts. Polymerization can also be effected by melting the acid with the metal acetate or oxide at 150°C.

The polymeric nature of these materials has been shown by measurements of solutions viscosity (Fig. 6.15). At low concentration in pyridine a zinc dicarboxylate has the same viscosity as a similar monocarboxylate but at high concentration it is much more viscous because of the increased degree of association.

6.12.2 Ionenes [20]

Tetramethyl-6, 3-ionene bromide is an example of an ionene:

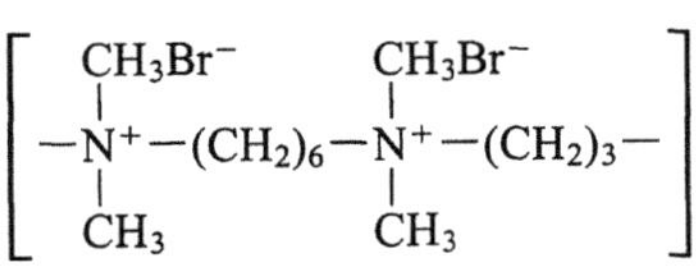

They are synthesized by the Menschutkin reaction:

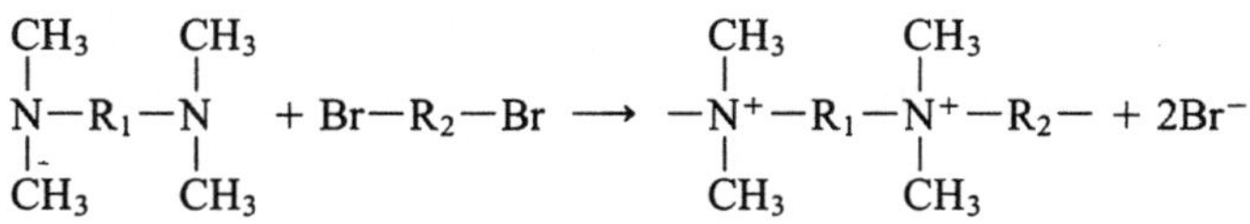

Some ionenes have been studied for their bacteriostatic and bactericidal activity. The tetramethyl ionene above has been evaluated for its pharmacological and antiheparin action. Ionenes with segments of polypropylene oxide in the backbone have been evaluated as thermoplastic elastomers. A type of redox polymer has been made by reacting 4,4-bipyridyl as the difunctional tertiary amine with various alkyl dibromides (polyviologens) [21]; the advantages of these over redox polymers based on hydroquinone is their simple synthesis and hydrophilic nature.

6.12.3 Polyethylenimine [22]

Polyethylenimine (PEI) is an integral polyelectrolyte which is available commercially, for example Polymin™ (BASF). It is formed by the ring-opening of ethylene imine (aziridine) to form a polyamine, 50% of which has the expected secondary amine functionality and 50% primary and tertiary amine functionality owing to branching:

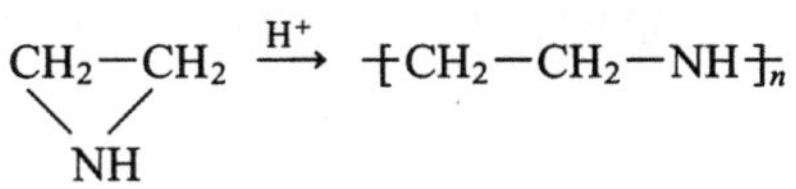

and

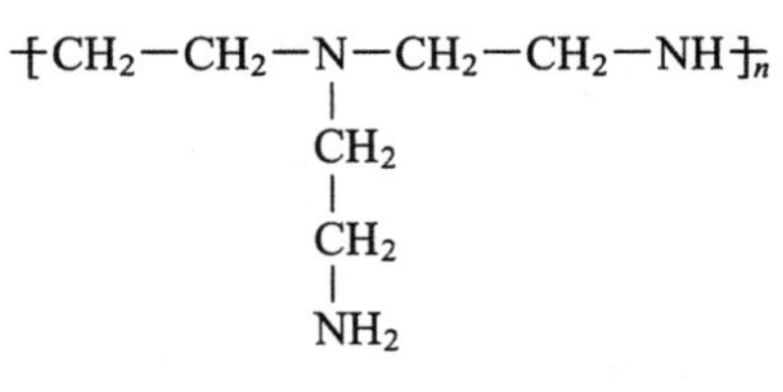

PEI has typical polyelectrolyte properties, for example water solubility, and is mainly used as a size, flocculating agent or protective colloid, notably in the paper and textile industries because of its ability to bind to cellulosic fibres. Membranes based on PEI were introduced in 1970 for use in reverse osmosis to desalinate water. These membranes, known as NS100 and NS101 [23, 24] are made by forming a PEI skin on polysulphone support and insolubilizing it by treatment with toluene di-isocyanate or phthaloyl chloride to produce a polyurea or polyamide:

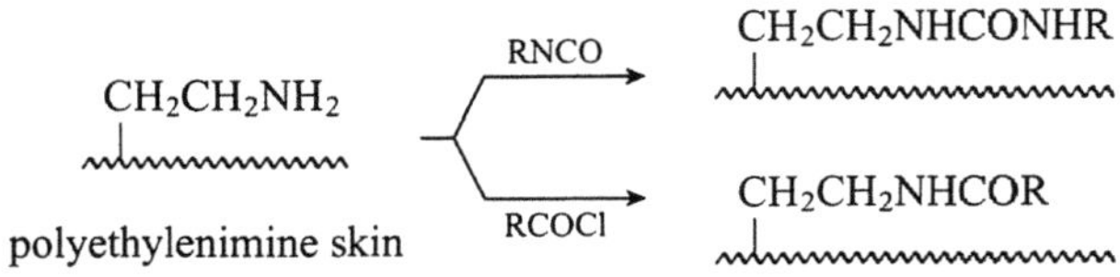

6.13 Polyelectrolyte complexes

When aqueous solutions of anionic and cationic polyelectrolytes are mixed a precipitate forms which is a complex of the two polymers with properties different from either. The reacting polymers may be strong or weak electrolytes and one of them may be of relatively low molecular weight. The complex may be **stoichiometric** or **non-stoichiometric** according to the composition of the mixture or the thermodynamics of its formation.

Generally, polyelectrolyte complexes (PECs) are insoluble and behave like cross-linked non-ionic polymers. They retain some of their parent polymers' hydrophilicity, particularly if they are non-stoichiometric. Those derived from strong electrolytes are hard, brittle and readily powdered but can be plasticized with salts into tough horn-like or leathery materials. A PEC's molecular structure depends upon the spatial distributions of ions along the two contributing polymer chains, but for simplicity it may be visualized as ladder-like:

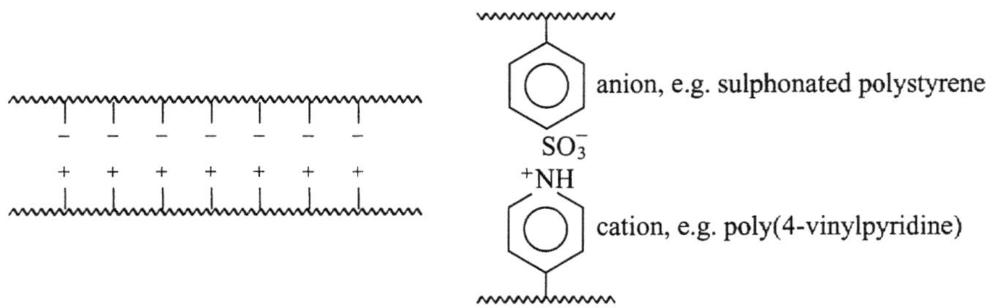

The complexes can be made by mixing as above, but careful control is necessary in choosing the concentrations. Another method is forcefully to mix concentrated solutions in a high-shear apparatus. Other techniques include *in situ* polymerization whereby an ionic monomer is polymerized in an aqueous solution of the complementary polyelectrolyte, that is a kind of template polymerization. In this way PECs may have some fundamental significance in prebiological evolution because their formation is analogous to the reactions between complementary pairs of biopolymers responsible for the self-assembly of biological structures.

PECs have minor applications of a varied nature but it is hard to pick out a common thread which stems from a unique characteristic of PECs. They have been used, for instance, in the micro-encapsulation of dyes for coating paper and have potential in membrane processes where special permselective properties may be required. Complexes having a small ion content (which strictly speaking are ionomer complexes rather than PECs) are being evaluated as a means of making immiscible polymers compatible, such as polystyrene and poly (ethyl acrylate). Thus, when these two polymers are mixed they normally form an opaque mass which has two T_g values representative of the individual polymers, but when a polyanionic, lightly sulphonated polystyrene (5 mol%) is mixed with an equivalent cationic poly (ethylacrylate-co-4-vinyl pyridine) the mixture is homogeneous and transparent with a single T_g. This compatibility is a consequence of the ionic interation and salt formation [25].

6.14 Blends of polymers and salts

Generally speaking, polymers containing chemically bound ions are very different from similar compositions which are only physical blends [26]. There are, however, some physical blends which show the characteristics of polymers with bound ions because a polymer/salt complex is formed. An example is polypropylene oxide containing zinc chloride [26].

These blends can be made by drying down mixed solutions in methanol. Their T_g values increase from −60°C to +30° as the proportion of zinc chloride is raised from 0 mol% to 30 mol%, but above 30% there is no further increase because the polypropylene oxide is saturated with zinc chloride. The blends are tough, transparent, easily moulded and absorb less water than expected from their salt content. X-ray diffraction shows no crystalline $ZnCl_2$ except at concentration above 30 mol%, but unlike the ionomers there is no narrow-angle scatter suggestive of an ionic microphase. Other salts behave like zinc chloride, for example, $FeCl_3$ and ZnI_2, but, strangely, $CuCl_2$ and LiCl do not and their blends are two-phase, opaque and water-sensitive. The complex is believed to be a cyclic structure involving the oxygens from adjoining polypropylene oxide units:

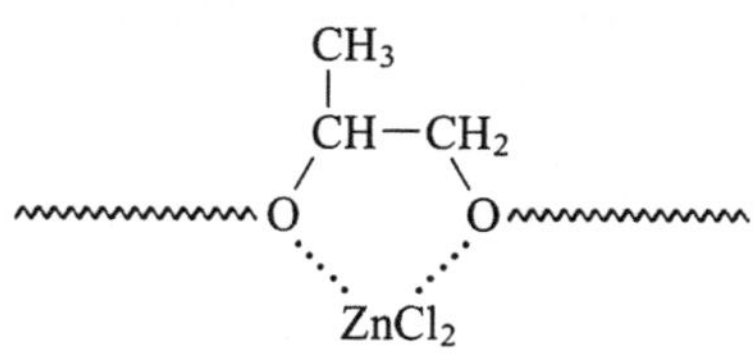

6.15 Biological and inorganic ionic polymers

Biological polymers are often ionic and it is useful to remind oneself that ionic polymers are not entirely an invention of humankind, it is only some of their applications that are. Biopolymers are a world apart from the synthetic materials in this chapter, but for completeness the ionic nature of the proteins, polysaccharides and nucleic acids should be pointed out.

The proteins are linear polyamides of α-amino carboxylic acids. Of the 20 or so acids which occur in nature, six are difunctional in either NH_2 or CO_2H and form polyamides in which the extra functional groups are left pendant. Thus glutamic acid forms pendant $-CH_2CH_2CO_2H$ and lysine $-CH_2CH_2CH_2NH_2$. Gelatin, derived from the protein collagen, is a polyanion containing 12% of glutamic acid. Amongst the polysaccharidies are gums such as gum arabic which contains units of glucuronic acid having pendant $-CO_2H$. Heparin is a poly(hexasaccharide) with four types of pendant anion, $-SO_3^-$, $-CH_2OSO_3^-$, $-NHSO_3^-$ and $-CO_2^-$. The nucleic acids are polyampholytes in which each nucleotide unit of the backbone has an integral phosphoric acid and a pendant amine base:

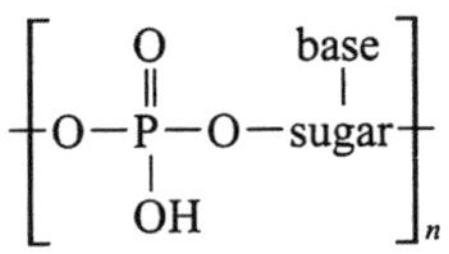

nucleotide

Inorganic ionic polymers arise in the from of mineral silicates, inorganic glasses and polyphosphates. They are a complex group of materials outside the considerations of ordinary polymer chemistry but they are ionic polymers of interest in their own right. In their simplest forms they are:

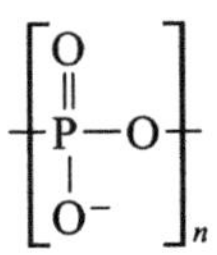

and

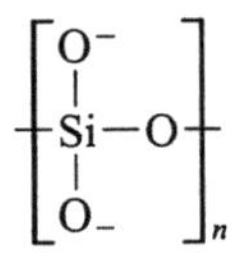

6.16 Conclusions

Synthetic ionic polymers span a vast range of physical properties. A few find applications as commodity plastics, a fair number find applications in small quantities as highly specialized materials while many are, at the moment, research curiosities which may find applications in the future. The position that any particular polymer occupies within this property range and the types of application for which it is suitable depend mainly upon the concentration of ions within it. While the type of ion and the type of covalent backbone matter, they are of secondary importance. Ions, whatever their type, impart to various polymers certain well-defined characteristics which make possible different applications depending on the amount of ion present. This is summarized in Table 6.8.

Synthetic ionic polymers are mostly relatively modern materials. However, as thermoplastics and thickeners they have probably reached a degree of maturity

Table 6.8 Applications for organic ionic polymer in relation to the amounts of ion present

Ion content (equivalent kg^{-1})	Designation	Dominant property	Application	Example
<0.5	Ionomer	Ionic cross-linking	Thermoplastic	Ethylene–acrylic acid copolymer
–1.5	Ionomer	Ion exchange	Membranes (electrodialysis, etc.)	PTFE copolymer
–1.5	Ionomer	Hydrophilicity	Membranes (reverse osmosis)	Sulphonated polyarylylene
>4	Polyelectrolyte	Water solubility	Thickeners, dispersants, flocculents and sizes	Polyacrylic acid or copolymer
>4	Polyelectrolyte (cross-linked)	Ion exchange	Ion-exchange resins	Sulphonated polystyrene
4.14	Polyelectrolyte	Ionic cross-linking	Dental cements	Polyacrylic acid

that may not allow much novel development in the future. As ion-exchange resins, where their ionic functionality is vital to the application, they probably achieve their greatest importance, although that too is now a mature industry. They do have a bright future, however, probably as specialist materials manufactured in small quantities for use in novel technologies.

The type of novel technologies which are becoming increasingly important are separation processes involving membranes. Undoubtedly these will demand organic ionic polymers tailor-made for their hydrophilicity and ion-exchange capability coupled with a high level of permselectivity. Associated with these processes will be the requirement of selective catalytic power, with the ionic polymer in membrane form or as an ion-exchange type granule.

References

1. Eisenberg, A., Farb, H. and Cool, L.G. (1966) *J. Polym. Sci.*, A–2, **4**, 855–68.
2. Bazuln, C.G and Eisenberg, A. (1981) *Ind. Eng. Chem. Prodn. Res. Devel.*, **20**, 271–86.
3. Jenkins, D.K. and Duck, E.W. (1975) In *Ionic polymers* (ed. L. Holliday), Applied Scieince Publishers, London, Ch. 3, p. 177.
4. Jenkins, D.K. and Duck, E.W. (1975) In *Ionic polymers* (ed. L. Holliday), Applied Science Publishers, London, Ch. 3, p. 185.
5. Rees, R.W. and Reinhandt, H.G. (1976) US Patent, 3997, 487.
6. Makowski, H.S., Lundberg, R.D., Westerman, L. and Bock, J. (1978) *Am. Chem. Soc. Divn. Polym. Chem. Polym., Preprints*, **19**(2), 292–7.
7. Maynard, T.G. and Johnson, R.R. (1963) *Rubber Chem. and Technol.*, **36**, 963–74.
8. Brockman, N.L. and Eisenberg, A.E. (1983) *J. Polym. Sci.: Polym. Chem. Edn.*, **21**, 3563–71.
9. Stinson, S.C. (1982) *Chem. and Engng. News*, 15 March, 22–25.
10. Kyu, T. (1985) *Am. Chem. Soc. Symp. Ser.*, **269**(4), 365–405.
11. Flett, D.S. (ed.) (1983) *Ion-Exchange Membranes*, Ellis Horwood, Ch. 8–11, various authors.
12. Chaudhari, D.D. (1983) *Chem. and Ind.*, 18, July p. 569

13. Noshay, A. and Robeson, L.M. (1976) J. Appl. Polym. Sci., **20**, 1885–1903.
14. Kimura, S.G. (1971) *Ind. Engng. Chem. Prodn. Res. Devel.*, **10**(3), 335–9.
15. Huang, R.Y.M. and Kim, J.J. (1984) *J. Appl. Polym. Sci.*, **29**, 4017–27.
16. Amberlyst 15, Rohm and Haas Co. (1980) Tech. Booklet, IE-94–65/78.
17. Klein, J. (1981) *Makromol. Chem., Suppl.* 5, 155–78.
18. Wilson, A.D. (1981) *Polym. Sci. Technol.*, 25, (*Biomed. Dent. Appl. Polym.*), **14**, 419–26.
19. Economy, J. and Mason, J.H. (1975) In *Ionic polymers* (ed. L. Holliday), Applied Science Publishers, Ch. 5.
20. Remboum, A., Baumgartner, W. and Eisenberg, A. (1968) *Polym. Lett.*, **6**, 159–71.
21. Factor, A. and Heinsohn, G.E. (1971) *Polym. Lett.*, **9**, 289–95.
22. Tomalia, D.A. and Killat, G.R. (1985) *Encyclopedia of Polymer Science and Engineering*, John Wiley, Chichester and New York, **1**, 680–739.
23. Cadotte, J.E. and Peterson, R.J. (1981) *Am. Chem. Soc. Symp. Ser.*, **153**, Ch. 21, pp. 305–325.
24. *Synthetic membranes, Vol. 1: Desalination*, (Ed. A.F. Turbac).
25. Eisenberg, A., Smith, P. and Zhou, Z-L. (1982) *Polym. Engng. and Sci.*, **22**(17), 1117–22.
26. James, D.B., Wetton, R.E. and Brown, D.S. (1978) *Am. Chem. Soc. Divn, Polym. Chem. Polym. Preprints*, **19**(2), 347–52.

7 Liquid crystal polymers

D. COATES and M. VERRALL

7.1 Introduction to thermotropic liquid crystals

Although thermotropic liquid crystal phases have been recognised since 1888 [1] the concept of liquid crystal polymers was not established until the mid-1970s. Therefore, to help understand liquid crystal polymers, which can in some instances be regarded as very viscous analogues of low molar mass systems, the important physical features of the more comprehensively studied low molar mass system is described.

Thermotropic liquid crystals, which are formed by the action of heat on a solid, are exhibited by organic compounds constituted from anisotropic molecules; for many years only rod-shaped anisotropic molecules (Table 7.1) were generally considered, and indeed today these are the only type that have found commercial applications. These types of liquid crystal phase are now referred to as calamitic to differentiate them from other recently discovered liquid crystal types derived from alternative anisotropic shaped molecules such as discs [2], for example molecules based on hexasubstituted truxenes and hexasubstituted triphenylenes:

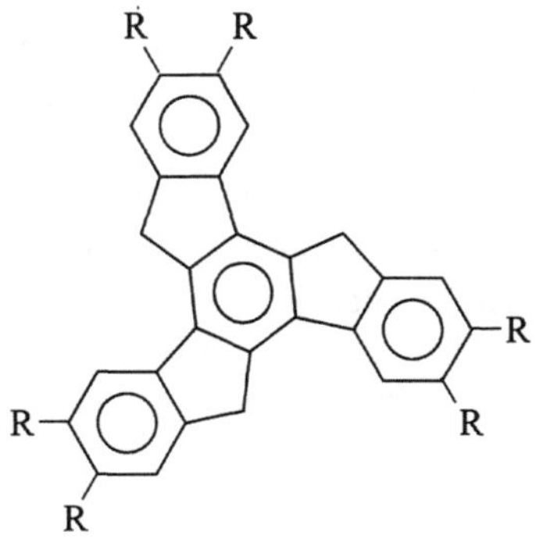

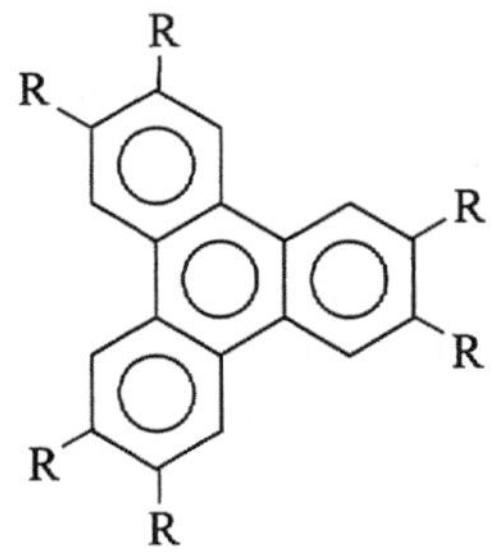

these form discotic liquid crystal phases in low molar mass and liquid crystal polymer system [3].

More recently, lath-shaped molecules and phasmids (a hybrid of calamitic and discotic phases) have also been shown to exhibit liquid crystal phases. In the future it is probable that liquid crystal polymers incorporating these units will also be developed.

Figure 7.1 shows the typical sequence of heating a solid, via liquid crystal phases to the isotropic liquid. Two general types of liquid crystal phase have been recognized and are referred to as smectic and nematic.

Table 7.1 Data for some typical thermotropic liquid crystals, showing transition temperatures between the crystalline (C), nematic (N), smectic (S) and isotropic (I) phases. S_C = tilted and disordered molecules within layers in the smectic phase; S_A = orthogonal and disordered molecules within layers in the smectic phase

	Transition temperature (°C)				
Structure	$C \rightarrow N$	$C \rightarrow S_C$	$S_C \rightarrow S_A$	$S_A \rightarrow N$	$N \rightarrow I$
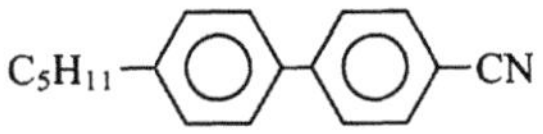	21	–	–	–	35
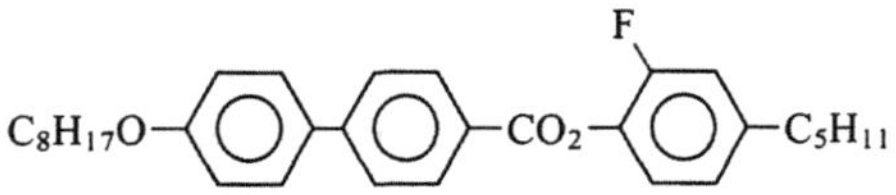	–	49	121	128	164
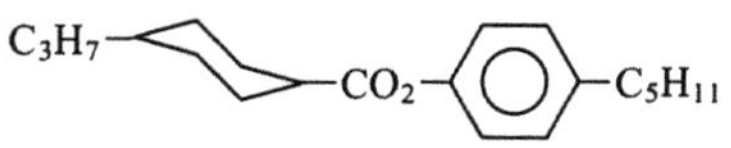	32	–	–	–	(29)
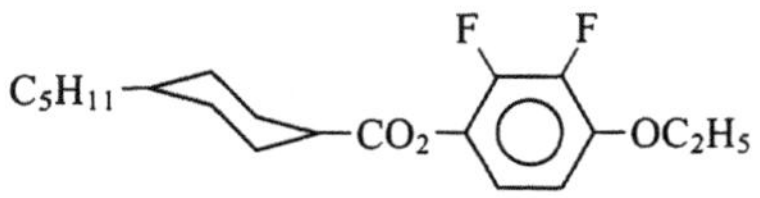	51	–	–	–	63

A variant of the nematic phase is the chiral nematic phase, often called the cholesteric phase, which replaces the nematic phase in liquid crystals which contain a chiral centre. Most early examples of this phase were exhibited by derivatives of cholesterol, hence its earlier trivial name.

Discotic liquid crystals exhibit analogous phases [2] known as columnar (in which the discs form columns and of which there are several types) and discotic nematic.

In liquid crystal polymers the same phase sequences are encountered, with the addition of a glass transition; as with other polymers some crystallinity can also occur. The phase transisitons can be detected as endotherms and exotherms on heating and cooling, respectively. A characteristic of liquid crystal transitions is that they occur at a similar temperature on cooling and heating cycles, unlike

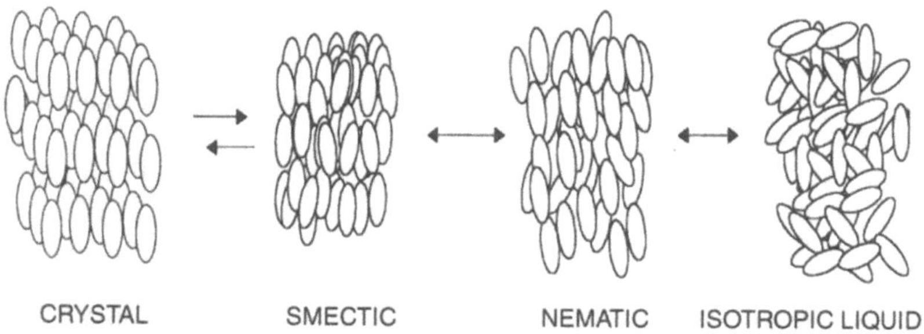

Figure 7.1 Nematic and smectic thermotropic liquid crystal phases are intermediate between a solid and the isotropic liquid.

melting and crystallization phenomena which usually occur at very different temperatures owing to supercooling of the melt prior to crystallization.

7.1.1 The nematic phase

The nematic (N) phase is by far the most studied and commercially important liquid crystal phase. The long axes of the molecules tend to lie parallel to a common axis, called the director; however, they have only short-range order; in the bulk, the director varies throughout the volume of the sample, giving rise to local changes in refractive index leading to a light-scattering, turbid liquid. In liquid crystal applications, which are usually based on thin films, it is critical that the director is unidirectionally aligned to provide a pseudo single-crystal orientation. In the nematic phase, the molecules undergo significant rotation about their long axis (10^{-10} s), end-over-end tumbling (10^{-5} s) and gyratory motions. Clearly, these motions and correlation times will be very restricted in liquid crystal polymers.

The nematic phase of liquid crystal polymers has the same microscopic textural appearances as in low molar mass systems and is used to identify the phase when examined as a thin film between crossed polarizers. For a polymer, the nematic phase is relatively fluid and has anisotropic flow properties, in main-chain liquid crystal polymers this has attractions in injection-moulding applications.

The nematic phase is used in nearly all liquid crystal displays; however, because a single component possessing all the required properties for an application has never been found, all applications involving the nematic phase use mixtures of many components; this may be indicative of what will have to be done in liquid crystal polymers.

7.1.2 The chiral nematic phase

The chiral nematic (N*) phase is a direct analogue of the nematic phase, formed when the molecules contain a chiral centre; it is characterized by a helical arrangement of the molecules (Fig. 7.2). The chiral nematic phase is particularly interesting for its unique optical properties [4, 5]. when light is shone on the helix 50% of certain wavelengths of light [$\lambda_{max} = \tilde{n}P$, where $\tilde{n}$ is the mean refractive index, P is the pitch of the helix and where the bandwidth is given by $\Delta\lambda = P\Delta n$, where $\Delta n = n_e - n_o$, subscripts e and o indicating light split into extraordinary and ordinary rays, respectively (subsection on optical properties in Section 7.1.4)] are reflected and circularly polarized with the same sense as the helical pitch sense; the other 50% of this light is transmitted with the opposite sense of circular polarization. All other wavelengths are unaffected unless polarized light is used in cases where a very large rotation of the polarization direction is found. In most compounds the pitch length has some small to modest temperature dependence (owing to changes in order parameter; Section

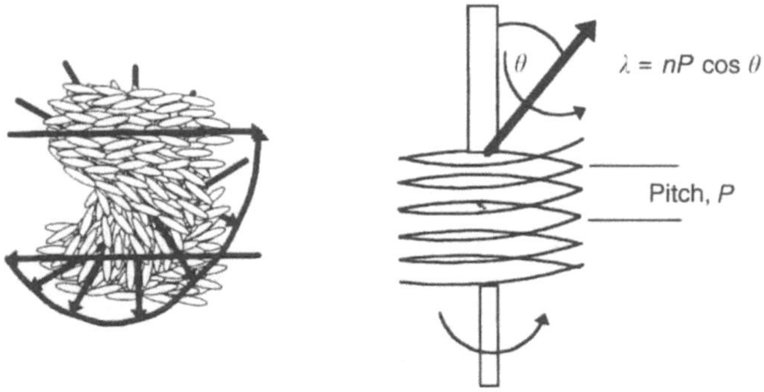

Figure 7.2 Diagrammatic depiction of the molecular orientation of the chiral nematic phase, showing that 50% of incident light of a specific wavelength, λ, is reflected and circularly polarized whereas the other 50% is transmitted. n = refractive index.

7.1.4 but in certain materials in which the N* phase is 'unwound' by an underlying S_A (or S_C^*) phase (Section 7.1.3 below) this can be extreme such that the film can, on cooling, change in appearance from blue to red (and infra-red); this effect has found use in thermography and thermometers.

7.1.3 The smectic phase

The smectic (S) phase is usually depicted as having molecules arranged in a regimented layer structure (Fig. 7.1), and although this simple picture may aid our understanding of the properties of the smectic phase it is more precisely described [6] as having molecules which are arranged to provide a single sinusoidal density wave.

In side-group liquid crystal polymers the polymer backbones prefer to lie together and between the smectic layers, therefore the majority of side-group liquid crystal polymers exhibit smectic rather than nematic phases (where the polymer backbone is less easily orientated). The stability of the smectic phase is somewhat greater than it would be for low molar mass analogues because the polymer backbone provides some constraint on the free movement of the side groups. However, too much constraint (such as found in methacrylate backbones) leads to a lowering of transition temperatures relative to more flexible backbones such as polysiloxanes. By contrast, main-chain liquid crystal polymers are invariably nematic.

Within the layers the molecules can be arranged in several ways, for example, orthogonal and disordered (known as smectic A, S_A), orthogonal and hexagonally ordered (known as smectic B, S_B) or tilted and disordered (known as smectic C, S_C), etc; in total at least 12 variants are known. However, in recent years, some of the more ordered variants, in which there is some correlation of molecular ordering from one layer to another, have been reclassified [7] as soft crystals.

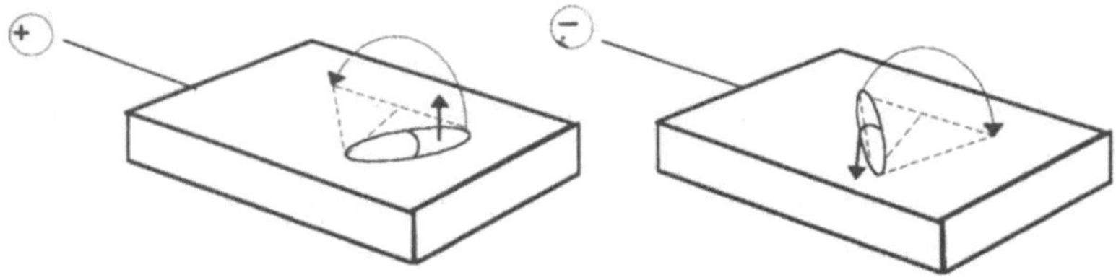

Figure 7.3 When an electric field is applied to a surface-stabilized chiral smectic C phase the molecules undergo a reorientation around a cone. The half angle of this cone is equal to the tilt angle of the molecules within the layers.

In low molar mass systems, the S_A and S_C phases are of particular importance; the S_A phase can be used in laser addressed optical data storage devices and bistable displays [8]. When composed of chiral molecules, the S_C^* phase exhibits an optically active liquid crystal phase (denoted by an asterisk superscript) in which the molecules, which are tilted within the layers, orientate themselves into a helix. This is additional to the normal optical activity associated with chiral molecules containing a chiral centre. Such materials are of interest because they can exhibit ferroelectric properties which can be utilized in fast switching (submillisecond) display devices [9]. The pitch of the helix and the magnitude of the ferroelectricity (spontaneous polarization, P_s) are strongly dependent on the nature and position of the chiral group.

When aligned in a thin film such that the long molecular axes are orientated in one direction (Fig. 7.3) there is a resultant dipole which can be arranged such that it is substantially perpendicular to the long molecular axes [10]. When a small direct current (d.c.) electric field is applied across the thin (1–3 μm) film, the dipole moves to realign in the applied field and the long molecular axes of the molecules move around a conical trajectory giving a change in angle and thereby change in transmission value when viewed between crossed polarizers. Reversing the d.c. electric field also reverses the orientation of the molecules. It is important that the effect of the inherent helix is nullified within the cell by using long-pitch-length materials and that the molecules stack in a bookshelf geometry between the two substrates. This is the basis of surface-stabilized ferroelectric displays [11]. Liquid crystal polymer based ferroelectric materials are the subject of much current research.

When the S_A phase contains chiral molecules it can exhibit an 'electroclinic' effect which is also of potential use in fast switching (<1 μs) electro-optic applications [8, 12, 13]. In this case the liquid crystal phase is not optically active (because the molecules are orthogonal to the smectic layers), hence an asterisk is not used. (New nomenclature to further define these phases is being devised.) In the electroclinic effect the molecules are induced to have a small overall tilt within the layers (which can be visualized as a biasing of the normal gyratory motion of the molecules about the long molecular axis and in the direction of the smectic layers). This is characterized by being proportional to the applied field; on removal of the field the tilt quickly relaxes back to normal.

Most liquid crystal polymers exhibiting an S_C^* phase also exhibit a chiral S_A phase such that the electroclinic effect usually accompanies ferroelectric properties; indeed, there is often a complex biphasic region present in which both features occur.

Smectic phases also exhibit characteristic microscopic textures when viewed as thin films between crossed polarizers; this is very useful as it provides a means of identification [7]. However, owing to the much higher viscosity of liquid crystal polymers this technique is more restricted and very often long anneal times are required before a smectic phase type can be microscopically identified; in some cases this has proven to be impossible. X-ray scattering studies can provide a more conclusive identification but even this technique often requires aligned samples which are again difficult to achieve. As a result many liquid crystal polymer phases remain unidentified.

7.1.4 Anisotropic properties

The anisotropic properties of the liquid crystal phase are key to its usefulness and success and can be changed by modification of the molecular structure. The same principles apply to both low molar mass liquid crystals and liquid crystal polymers.

Transition temperatures. In general, longer molecules have higher liquid crystal transition temperatures that are significantly influenced by the additional effect of terminal groups; polar terminal groups which extend the conjugation of the molecule tend to provide higher transition temperatures. However, these positive features are often offset by the higher melting points that they provide. Lateral substituents broaden the molecule and thus reduce intermolecular attractiveness forces which result in reduction of transition temperatures, but can also reduce melting points. Longer terminal alkyl chains increase lateral attractions and therefore layered structures (smectic phases) are favoured (Table 7.1). Detailed molecular engineering can be used to produce selectively particular phases and to tailor transition temperatures and other physical properties [14, 15].

In the case of liquid crystal polymers the transition temperatures are usually spread over several degrees rather than being very sharp (ranging over more than 1°C). A biphasic region occurs in which both upper and lower temperature phases exist, this is explained by the existence in a liquid crystal polymer sample of many molecular weight fragments each of which exhibits slightly different transition temperatures (Section 7.3.10).

Order parameter. The degree to which the long molecular axes are aligned with respect to a director which describes their orientation is called the order parameter (S). Perfect ordering, where all the molecules point in one direction, would have $S = 1$ (practically, this is never achieved) whereas complete disorder is when $S = 0$ (e.g. the isotropic liquid). Typical values of S are 0.4 near the nematic-to-isotropic transition (T_{N-I}) and about 0.7 when at a low reduced temperature ($T_{red} = T_{exp}/T_{clpt}$, where T_{exp} is the temperature at which the experiment is carried out, and T_{clpt} is the clearing point temperature), e.g. some 50°C or so below T_{N-I}. S is also very dependent on the shape and polarizability of the molecules.

Dielectric properties. In most, but not all, cases the dielectric permittivity (ε) parallel ($\varepsilon_{\parallel}$) to the long molecular axis is higher than across it ($\varepsilon_{\perp}$), thus the resultant dielectric anisotropy is positive ($\Delta\varepsilon = \varepsilon_{\parallel} - \varepsilon_{\perp}$); in this case, the molecules align parallel to an applied electric field.

Incorporating a polar terminal group (e.g. nitrile is often used) increases the electron polarizability along the long molecular axis, which results in a higher $\varepsilon_{\parallel}$ Polar lateral substituents and especially small ones,. for instance fluorine, increase $\varepsilon_{\perp}$ such that materials of negative dielectric anisotropy can be produced, of particular importance is the 2,3-difluoro phenyl group (Table 7.1).

Optical properties [16, 17]. When light is incident on an aligned sample of liquid crystal it is split into it two component rays knows as the ordinary (refractive index n_0 or $n_{\perp}$) and extraordinary (refractive index n_e or $n_{\parallel}$) rays; these are polarized perpendicular or parallel to the liquid crystal (nematic) director, respectively. The speed of the two rays through the sample is different owing to the aligned anisotropic molecules having different, direction-dependent refractive index values, thus the two beams are retarded and therefore deflected by different amounts. Liquid crystals exhibit a positive birefringence ($\Delta n = n_e - n_0$).

The refractive index values are influenced by the polarizability of the individual molecules and by how well ordered they are, an approximate relationship being

$$S \approx \frac{n_e^2 - n_o^2}{\tilde{n}^2 - 1}$$

where $\tilde{n}$ is the mean refractive index, $(2n_e^2 + n_o^2)/3$. Thus aromatic compounds with a polar terminal group have high n_e values and high Δn values; conversely, cyclohexyl-based compounds have much lower Δn values. At lower reduced temperatures, Δn increases because S and density increase.

Therefore, the molecules of the liquid crystal phase can be reorientated by an electric field acting upon an induced dipole moment (as above) or on a permanent dipole (ferroelectric S_C^*) to produce a change in refractive index which can be observed between crossed polarizers. Electro-optic displays usually consist of aligned thin films of liquid crystal in which this effect is optimized.

Side-group liquid crystal polymers can also be reorientated in the same way in electric fields; in the case of nematic liquid crystal polymers these orientations slowly relax when the field is removed; however, in smectic phases relaxation occurs to only a small extent.

Magnetic properties. Most liquid crystals are diamagnetic and in an analogous way to the dielectric anisotropy exhibit a diamagnetic anisotropy ($\Delta\chi$). This is usually positive and relatively weak. Thus the long molecular axis can be aligned parallel to an applied magnetic field and although this is of little use in devices it does provide an alignment technique for thick films required in some physical studies.

7.1.5 Lyotropic liquid crystals

Certain elongated molecules which contain a hydrophilic terminal group (e.g. phosphate, etc.) dissolve in water to produce layered structures of molecules (lamellar phases) or elongated miscelles which in turn orientate themselves; these are referred to as lyotropic liquid crystal phases and are formed by the action of a solvent on a solid. Lyotropic liquid crystals are important in the cosmetic and detergent industry.

7.2 Liquid crystal polymers

The traditional molecular requirements for low molar mass liquid crystals are also largely necessary for polymeric liquid crystals, where the monomeric units will typically be composed of a rigid core linked by a spacer group to the polymeric backbone, or to another mesogenic group.

Liquid crystal polymers are most commonly grouped into two categories – 'main-chain' liquid crystal polymers in which the mesogenic groups are linked to each other in a head-to-tail arrangement, and 'side-group' liquid crystal polymers in which the mesogenic groups are attached via a linking group to a polymer backbone. The structural diversity of materials in these categories is ever increasing and the distinction between the two fields often becomes blurred, most notably with combination polymers bearing mesogenic groups in both the polymer backbone and as side groups [18, 19]. A selection of typical liquid crystalline polymer structures is shown in Fig. 7.4.

7.2.1 Main-chain liquid crystal polymers

The field of main-chain liquid crystal polymers is a subject which has received intensive study since the first commercial examples of such materials were produced by Du Pont in the 1970s. Prior to this the occurrence of liquid crystal phases in solutions of rigid, rod-like polymer molecules had been predicted [20–22] and was first experimentally confirmed in 1956 [23]. Main-chain liquid crystalline polymers have taken a position as commodity polymers with exceptional properties and the field has been widely reviewed [24–26]. In this chapter we will concentrate on side-group liquid crystal polymers.

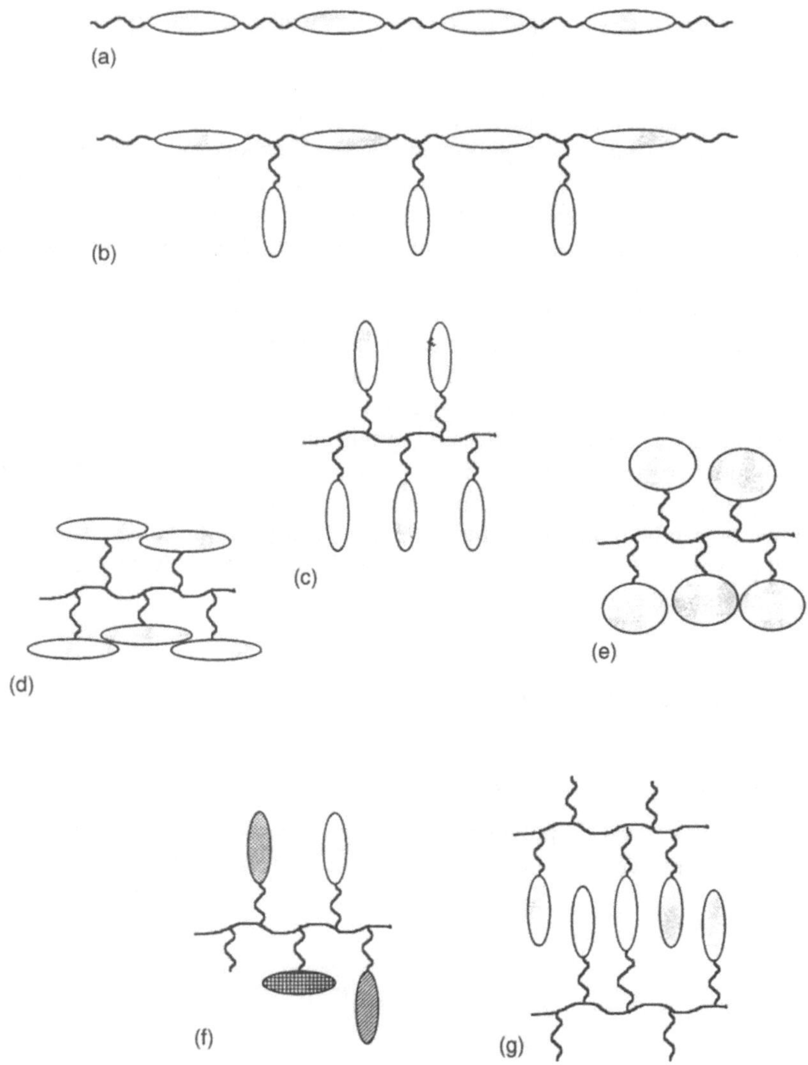

Figure 7.4 Various types of liquid crystal polymer: (a) main-chain; (b) mixed main-chain and side-group; (c) side-group (terminally attached); (d) side-group (laterally attached); (e) discotic side-group; (f) mixed side-group; and (g) liquid crystal elastomer.

7.2.2 Side-group liquid crystal polymers

The first attempts at combining mesogenic behaviour with macromolecular properties involved rigid core groups of the type well known in low molar mass liquid crystals being attached directly to a polymer backbone [27, 28]. This work did not generally produce materials displaying the desired broad temperature range liquid crystal phases, as the direct bonding of the side groups to the backbone resulted in very rigid materials which tended toward crystallinity and high glass transition temperatures (T_g).

In 1978 a breakthrough occurred which sparked an explosion of work in this field; Finkelmann *et al.* [29] proposed that liquid crystallinity could be promoted by inserting a flexible spacer group between the backbone and core group, and they went on to prepare the first examples of liquid crystal polymers in which the mesogenic groups were decoupled from the polymer backbone by alkylene spacer groups. This approach allowed the mesogenic side groups more readily to form the liquid crystal phases associated with low molar mass liquid crystals, whilst the polymeric backbone can impart some of the desirable characteristics of macromolecular materials. This opened up a wide range of possible applications where the anisotropic nature of liquid crystal materials could be combined with the physical and mechanical properties of polymers.

The majority of materials disscussed in this chapter have rod-shaped or lath-shaped liquid crystal side groups of the terminally attached variety and, as such little explanation of the behaviour of other strucural types will be given. Further information can be found in the literature on laterally attached side-chain liquid crystal polymers [30–32], discotic side-group liquid crystal polymers [3, 33, 34] and numerous other conformations [35, 36].

7.3 The structure of side-group liquid crystal polymers

7.3.1 General structural features

Side-group liquid crystal polymers of the type proposed by Finkelmann *et al.* [29] may be considered as consisting of four main constituent parts:

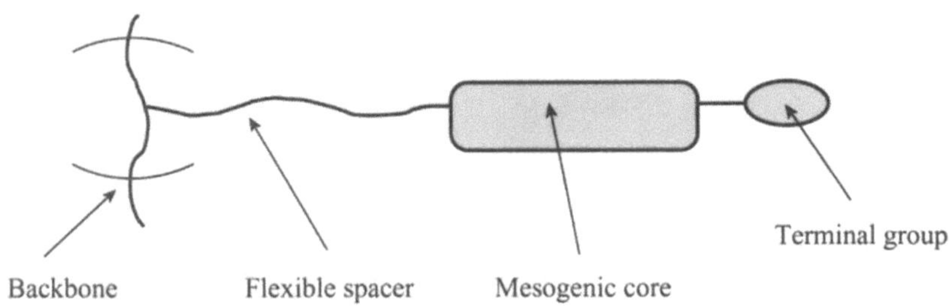

The repeat unit of this class of liquid crystal polymer typically consists of a polymeric backbone or main chain, a (more or less) flexible spacer group, a mesogenic core and a terminal group; the chemical nature of each of these parts can be varied to encompass a vast range of materials. The mesogenic core and terminal group can be of almost any chemical structure known in low molar mass liquid crystals, of which there are many reviews [14, 15]. Likewise the backbone or main chain is limited only by the requirement that it may be functionalized to incorporate the side-group spacer. Within these broad constraints a wide variety of side groups and polymeric backbones have been reported.

7.3.2 *Liquid crystal poly(acrylate)s*

Probably the most widely studied of the side-group liquid crystal polymers are those based upon poly(acrylate) and poly(methacrylate) backbones. These have the advantage of being highly accessible in terms of their synthesis and polymerization in solution and are typically prepared by radical or ionic polymerization using standard thermal initiators.

Consider an example based upon commercially available 4′-cyano-4-hydroxybiphenyl. Alkylation with a suitable ω-bromoalkanol followed by esterification with acryloyl or methacryloyl chloride will yield a monomer which can be polymerized using standard free radical techniques (Scheme 7.1). In this way one of the simplest series of liquid crystal polymers can be prepared, and numerous examples can be found in the literature. A selection is shown in Table 7.2. The materials tabulated in Table 7.2 highlight two features of

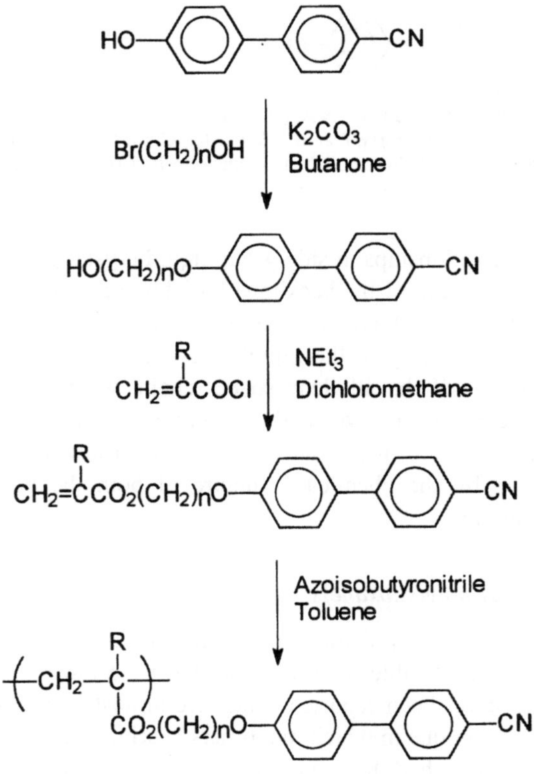

Scheme 7.1 Typical synthetic route for poly(acrylate)s.

Table 7.2 Transition temperatures for some liquid crystalline poly(acrylate)s[a]. Phases: N, N_{re} = nematic and re-entrant nematic, respectively; S_A, S_C = smectic A and smectic C, respectively; I = isotropic

n[a]	T_g (°C)[b]	Transition temperature[b]	Ref.
R=H			
2	50	$N \xrightarrow{112°C} I$	Shibaev *et al.* (1982)[c]
3	47	$S_A \xrightarrow{83°C} I$	Merck Ltd, LCP95
4	42	$N \xrightarrow{229°C} I$	Dubois *et al.* (1986)[d]
5	35	$S_A \xrightarrow{120°C} N \xrightarrow{124°C} I$	Dubois *et al.* (1986)[d]
6	32	$N_{re} \xrightarrow{80°C} S_A \xrightarrow{124°C} N \xrightarrow{132°C} I$	Dubois *et al.* (1986)[d]
11	25	$S_C \xrightarrow{30°C} S_A \xrightarrow{145°C} I$	Kostromin *et al.* (1984)[e]
R=CH_3			
2	95	–	Shibaev *et al.* (1982)[c]
3	84	–	Merck Ltd, LCP96
6	55	$S \xrightarrow{100°C} I$	Shibaev and Platé (1985)[f]
11	40	$S_A \xrightarrow{121°C} I$	Shibaev *et al.* (1982)[c]

a 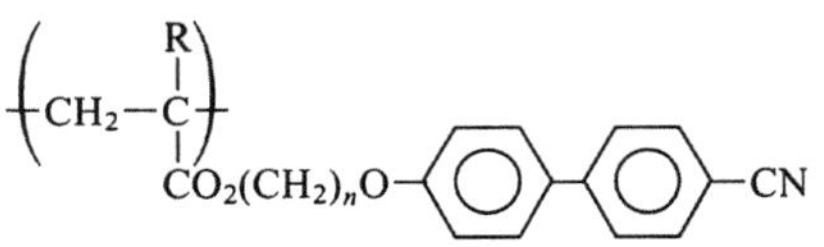

[b] Glass transition temperature.
[c] Shibaev, V.P., Kostromin, S.G. and Platé, N.A. (1982) *Eur. Polym. J.*, **18**, 651.
[d] Dubois, J.C., Decobert, G., Le Barney, P. *et al.* (1986) *Mol. Cryst. Liq. Cryst.*, **137**, 349.
[e] Kostromin, S.G., Simitsyn, V.V., Talroze, R.V. and Shibaev, V.P. (1984) *Polym. Sci. USSR*, **26**, 370.
[f] Shibaev, V.P. and Platé, N.A. (1985) *Pure Appl. Chem.*, **57**, 1589.

structure property relationships in side-group liquid crystal polymers: firstly, the effect of hindered rotation of the backbone by adding a methyl substituent onto the acrylate backbone, which not only causes a significant increase in T_g but also reduces the thermal stability of the liquid crystal phases; second, the effect of increasing the length of the alkylene spacer chain, which has a considerable impact on both the mesophase type and the thermal stability of the liquid crystal phases. The precise nature of this latter effect cannot always be predicted and will be influenced by the chemical structure of both the backbone and the mesogenic side group.

7.3.3 Liquid crystal poly(siloxane)s

The second major group of polymers in this field are the poly(siloxane)s, which are probably the most flexible (and therefore have the lowest T_g) liquid crystal polymers that have yet been reported. They are almost invariably prepared by the hydrosilation technique in which monomers bearing a terminal alkenyl group in the spacer are attached to a poly(hydrogenmethylsiloxane) backbone in solution using a platinum catalyst. The example synthesis [37] shown in Scheme

7.2 uses cholesterol and cholestanol to create chiral mesogenic side groups which are attached to a poly(hydrogenmethylsiloxane) backbone by this technique:

7.3.4 Liquid crystal poly(styrene)s

For many applications of liquid crystal side-group polymers a high T_g is desirable, and the poly(styrene) backbone with it bulky aromatic ring attached directly to the backbone can offer relatively high glass transitions temperatures without compromising the thermal stability of the liquid crystal phases. In the following examples [38] a polymer analogous reaction is used to functionalize a preformed backbone with mesogenic side groups chosen for their potentially

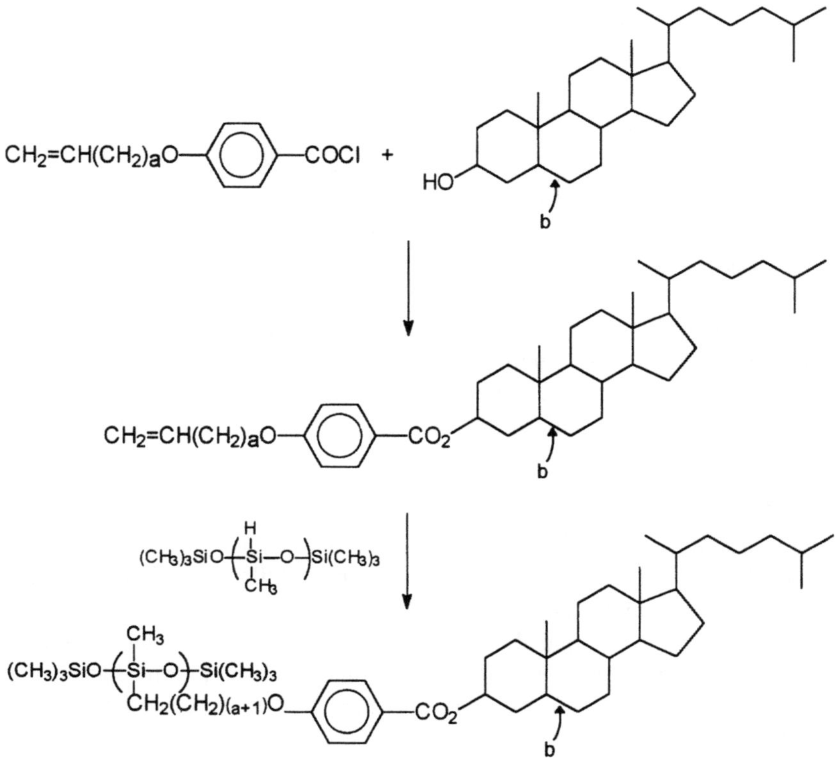

a	*b*	Monomer phase behaviour	Polymer phase behaviour
1	C=C	K $\xrightarrow{124°C}$ N* $\xrightarrow{226°C}$ I	K $\xrightarrow{95°C}$ S $\xrightarrow{272°C}$ I
1	sat.	K $\xrightarrow{100°C}$ N* $\xrightarrow{226°C}$ I	g $\xrightarrow{101°C}$ S $\xrightarrow{230°C}$ I
9	sat.	K $\xrightarrow{108°C}$ N* $\xrightarrow{183°C}$ I	g $\xrightarrow{94°C}$ S $\xrightarrow{206°C}$ I

Scheme 7.2 Typical synthetic route for some liquid crystalline poly(siloxane)s.

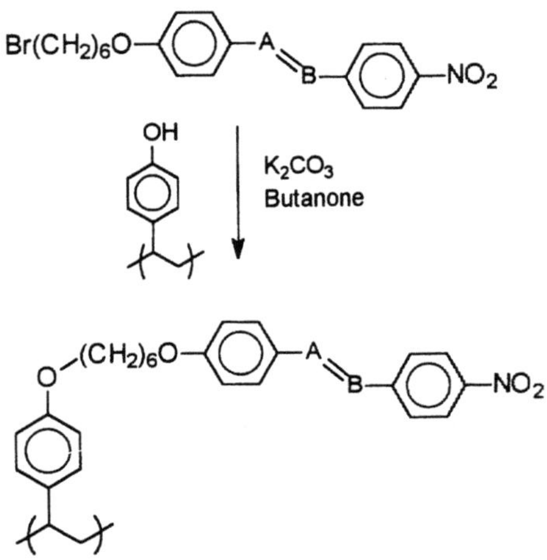

A	B	T_g (°C)	Liquid crystal phases
CH	CH	85	$S_A \xrightarrow{165°C} I$
N	CH	75	$S_A \xrightarrow{118°C} I$
N	N	85	$S_A \xrightarrow{151°C} I$

Scheme 7.3 Typical synthetic route for some liquid crystalline poly(styrene)s.

non-linear optical properties (Scheme 7.3). In this instance a high T_g would be desirable in order to allow the liquid crystalline alignment of the side groups to be permanently 'frozen in' at ambient temperatures, thereby retaining any non-linear optical properties indefinitely.

7.3.5 Liquid crystal poly(ester)s

Poly(ester)s form a highly important group in the field of commercial and commodity polymers but have been surprisingly little studied as side-group liquid crystal polymers. The potential structural diversity of poly(ester)s is of course extremely wide, and the commercial availability of suitable diols and diacids is good, although the preparation of functionalized intermediates can be a challenge. The following example polymers [39] are prepared by a trans-esterification process catalysed with titanium(IV) isopropoxide in diglyme (Scheme 7.4). The technique clearly shows the potential for preparation of copolymers with an alternating ABAB type structure which can strongly influence the physical properties of the polymer.

7.3.6 Liquid crystal poly(vinyl ether)s

The formation of polymers by cationic polymerization is well known, for instance the polymerization of epoxide monomers to yield poly(oxyethylene)s,

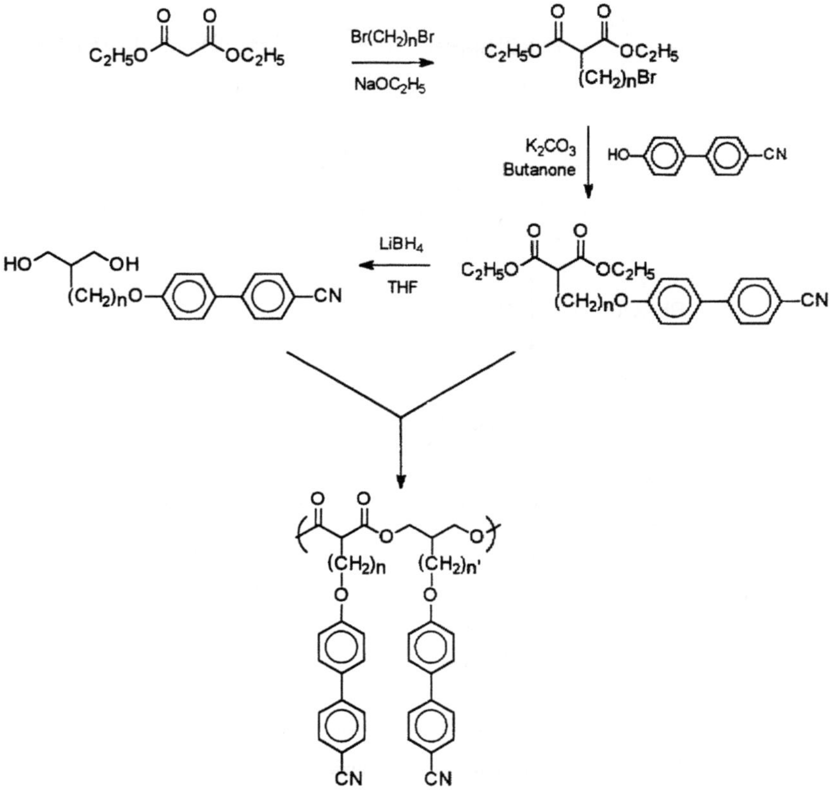

Scheme 7.4 Typical synthetic route for some liquid crystalline poly(ester)s.

or the polymerization of vinyl-based monomers to give pol(vinyl)s. These latter materials are particularly suited to liquid crystal phases in the monomer. Relatively simple liquid crystalline vinyl ether monomers can therefore be prepared and polymerized within a nematic or smectic phase, and this effect will be discussed in detail later in this chapter. Scheme 7.5 shows a series [40] of vinyl ether monomers with differing alkyl spacer groups which were polymerized using $BF_3.O(C_2H_5)_2$ as the cationic initiator.

7.3.7 Other classes of liquid crystal side-group polymer

It should already be clear that the scope for liquid crystal side-group polymers is very broad, and it is of course also possible to combine many of the polymeric backbones described above to form mixed copolymers, block copolymers and so on. Indeed it would be possible to continue exemplifying different structural types of these materials far beyond the space permitted in this chapter.

Having considered radical polymerization, ionic polymerization and polymer analogous reactions, it is perhaps useful to consider one final example in an

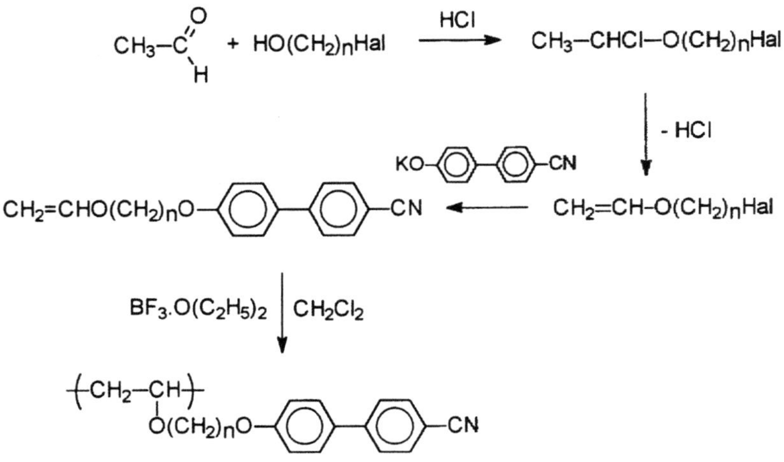

n	Monomer transition temperatures	Polymer transition temperatures
2	$K \xrightarrow{116^\circ C} I$	No mesophase
3	$K \xrightarrow{61^\circ C} I$	$N \xrightarrow{78^\circ C} I$
4	$K \xrightarrow{71^\circ C} N \xrightarrow{77^\circ C} I$	$N \xrightarrow{73^\circ C} I$
5	$K \xrightarrow{158^\circ C} N \xrightarrow{41^\circ C} I$	$S_A \xrightarrow{65^\circ C} N \xrightarrow{98^\circ C} I$
6	$K \xrightarrow{70^\circ C} N \xrightarrow{75^\circ C} I$	$S_A \xrightarrow{98^\circ C} N \xrightarrow{100^\circ C} I$

Scheme 7.5 Typical synthetic route for some liquid crystalline poly(vinyl ether)s.

attempt to demonstrate that liquid crystal polymers can be prepared with even somewhat exotic backbones and to show further that the limits to structural variations in these materials exists only in the imagination and abilities of the chemists within the field. In Scheme 7.6 the living ring-opening metathesis polymerization of a norbornene derivative [41] is shown, and the liquid crystal behaviour of the resultant polymer exemplified for various number average molecular weights ($\bar{M}_n$) (the weight average molecular weight, $\bar{M}_w$, as a proportion of $\bar{M}_n$ is also shown).

7.3.8 *The mesogenic core group*

From the structural variations in the preceding examples it is clear that the mesogenic core will have a major influence on the liquid crystalline properties of these side-group polymers. It is well known in low molar mass liquid crystals that one of the most important structural features of a core group in determining the thermal stability is the length-to-breadth ratio, and this is also true for liquid crystal polymers. By extending the aromatic core group in a linear manner this ratio will increase and the clearing temperature of the material rises dramatically. This effect can be seen clearly in the series of poly(siloxane)s (Table 7.3).

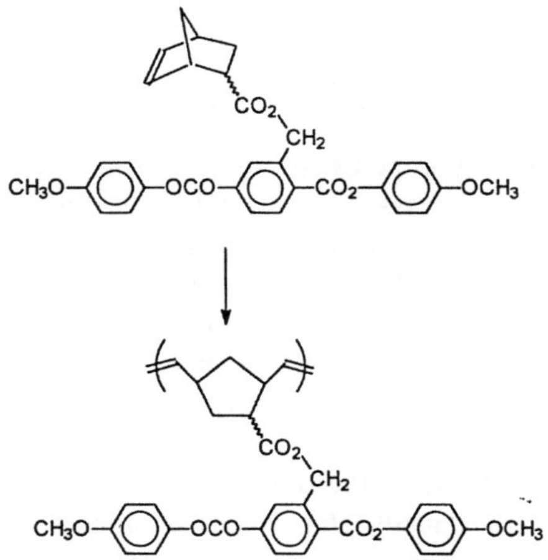

$\overline{M}_n$	$\overline{M}_w/\overline{M}_n$	Transition temperatures (°C)
2670	1.16	$g \xrightarrow{79^\circ C} N \xrightarrow{131^\circ C} I$
4290	1.27	$g \xrightarrow{85^\circ C} N \xrightarrow{145^\circ C} I$
7221	1.12	$g \xrightarrow{90^\circ C} N \xrightarrow{146^\circ C} I$
24 050	1.13	$g \xrightarrow{98^\circ C} N \xrightarrow{164^\circ C} I$
52 630	1.16	$g \xrightarrow{97^\circ C} N \xrightarrow{163^\circ C} I$

Scheme 7.6 Ring-opening reaction off norbornene to provide liquid crystal polymers.

Table 7.3 Transition temperatures of a series of poly(siloxane)s (liquid crystal polymers with longer molecular cores have higher transition temperatures). Phases: g = glass; N = nematic; I = isotropic

Structure	Transition temperature
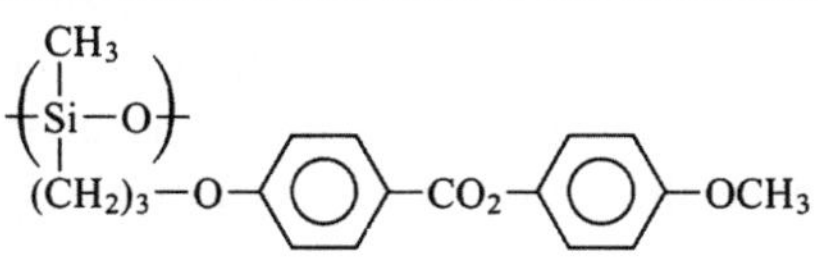	$g \xrightarrow{15^\circ C} N \xrightarrow{61^\circ C} I$
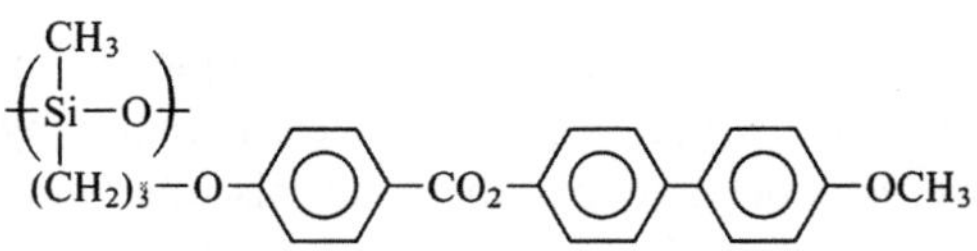	$K \xrightarrow{139^\circ C} N \xrightarrow{319^\circ C} I$
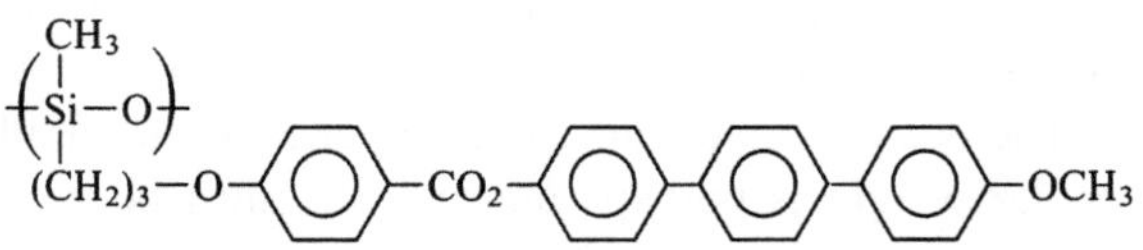	$K \xrightarrow{200^\circ C} N \xrightarrow{360^\circ C} I$

Further reviews of the structure of liquid crystal polymers are available in the literature [42–46].

7.3.9 Glass transitions in liquid crystal polymers

One of the most interesting properties of liquid crystal polymers which separates them from most low molar mass liquid crystals is the ability to form a glass. In liquid crystal polymers the T_g can be of great significance since a material might be heated to above the T_g and then exposed to some external aligning force such as an electrical or a magnetic field. When a desired bulk alignment of some or all of a sample has been achieved the sample can then be cooled to a temperature below T_g, whereupon the alignment is 'frozen' into the sample and cannot be greatly altered without once again heating to above T_g. This feature of liquid crystal polymers provides the scope for numerous applications in which samples of a thin mechanically stable film of anisotropic material on one substrate provides a permanence of anisotropic properties that are not possible using low molar mass liquid crystals. An example of such an application would be for optical or digital data storage where information could be stored indefinitely in a thin film of liquid crystal polymer kept below its T_g.

7.3.10 Molecular weight of liquid crystal polymers

It is important when discussing the properties of liquid crystal polymers to consider the role of the degree of polymerization (DP) or molecular weight. Two main factors must be considered when discussing the molecular weight of polymers – the average molecular weight (usually quoted as the number average $\bar{M}_n$), and the distribution of molecular weight or polydispersity (quoted as the quotient of weight average and number average, $\bar{M}_w/\bar{M}_n$). Most liquid crystal polymers throughout the literature are relatively low molecular weight, particularly those prepared by step or chain growth polymerization of a mesogenic monomer rather than polymer analogous reactions with a preformed backbone. Polymers reported with a degree of polymerization in excess of 100 repeat units are unusual, with 10–40 repeat units being typical. To the research scientist the study of polymers (or more accurately oligomers) within this range of molecular weights has many advantages in the physical handling and analysis of the materials, but causes difficulty in accurate comparison of materials since, within this range, the physical properties are highly dependent upon the molecular weight.

Consider first the effect of very low molecular weight using an example [47] in which poly(vinyl ether)s [40, 48] with narrow polydispersities were prepared by cationic polymerization. The liquid crystal behaviour of the materials was found to vary considerably according to the degree of polymerization, with nematic, smetic A and smectic C phases being observed depending on the molecular weight, as shown in Fig. 7.5. The occurrence of this extreme variation in liquid crystal behaviour with molecular weight is perhaps unusual but must be considered when dealing with oligomeric structures such as these. More commonly

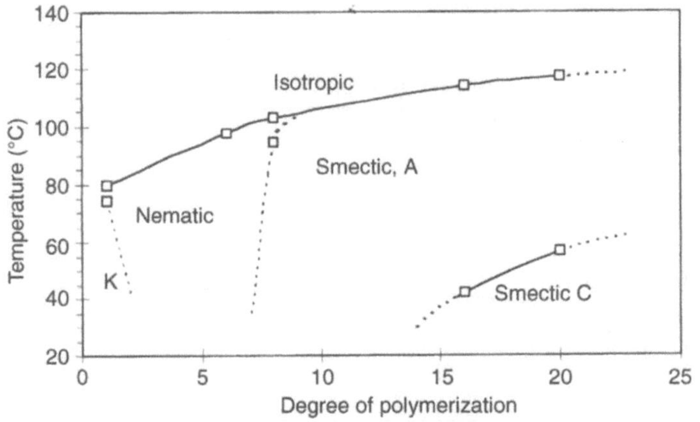

Figure 7.5 Liquid crystal phase types and transition temperatures by degree of polymerization for

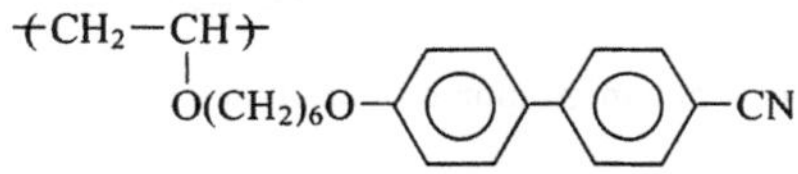

the phase sequence of a side-group liquid crystal polymer is fairly constant across a range of molecular weights but the temperature stability increases with increasing molecular weight until reaching a plateau, after which further increases in $\bar{M}_n$ cause no further change in the thermal properties of the material. The DP at which this plateau begins is dependent upon the chemical structure of both the backbone and the side groups, but will generally fall in the range of 50–100 repeat units. The behaviour of some norbornene-based polymers is shown in Fig. 7.6. Clearly care must be taken when considering the thermal properties of liquid crystal polymers with a DP which falls below this critical level.

7.4 Elastomeric liquid crystal polymers

In 1975 it was proposed by De Gennes [49] that in theory liquid crystal polymers might be cross-linked to form elastomers in which properties arising from the liquid crystalline behavior could be greatly stabilized to give anisotropic solids with unusual properties. In 1981 the first examples of liquid crystalline elastomers were reported by Finkelmann *et al.* [50], based upon a lightly cross-linked liquid crystalline poly(siloxane). These early elastomers were prepared by a one-step process in which monofunctional liquid crystal side groups and bifunctional cross-linking groups were simultaneously attached to a polymer backbone. Later work has shown the advantages to cross-linking a linear side-group liquid crystal polymer by ultraviolet or thermally initiated reaction after the polymer has formed a liquid crystalline phase. This technique allows the preparation of interesting materials which possess a molecular 'memory' of the anisotropic state in which the cross-linking took place. More recently, a great deal of research has been

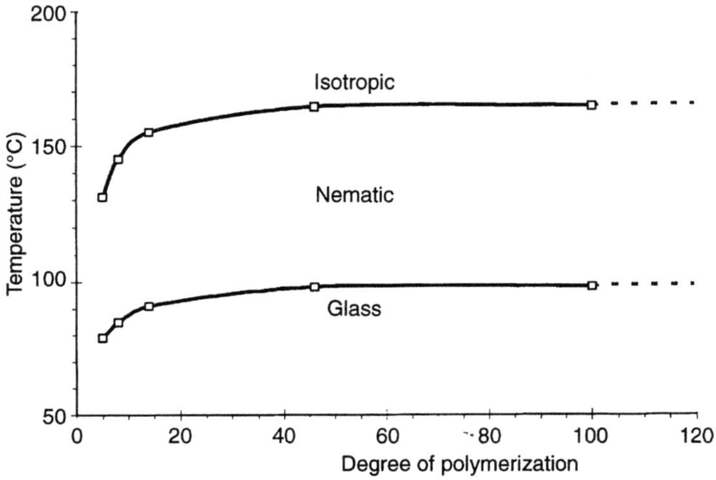

Figure 7.6 Effect of degree of polymerization on transition temperatures of liquid crystal polymers – norbornene-based polymers:

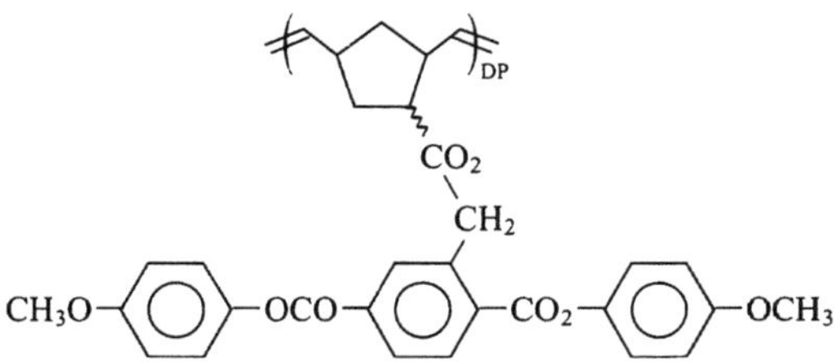

undertaken into the formation of the polymer networks by *in situ* polymerization of aligned low molar mass liquid crystalline monomers, and this work will be more fully discussed in the Section 8.

7.5 Laterally attached liquid crystal side-group polymers

In liquid crystal polymers with the mesogenic side group terminally attached to the backbone the mesophase types are predominantly smectic, even when the equivalent low molar mass liquid crystals would be strongly nematic. This is not surprising given the smectic lamellar structure will be greatly stabilized by the anchoring effect of the polymer backbone, which will essentially define the layer planes within the phase. This is not the case, however, when the side groups are attached laterally via a linking group to the polymer backbone. In this instance it is difficult to perceive of an arrangement of the mesogenic side groups in which the backbone could define or even fall within the layer planes of a smectic liquid crystal phase, and, indeed, experimentally, polymers of this type are almost exclusively nematic.

Laterally attached side-group liquid crystal polymers were first prepared with a poly(acrylate) backbone [51] and subsequently with poly(siloxane) [52] as well as other backbones; even when the mesogens contain long alkyl end groups and long alkyl spacers (both traditionally good promoters of smectic phases) the nematic phase is almost invariably found. The effect of lateral substitution in promoting nematic phases is in fact so strong that copolymerization of a terminally attached side-group polymer (normally smectic) with as little as 16% of a laterally attached side-group polymer will cause complete suppression of smectic behaviour and only a nematic phase will be found. An example [53] is given in Table 7.4.

7.6 Cyclic polysiloxanes

Cyclic liquid crystal polymers have been known since 1981 [54]. They are characterized by having rather high T_g values (for a siloxane) and when they pass beyond this point they provide fluid liquid crystal phases more reminiscent of low molar mass materials than of liquid crystal polymers. They are synthesized by hydrosilylation of well-defined cyclic siloxanes with ω-alkene mesogens using a platinum catalyst. Although it might be expected that these compounds should exhibit discotic phases they typically exhibit calamitic liquid crystal phases, and because of this they are drawn as shown in Fig. 7.7. The side groups are statistically arranged about the cyclic core, which can consist of 3–8 units, and typically 5 or 6 units.

Table 7.4 Properties of a typical laterally attached liquid crystal polymer:

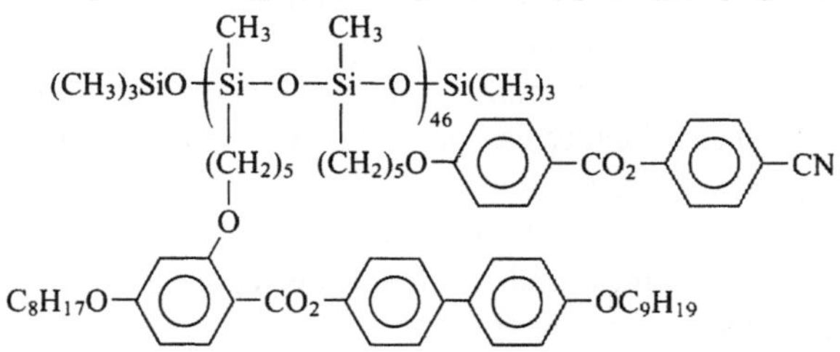

Terminal side-group polymer (%)	Lateral side-group polymer (%)	Phase behaviour
100	0	K $\xrightarrow{107°C}$ S_A $\xrightarrow{184°C}$ I
95	5	K $\xrightarrow{62°C}$ S_A $\xrightarrow{140°C}$ I
86	14	g $\xrightarrow{14°C}$ S_A $\xrightarrow{131°C}$ N $\xrightarrow{134°C}$ I
84	16	g $\xrightarrow{11°C}$ N $\xrightarrow{108°C}$ I
80	20	g $\xrightarrow{11°C}$ N $\xrightarrow{108°C}$ I
58	42	g $\xrightarrow{6°C}$ N $\xrightarrow{69°C}$ I
19	81	g $\xrightarrow{10°C}$ N $\xrightarrow{59°C}$ I
0	100	g $\xrightarrow{5°C}$ N $\xrightarrow{58°C}$ I

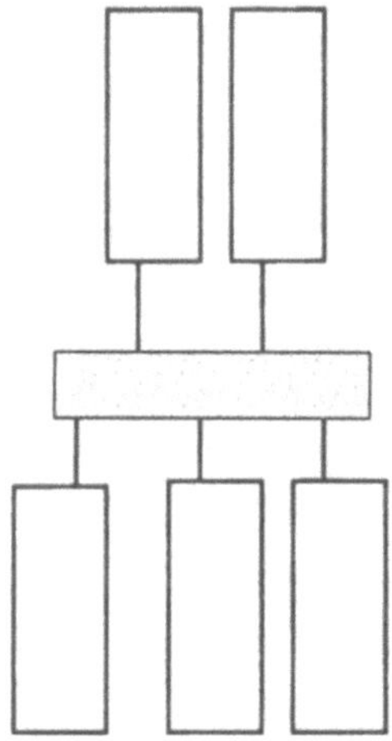

Figure 7.7 The side groups on cyclic polysiloxanes lie such that calamitic liquid crystal phases form.

The separation of the isomers is possible but very difficult; it is therefore usual to use mixtures of the isomers. In systems which have relatively small and flexible side groups, for example phenyl benzoates, the size of ring has little effect on the T_g or transition temperatures, but when a bulky side group is used, for example cholesterol, T_g and transition temperatures increase with increasing ring size. As the alkylene chain spacer is increased (from propylene to hexylene), T_g and smectic phase values tend to decrease, but nematic phase stability increases; this is the opposite situation to that of linear systems.

The most interesting compounds are copolymers incorporating cholesterol; cholesteryl homopolymers are purely smectic, but the addition of other small groups, such as phenylbenzoates, promotes a chiral nematic phase. These additional groups also serve to allow modification of the wavelength of reflected light (i.e. the helical pitch length). To produce a green colour, about half the side groups need to be cholesterol; red requires fewer cholesterol side groups. Some examples of cyclic siloxanes are [55]:

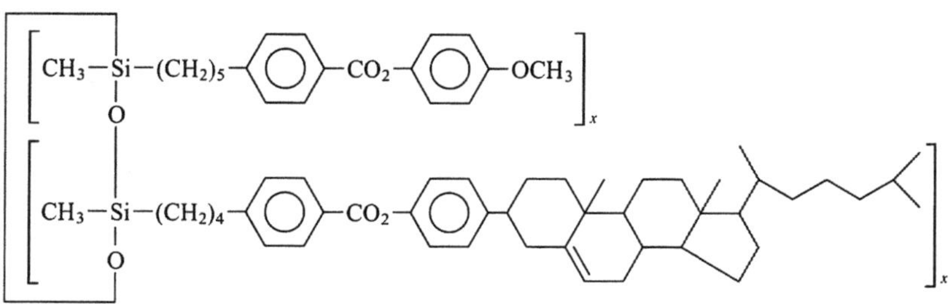

with a T_g of 38°C and a chiral nematic-to-isotropic phase-change temperature of 176°C. These compounds are of interest as an artists' paint [56] where they are applied in the liquid crystal phase (and allowed to cool below T_g), as cholesteric colour filters and as decorative or security films or flakes [57]. For some applications where their low viscosity above T_g is a problem intermolecular cross-linking of the side groups has been used to increase the molecular weight and viscosity.

7.7 Applications of side-group liquid crystal polymers

7.7.1 Display applications

In a traditional nematic liquid crystal display the performance of a material will be dependent upon many factors, not least of these being the viscosity, which has a strong influence on the speed with which the liquid crystal can reorientate under the application of a field. Depending upon molecular weight, the bulk viscosity of side-group liquid crystal polymers has been shown to be four to five orders of magnitude higher than their low molar mass equivalents [58]. This would suggest that the electrooptic behaviour of these materials might not readily be exploited in active applications, as their very high viscosity could result in extremely slow reorientation of the mesogenic groups under an electric field. This is more fully understood by considering the role of the backbone in the switching process of a hypothetical nematic side-chain polymer.

In nematic liquid crystal displays the director of the liquid crystal is required to reorientate through an angle of up to 90°. In terminally attached side-chain liquid crystal polymers the backbone is known to lie preferentially normal to the director within the sample [59]. Thus if the director were to be reorientated through 90° then the backbone would also be required to reorientate by a similar angle, and this would clearly be a greatly hindered process. The reorientation is depicted graphically in Fig. 7.8, with only the mesogenic cores and the backbone depicted for reasons of clarity.

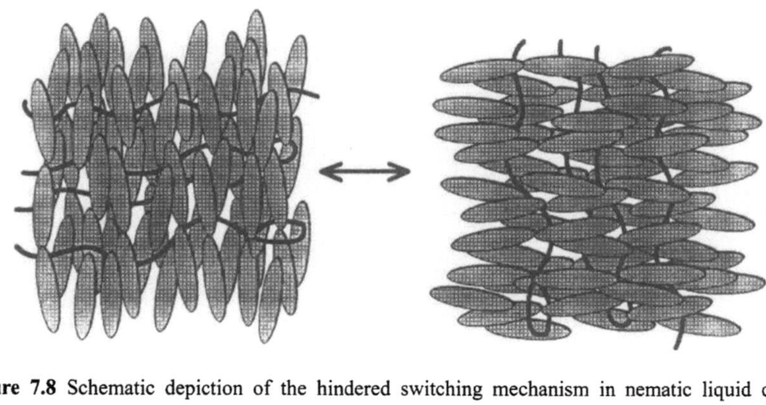

Figure 7.8 Schematic depiction of the hindered switching mechanism in nematic liquid crystal polymers.

Clearly, if a rapid electrooptic response is to be observed in liquid crystal polymers then the backbone must take no part in the reorientation of the mesogenic groups, and indeed this can occur by exploiting the ferrolectric response in the S_C^* phase.

In a side-group liquid crystal polymer smectic phase, the backbone will lie preferentially (although not exclusively) within the layer planes. In liquid crystal polymers with non-hydrocarbon backbones (e.g. polysiloxanes) this is sometimes described in terms of microphase separation and this may play a part in the formation of the smectic phase. This must, of course, be balanced against the physical constraints of the backbone to act as anything other than a layer boundary, particularly in stiffer backbones where a greater persistence length (i.e. backbone linearity) may be found. Within the S_C^* phase of a side-group liquid crystal polymer the backbone will therefore be lying predominantly within the layer planes at an angle to the director (dictated by the S_C^* phase tilt angle, θ). If the mesogenic side groups are suitably decoupled from the backbone by a flexible spacer group then they might rotate around a cone whilst maintaining the same angle with the layer planes and therefore also with the backbone, as shown in Fig. 7.9.

If such a material is placed in a suitable electrooptic cell then two stable positions of the mesogenic side groups around the cone can be induced, and the director can be selectively switched between the two by application of an electric field. Since theoretically this could occur without reorientation of the backbone it can be envisaged that a rapid response might be possible despite a very high bulk viscosity. This is depicted in Fig. 7.10 with the spacer groups again removed for clarity.

The first side-group liquid crystal polymer which demonstrated a S_C^* phase was reported in 1984 [60, 61] but no electrooptic behaviour was observed.

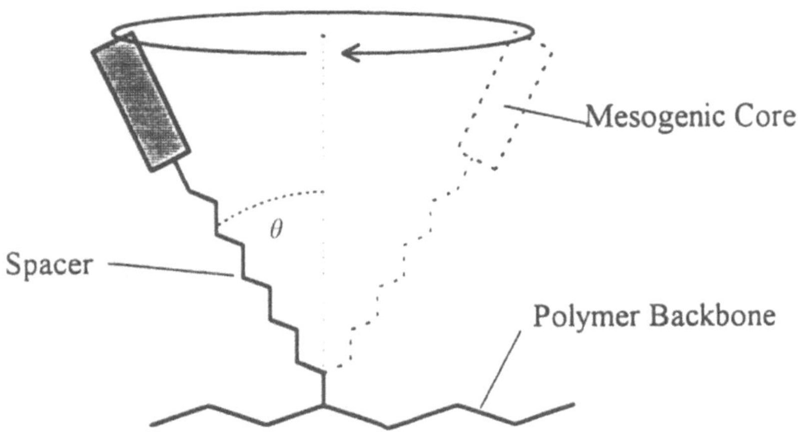

Figure 7.9 Side groups of S_C^* liquid crystal polymers can freely rotate around a cone.

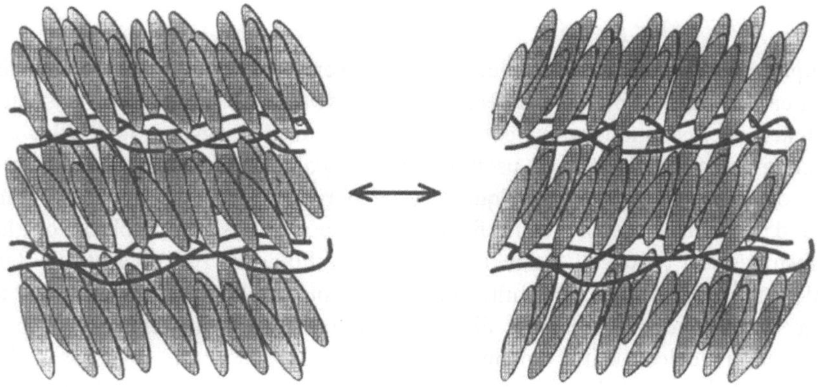

Figure 7.10 Schematic drawing of the possible switching mechanism-in a smectic liquid crystal polymer.

Despite many further reports of polymers with S_C and other tilted smectic phases [62, 63] it was not until 1988 that a ferroelectric response was measured in a switching liquid crystal polymer [64]. The ideal switching of these materials described above is probably far from the truth and the electrical response of the side groups will never be fully decoupled from the backbone. This is highlighted by considering the effect of molecular weight upon the response speed of the ferroelectric polymer [65] shown in Figure 7.11; increasing DP clearly lengthens response times. The response speeds shown are for complete switching in a 2 µm cell with 10 V μm^{-1} electric field applied.

Flexible films of S_C^* liquid crystal polymer have been made between two indium tin oxide coated plastic films; alignment of the liquid crystal phase was accomplished by shearing the sandwich of films over heated rollers [66, 67]. In order to increase response speed at ambient temperatures a composite of liquid crystal polymer and low molar mass liquid crystal was used. In this way the first flexible liquid crystal polymer based displays have been demonstrated; in present films, over a period of time the alignment deteriorates and contrast is lost.

Liquid crystal polymers exhibiting an S_A phase have been explored for display applications [68, 69]. Response times from a light-scattering focal conic texture [7] to a homeotropic texture, in which the mesogenic side groups are aligned parallel to the applied field of typically 10 V μm^{-1}, can be less than 1 s. However, such response times are only found close (within 5°C) to the clearing temperature. On cooling, both textures are stable and can be stored almost indefinitely. The addition of low molar mass liquid crystals provided improvements to the response times and temperature range. While this effect has not found use in displays it has been further explored in non-linear optic applications. Dichroic dyes have also been added to provide coloured displays [70].

7.7.2 Optical data storage

Liquid crystal polymers have been used in several ways to provide films having optical storage or memory effects. In early devices [71]] the film was sandwiched between transparent conducting plates and an electric field which, in conjuction with a laser, was used to change locally the orientation and thereby the optical properties of the liquid crystal polymer. Later devices have eliminated the need to use an electric field such that open films cast on plastic or glass substrates can be used. A thin film of smectic liquid crystal polymer (5–8 μm thick) is slowly cooled (1°C min^{-1}) from the isotropic liquid phase such that the director is random within the film and thus creates a light-scattering film; in projection devices using narrow collection angle optics this provides a dark background. A dye dissolved in the film is used to absorb laser light such that the laser (usually a low-power He–Ne laser) locally heats the liquid crystal polymer film to the isotropic liquid phase, which is optically clear, producing a transparent line. On fast cooling the film below the T_g the transparent line is stored on the film (Fig. 7.12). Polyacrylates and methacrylates, with high birefringence molecular cores attached, are usually chosen so as to maximise the light scattering effect (caused by local micron sized changes in director orientation giving rise to mismatches in refractive index). Line widths of 1–15 μm have been achieved with indefinite storage to 60°C. Local or total erasure is possible

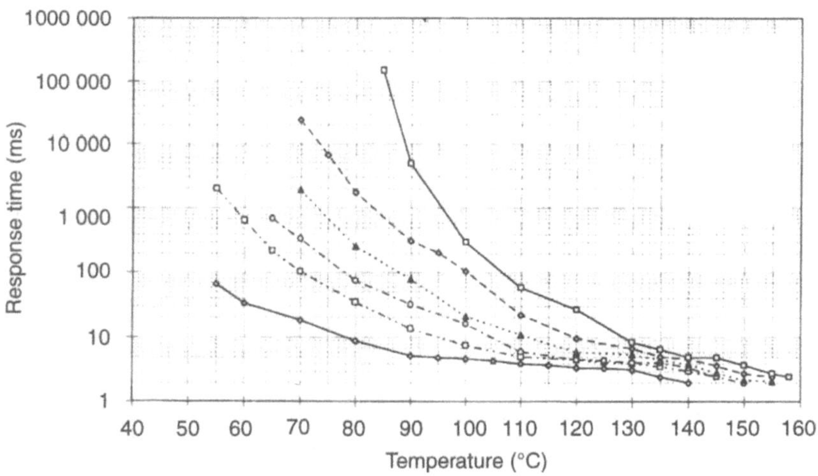

Figure 7.11 Effect of degree of polymerization (DP) on response times of an S_C^* liquid crystal polymer:

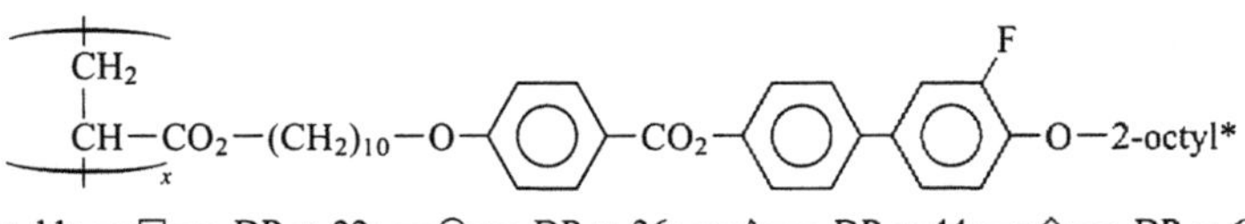

–◇–, DP = 11; --□--, DP = 22; --○--, DP = 36; --△--, DP = 44; --◇--, DP = 69; –⊟–, DP ≃ 150.

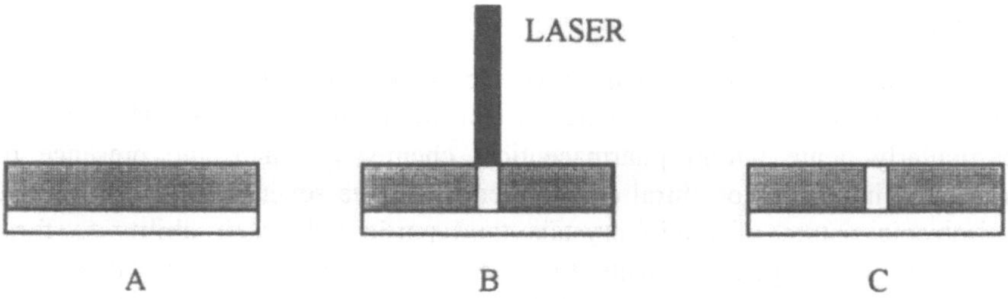

Figure 7.12 A side-group liquid crystal polymer in a light-scattering texture is addressed using a laser to provide stored images.

and films can be easily updated or re-used. Their use in microfiche is possible as they can be copied onto diazo film giving contrast ratios of over 10 : 1 in the copied film. At present, writing times are typically in the region of 0.5–1 m^{-1}; unfortunately, this is insufficient for many computer-based optical-storage applications.

A second type of optical storage device uses photosensitive groups. A linear *trans* azo group is incorporated into the molecular cores of a liquid crystal polymer such that a liquid crystal phase is exhibited. Upon irradiation with UV light, through a mask or locally by a laser, the *trans* isomers isomerize to the 'bent' *cis* form, which is much less conducive to liquid crystal phases, and a lower clearing point liquid crystal polymer is produced (Scheme 7.7). This gives rise to a lowering of Δn or, in the extreme case, an isotropic liquid is formed, in either case there is a detectable optical change. Reversal is accomplished by irradiation with visible light.

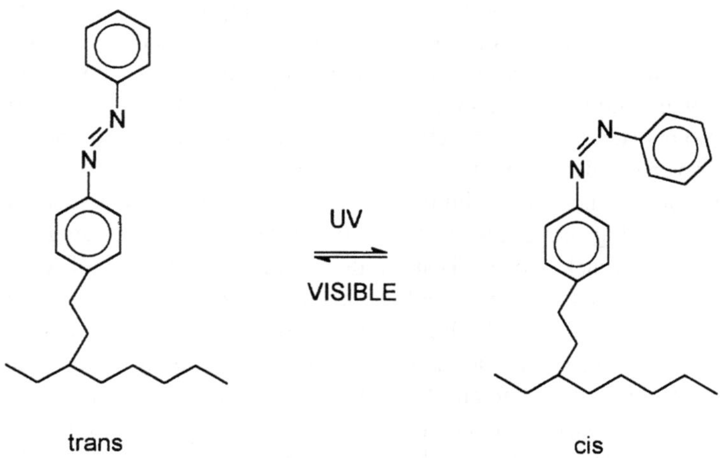

Scheme 7.7 *Trans* and *cis* isomerism of azo groups on irradiation with UV and visible light.

7.7.3 Chromatography

There is a constant drive to improve the separation efficiency of chromatographic columns on both the analytical and the production scale. This drive is particularly acute within pharmaceutical chemistry, where the presence of unwanted impurities or chiral enantiomers can have severe repercussions. The anisotropic nature of liquid crystals (and particularly their ability to form organized helical phases) means that they could prove to be very interesting media for chemical separation and, for some years, liquid crystals have been studied as stationary phases in gas chromatography.

The first use of liquid crystals in gas chromatography was reported in 1963 [72, 73] and since then many low molar mass liquid crystals and liquid crystal mixtures have been examined for this application – indeed a number of analytical columns using liquid crystal stationary phases have also become commercially available. The columns typically utilize the nematic phase rather than smectic phase, this being largely because of the availability of suitable liquid crystal mixtures with a wide nematic mesophase. As a general rule the comparative separation efficiency of nematic and smectic mesophases decreases as follows:

nematics
low-order smectics ↓ decreasing separation efficiency
higher-order smectics

There are some materials, however, which separate better on smectic liquid crystal stationary phases than on nematic ones, including alkylbenzene isomers, *n*-alkene isomers and phenanthrene/anthracene mixtures. Chiral nematic phases also display different separation characteristics to achiral ones, showing greater selectivity, for example, towards steroids and *cis* or *trans* isomers of *n*-alkenes, but poorer separation of alkylbenzene isomers. Not surprisingly, the efficiency of liquid crystals as separation media is also strongly influenced by the length-to-breadth ratio of the mesogenic cores, particularly for separation of isomers where this value will vary according to substitution patterns. For example on conventional columns *para*-xylene and *meta*-xylene elute in order of their boiling points (i.e. *para*-xylene elutes first), but on liquid crystal columns they elute in reverse order of length-to-breadth ratio (i.e. *meta*-xylene elutes first).

Studies of low molar mass liquid crystals have confirmed that such materials possess interesting properties as stationary phases in gas chromatography, but one major drawback exists. That is the relatively high volatility of these monomeric compounds which leads to a limited operating-temperature range and useful column life as the liquid crystal coating bleeds out of the column during use. It is therefore reasonable to presume that liquid crystal side-group polymers might demonstrate all of the interesting chromatographic properties of their low molar mass counterparts but with greatly enhanced stability and operating-temperature range. Although literature examples of liquid crystal polymer stationary phases are relatively few, the material illustrated in Table 7.5

was reported [74] to show considerably improved separation of 1- and 2-chloronapthalene and 1- and 2-bromonapthalene compared with a conventional column (SE-54). Some selected results are highlighted in Table 7.5.

7.7.4 *Non-linear optical applications*

The optical properties of a medium can change when the intensity of light propagating in it is changed or when another lightwave is introduced; such properties are material-dependent and give rise to non-linear optical effects. Of particular interest are the second-order and third-order macroscopic non-linear susceptibility terms (χ^2 and χ^3) which are dependent on the molecular susceptibilities β and γ, respectively, the packing density and on how well-ordered the active groups are. Liquid crystals and liquid crystal polymers in particular have been extensively explored and reviewed [75, 76] for these effects. The elongated molecules required for liquid crystal properties are also suitable for non-linear optical (NLO) properties and in principle, liquid crystal phases (and particularly the nematic phase) can be poled to give good alignment of the molecules which can be quenched by cooling into a glassy phase.

Second-order effects are of particular relevance because the effect is larger and the desired non-centrosymmetric orientation of the active groups can be achieved by poling of a liquid crystal polymer. Of particular interest is the doubling of the wavelength of laser light by passing through an NLO film.

Typical NLO active groups have been either doped as guests into liquid crystal polymers or bound into polymers as side groups, often as copolymers, this being particularly useful as more active groups can be incorporated without phase separation or crystallization of the guest occurring. Very often high T_g backbones such as methacrylates are desirable, but polysiloxanes have also been

Table 7.5 Relative retention times for some naphthalene compounds on a conventional column and a liquid crystal polymer (LCP) coated column. Liquid crystal polymer:

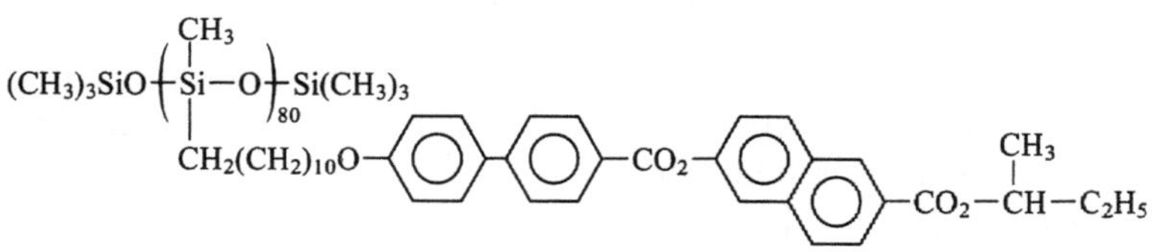

Compound	Boiling point (°C)	Length : breadth ratio	Relative retention times SE-54	Relative retention times LCP
1-chloronaphthalene	259	1.16	1.02	0.89
2-chloronaphthalene	261	1.40	1.00	1.00
1-bromonaphthalene	279–281	1.17	1.02	0.91
2-bromonaphthalene	281–283	1.40	1.00	1.00

Transition behaviour: g $\xrightarrow{12°C}$ N* $\xrightarrow{319°C}$ I used. This research has continued for many years, battling against long-term alignment and light-scattering problems in the films, with the result that commercial devices are not yet available with this technology.

7.8 *In situ* photopolymerization

The conventional methods to make side-group liquid crystal polymers are solvent-based and provide polymers of rather low molecular weight, typically a DP of 10–40. Such polymers are often used as aligned thin films (<100 µm thick) which take advantage of the short and induced long-range orientational ordering (caused by alignment) of the liquid crystal polymer which leads to anisotropic mechanical, optical, magnetic and electrical properties. However, it is relatively slow and difficult to create aligned liquid crystal polymer films compared with low molar mass liquid crystals, even under the influence of electric or magnetic fields, surface alignment or shearing. Alignment becomes increasingly difficult as the DP increases and becomes even more difficult, or impossible, as the liquid crystal polymer is cross-linked.

These problems can be overcome elegantly by aligning a low molar mass liquid crystal which carries one or more polymerizable groups and then polymerizing it *in situ* to form a thin, aligned polymer film. The concept of *in situ* photopolymerization of a photoreactive liquid crystal monomer (often referred to as a reactive liquid crystal or reactive mesogen) was reported [77] in the late 1980s and has now been extended to many systems and explored for many applications. Usually, photopolymerization is chosen rather than thermal polymerization because in this way the polymerization action is divorced from the thermal properties of the material which of necessity are limited by the finite temperature range over which the liquid crystal phases can exist. Standard UV-excited photo-initiators can be used to induce polymerization.

This technique has many advantages and unique features:

- processing of the relatively low-viscosity and soluble reactive liquid crystal can be achieved in many ways such as spin coating or deposition from solution to form a thin fluid film;
- alignment of the thin fluid film of reactive liquid crystal can be achieved quickly by standard methods used for liquid crystals; it is also possible selectively to align some areas in one orientation and adjacent areas in a different orientation by using rubbed surface aligning layers, polarized photopolymerization of cinnamate aligning agents, electric or magnetic fields or combination of these;
- irradiation can be carried out over all the film or in selected areas by use of a photomask;
- irradiation can be carried out in selected areas as above but also at selected temperatures, thus allowing curing in different liquid crystal phases, or, in the

case of a chiral nematic material, at different pitch lengths to provide multicoloured films;

- photopolymerization can be very fast if irradiated with higher-power UV lamps, which thus reduces kinetically-controlled phase separation when the polymer formed is insoluble in the monomer;
- in reactive liquid crystals with a functionality of two or more, a thin film of highly cross-linked polymer network is formed which cannot be otherwise obtained.

In the solvent-based methods of synthesis of liquid crystal polymers it is irrelevant whether the monomer exhibits a liquid crystal phase or not. However, in this technique, the monomer (or mixture of monomers) must necessarily exhibit a liquid crystal phase so that it can be aligned prior to polymerization. If polymerization is carried out outside the liquid crystal phase region then there is no preferred ordering of the side groups. This requirement places some constraints on the materials that can be used, because attaching a polymerizable group at the end of linear molecule tends to lower liquid crystal stability.

In principle, any polymerizable group could be used; however, those undergoing addition polymerization are favoured. Free radical photoinitiators acting on acrylates are most common, although the cationic photoinitiation of liquid crystal epoxies [78] and vinyl ethers [79] is also known.

7.8.1 Monoreactive liquid crystals

Upon polymerization the ordering of the liquid crystal phase is increased, which increases the absolute thermal stability of the liquid crystal phase and also tends to provide a higher-order phase. The exception to this is in cases in which the mesogenic cores cannot align in the polymer as well as they did in the monomer phase. Side-group liquid crystal polymers are predominantly smectic rather than nematic; however, the reactive liquid crystal could exhibit nematic and perhaps one or more smectic phases and it is possible to polymerize the reactive mesogen in any of these liquid crystal phases. Thus a reactive mesogen polymerized while in its nematic phase can provide a polymer which, at the polymerization temperature, is smectic. This interesting situation provides a thin polymer film that has the microscopic texture of the nematic phase because the fast polymerization locks in this ordering which is maintained in the polymer, especially near the surfaces. However, the bulk of the film has the elecrooptic properties of the smectic phase. The phase transitions of a reactive liquid crystal and also those of the polymer derived from it by *in situ* photopolymerization of the monomer in the nematic phase (at about 90°C) are shown in Table 7.6. When assessed at the same reduced temperature, the birefringence of the polymer film is higher than that of the initial monomer film.

The transition temperatures of the photopolymerized sample are slightly higher than those of the same polymer formed by a solvent-based method (thermal polymerization) owing to the higher DP values obtained with *in situ* photopolymerization techniques (typically, DP.150).

Typically over 95% of the monomer is polymerized when polymerization takes place in the nematic phase; in the smectic phase longer UV exposure followed by an increase in temperature is required to achieve complete polymerization; this feature is especially prominent if the polymer being formed exhibits a glassy phase at the polymerization temperature being used.

The polymerization rate is usually faster in the nematic phase than it is in the isotropic phase because the reactive groups are more ideally aligned for the reaction to occur. With use of a low irradiance of UV (1–2 mW cm^{-2}) complete reaction occurs in 10–30 s.

This technique has not yet been explored extensively for monoreactive liquid crystal systems, although the use of reactive liquid crystals to form ferroelectric liquid crystal polymers is also being explored as an alternative method to the traditional shear method of aligning ferroelectric liquid crystal polymers to form large-area films for use in displays.

7.8.2 Direactive liquid crystals – anisotropic networks

The concept of *in situ* photopolymerization with the incorporation of multireactive mesogens is opening a new and exciting area based on anisotropic networks. For many purposes side-group liquid crystal polymers can be conveniently regarded as very viscous liquid crystals whose fluidity increases with temperature, especially at low DP values. Although this is useful for some electrooptic applications of liquid crystal polymers it is not so desirable if they are to be used as coherent, mechanical self-supporting films. To some extent,

Table 7.6 Transition temperature for a reactive liquid crystal:

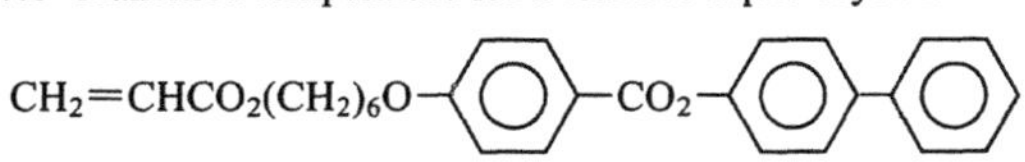

and the polymer derived from it:

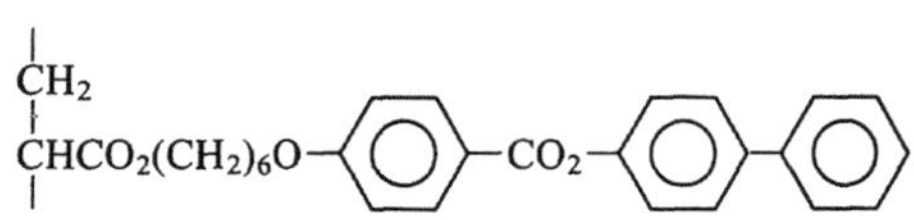

Birefringence values, Δn, are measured at a reduced temperature of 0.90

	Transition temperature (°C)			
	C → S/N	S → I	N → I	Δn
Reactive liquid crystal	88	76[a]	98	0.195
Polymer	87[b]	172	190	0.205

[a] Monotropic phase transition.
[b] Transition from a partially crystalline solid.

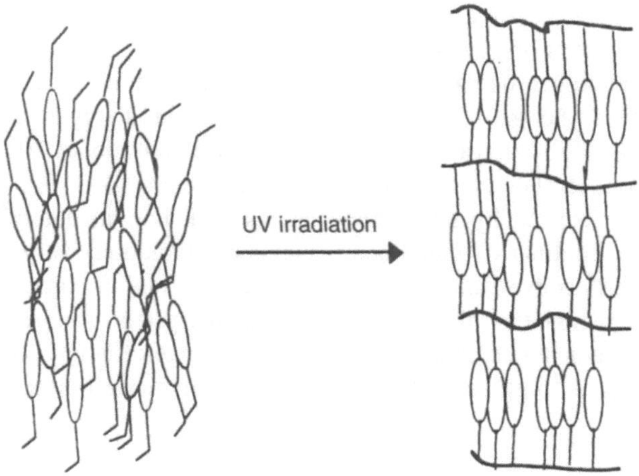

Figure 7.13 Anisotropic network polymer films synthesized by *in situ* photopolymerization of reactive liquid crystals by means of UV light.

cross-linking of the backbones to form elastomers partially overcomes this feature, but a more comprehensive solution is to increase drastically the cross-linking by the use of multireactive liquid crystals.

Figure 7.13 illustrates a direactive liquid crystal whose molecules are oriented to represent an aligned nematic phase. In a thin film this undirectional order is relatively easy to produce by surface-alignment techniques. Upon photo-polymerization, this order is locked into a three-dimensional anisotropic network [80].

Such polymers have very high cross-link densities and, as such, their optical and mechanical properties are very stable to high temperatures (>200°C); by differential scanning calorimetry (DSC) only a T_g can be detected (often with difficulty, in the 70°C–90°C region); there are no observable phase transitions and they are insoluble in common solvents, though they may absorb relatively small amounts of solvent, depending on the cross-link density.

Within these networks the individual molecular cores have liquid-crystalline-like order which is frozen in during polymerization, but they do not have the freedom of movement normally associated with, and characteristic of, a liquid crystal phase, nor do the films change state on heating or cooling in the same way that, for example, a side-group liquid crystal polymer does. Thus, these films are probably better described as anisotropic networks whose constituent molecules have liquid-crystal-like ordering rather than as true liquid crystal polymers.

Such films can be cast on complex surfaces as a liquid crystal and, when aligned and polymerized, provide a coating of anisotropic polymer film. If they are not aligned, the film consists of many domains, each domain having liquid crystal ordering (e.g. nematic), but the director configuration of the domains is random and therefore, on a macroscopic scale, an averaging of the physical and optical properties is observed.

Polymerization can be carried out in the isotropic liquid, nematic or smectic phase. Because the gel point is reached earlier in diacrylate systems, the maximum rate of reaction occurs at a lower conversion (about 20%) compared with the monoacrylate system (which gel at about 25–35%). About 80% conversion is detected by DSC; however, as with normal diacrylates, photopolmerization continues (under continued UV irradiation) very slowly until about 90%–95% conversion is finally achieved. The rate of conversion is much slower for direactive than for monoreactive liquid crystals. Temperture has little effect on free radical reactions unless there is a change of liquid crystal phase or the reaction is at high temperatures, when depolymerization competes with polymerization.

Optical properties. Although the polarizability of the reactive liquid crystal (and therefore the mean refractive index $\tilde{n}$) owing to the elimination of the vinyl groups, the increase in covalent bonds owing to polymerization is lowered upon polymerization significantly increases the density of the resultant polymer film and thus, overall, increases $\tilde{n}$.

In reactive liquid crystals which exhibit a high-order parameter and are polymerized at low temperatures, the order parameter changes very little when the sample is polymerized (Table 7.7). However, in lower-order parameter samples such as the same material at high reduced temperatures, polymerization can increase the order parameter and a highly ordered network is still produced.

In systems in which the monomers can pack together quite well but where because of the constrained orientation of the molecules in the polymer network the close packing is inhibited, for example again, lateral methyl compounds (Table 7.7), there can be a reduction in the higher-order parameters upon

Table 7.7 Order parameter (S) values for reactive liquid crystals:

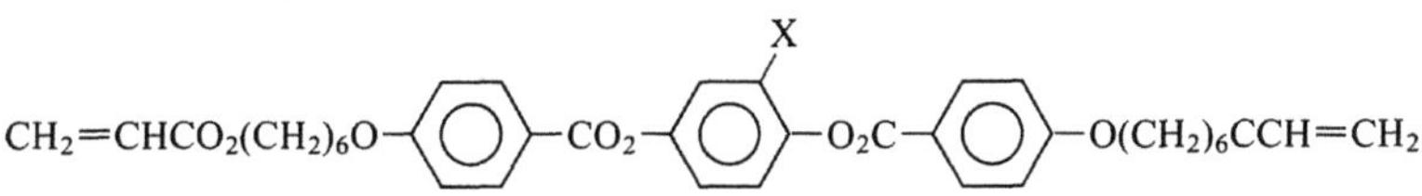

and their polymers, for X = H or X = CH_3 and for polymerization at different reduced temperatures ($T_{red} = T_{exp}/T_{N-1}$)

	Order parameter			
Reduced temperature	monomer (X = H)	polymer (X = H)	monomer (X = CH_3)	polymer (X = CH_3)
0.86	0.69	0.69	0.69	0.61
0.98	0.46	0.57	0.46	0.50

polymerization, but at lower-order parameter values (higher temperatures) an increase in S is seen, but not by as much as for systems which allow better packing.

Therefore, there is a relationship between the temperature of cure and the order parameter of the polymer film, which in turn leads to a relationship between the temperature of cure and the birefringence (Table 7.8); thus it is possible to control the birefringence of the film by controlling the temperature of cure as well as by the choice of monomers [80].

When the reactive liquid crystal molecules contain very short terminal alkylene chains the order parameter measured in the final polymer can be lower than the initial reactive liquid crystal because the cores are more constrained by the stereochemistry of the reacting acrylate groups. In compounds with alkylene chains above hexylene this effect is negligible.

Physical properties. Owing to the highly anisotropic ordering of the molecules in anisotropic network polymers, physical properties such as tensile strength are also anisotropic [81]. In well-orientated films of polymer the tensile moduli (E) in the direction of the nematic director ($E_{\parallel}$) is about 2–2.2 GPa, which is twice that of $E_{\perp}$ and that of isotropic acrylate polymers.

The expansion coefficient in films polymerized in the nematic phase [82] overall show a small shrinkage in the z direction (along the direction of the long molecular axis – the nematic director) when heated from 20°C to 150°C. The shrinkage is larger (2%) for materials with longer alkylene spacer groups between the aromatic cores and the polymer backbones; typical values are between 0.2% and 2%. On cooling to −50°C very little change in size occurs, and the change is slighty negative. In the other two directions (x and y) the expansion is similar and again depends on the alkyl spacer length, typical values being 2% (for short chains) and 4% (for long chains).

In chiral nematic films the major expansion (about 5%) is in the direction of the helix; within the plane of the helix the expansion or shrinkage is, as expected, isotropic and very small (<0.1%).

Table 7.8 The birefringence of a polymer network based on

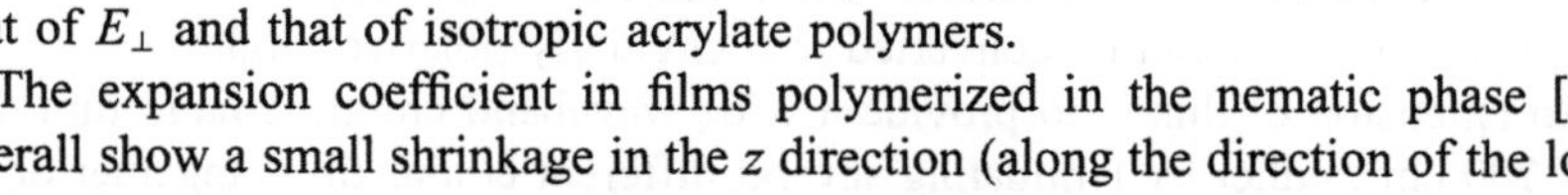

as it varies with the temperature at which the film was cured

Temperature of cure (°C)	Birefringence of aligned polymer
80	0.164
100	0.15
120	0.132
140	0.104

7.8.3 Applications of anisotropic networks

Polarizing beam splitter. A thin film of reactive mesogen is aligned and then polymerized between two glass or quartz prisms. One refractive index of the anisotropic network (n_o) is selected to be similar to that of the prisms, and the other (n_e) is higher. Light entering the prism impacts against the anisotropic network; if the wave vector of the light is orientated parallel to n_o it passes through the prism, whereas if it is orientated perpendicular to n_o the higher refractive index n_e is observed and the light follows a different path out of the prism block. Thus incident light is split into two linearly polarized light beams. One beam can be rotated using a $\lambda/4$ plate to provide, in principle, 100% conversion of randomly polarized light to linearly polarized light. Such a device has uses in liquid crystal display (LCD) projection televisions and in other applications which require linearly polarized light [83].

Circular polarizers. Circularly polarized light can be conveniently produced by using a chiral nematic film (Section 7.1.2). By mixing a nematic reactive mesogen, it is possible to generate copolymer films having more or less any desired pitch length and therefore, over a narrow bandwidth ($\Delta\lambda = P\Delta n$), any wavelength of light can be converted into circularly polarized light.

An extension of this is to provide a wide waveband chiral nematic film ($\Delta\lambda$ 400–700 nm), either by laminating several different colour films together or by having a film with a helix gradient [84] which can be used to reflect all visible light. Such films have been called ‘reflective polarizers’ and, in conjunction with a quarter wave plate to convert circularly polarized to linearly polarized light, can be used in backlit LCDs to produce polarized light more efficiently than is achieved by using conventional linear polarizers which absorb about 50% of the incident light. In this way LCDs can be made up to 80% brighter [85].

Non-Linear optics. By using the highly-ordered polymer network to align a dichroic chromophore, high non-linear optic coefficients can be produced. If the dichroic chromphore is also copolymerized into the network, more of it can be used without fear of phase separation.

Compensation films. LCDs generally use linearly polarized light, but this becomes elliptically polarized within the liquid crystal cell. The result of this are a number of undesirable optical properties such as reduced viewing angle, colour inversion and grey scale inversion in twisted nematic displays or the formation of unwanted birefringent colours in supertwisted nematics (STN) displays. These problems can be alleviated by incorporating a compensation film between the liquid crystal cell and the polarizers whose molecular orientation is similar but opposite to the structure of the liquid crystal in the liquid crystal cell. This can be accomplished by generating such a structure in the reactive monomer, followed by polymerization to provide a polymer film. Such films can be chiral nematic, nematic or smectic.

Displays. LCDs which incorporate a nematic or chiral nematic low molar mass liquid crystal are usually reversible when the electric field is removed; however, by incorporating a few per cent of a diacrylate crystal and polymerizing this within the display cell it has been possible to produce polymer stabilized liquid crystal displays that are bistable [86, 87].

Other applications. This technology has been known only since the early 1990s and many other applications are being discussed. For example, chiral nematic film could be used in security applications, as pigment flakes in various inks and as colour filters for LCD displays. They may also be used in adhesives. Nematic, and perhaps smectic, films could be used in holographic films, optical fibre claddings and dental fillings.

References

1. Reinitzer, F. (1888) *Montash Chem.*, **9**, 421.
2. Chandrasekar, S. Shadashiva, B.K. and Suresh, K.A. (1977) *Pramana*, **9**, 471.
3. Kreuder, W. and Ringsdorf, H. (1983) *Makromol. Chem. Rapid Commun.*, **4**, 807.
4. Fergason, J. (1966) *Molec. Cryst.*, **1**, 293.
5. Sage, I. (1990) In *Liquid Crystals, Applications and Uses*, vol. 3 (ed. B. Bahahur, World Scientific, Singapore.
6. Leadbetter, A.J. (1987) In *Thermotropic Liquid Crystals* (ed. G.W. Gray), John Wiley, Chichester, Sussex and New York.
7. Gray, G.W. and Goodby, J.W. (1984) In *Smectic Liquid Crystals*, Leonard Hill, London.
8. Coates, D. (1990) In *Liquid Crystals, Applications and Uses*, vol. 1, (ed. B. Bahahur), World Scientific, Singapore.
9. Meyer, R.B., Lieber, L., Strzelecki, L. and Keller, P., (1975) *J. Phys.* (*Paris*), **36**, L-69.
10. Dijon, J. (1990) In *Liquid Crystals, Applications and Uses*, vol. 1, (ed. B. Bahahur), World Scientific, Singapore.
11. Clark, N.A. and Lagerwall, S.T. (1980) *Appl. Phys. Lett.*, **36**, 899.
12. Bahr, C.H. and Heppke, G. (1987) *Liquid Crystals*, 2, 825.
13. Anderson, G., Dahl, I., Keller, P., *et al.* (1987) *Appl. Phys. Lett.*, **51**, 640.
14. Coates, D., (1990) In *Liquid Crystals, Applications and Uses*, vol. 1, (ed. B. Bahahur), World Scientific, Singapore.
15. Toyne, K. (1987) In *Thermotropic Liquid Crystals* (ed. G.W. Gray), John Wiley, Chichester, Sussex and New York.
16. De Jeu, W.H. (1980) In *Physical Properties of Liquid Crystal Materials* (ed. G.W. Gray), Gordon and Breach, London and New York.
17. Sage, I. (1990) In *Ullman's Encycolpedia of Industrial Chemistry*, vol. A15, VCH Publishers, New York.
18. Reck, B. and Ringsdorf, H. (1985) *Makromol. Chem. Rapid Commun.*, **6**, 291.
19. Reck, B. and Ringsdorf, H. (1986) *Makromol. Chem. Rapid Commun.*, **7**, 389.
20. Onsager, L. (1949) *Proc. New York Acad. Sci.*, **51a**, 627.
21. Ishihara, A., (1951) *J. Chem. Phys.*, **19**, 1142.
22. Flory, P.J. (1956) *Proc. R. Soc. London, A*, **234**, 60.
23. Robinson, C. (1956) *Trans. Faraday Soc.*, **52**, 571.
24. Plate, N.A. (1993) In *Liquid-crystal Polymers*, Plenum, New York.
25. Cox, M. (1987) In *Liquid-crystal Polymers* (ed. R. Meredith), RAPRA report number 4, Pergamon Press, Oxford, Sydney, and Elmsford, NY.
26. Blumstein, A. (1985) In *Polymeric Liquid Crystals*, Plenum, New York.
27. Wendorff, J.H. (1978) In *Liquid Crystalline Order in Polymers* (ed. A. Blumstein), Academic Press, London, New York, Orlando, Fl, and San Diego, CA.

28. Blumstein, A. (1978) In *Mesomorphic Order in Polymers and Polymerisation in Liquid Crystalline Media*, ACS Symp. Ser. 74, American Chemical Society, Washington, DC.
29. Finkelmann, H., Happ, H., Portugall, M. and Ringsdorff, H. (1978) *Makromol. Chem.*, **179**, 2541.
30. Hessel, V. and Finkelmann, H. (1985) *Polym. Bull.*, **14**, 3751.
31. Hessel, V., Herr, R-P. and Finkelmann, H. (1987) *Makromol Chem.*, **188**, 1597.
32. Keller, P. (1987) *Int. Conf. On Liq. Cryst. Polymers, Bordeaux*, abstr. 20P1.
33. Ringsdorf, H. and Wustefeld, R. (1987) *Int. Conf. on Liq. Cryst. Polymers, Bordeaux*, abstr. 10P1.
34. Sirlin, C., Bosio, C. and Simon, J. (1988) *Mol. Cryst. Liq. Cryst. Bull*, **3**, 49.
35. McArdle, C.B. (1989) In *Side Chain Liquid Crystal Polymers* (ed. C.B. McCardle), Blackie, Glasgow and London.
36. Plate, N. and Shibaev, V. (1988) In *Comb Shaped Polymers and Liquid Crystals*, Plenum, New York.
37. Adams, N.W., Bradshaw, J.S., Bayona, J.-M., *et al.* (1987) *Mol. Cryst. Liq. Cryst*, **147**, 43.
38. McCulloch, I.A. and Bailey, R.T. (1991) *Mol. Cryst. Liq. Cryst.*, **200**, 157.
39. McRoberts, A.M., Denman, R., Gray, G.W. and Scrowston, R.M. (1990). *Makromol. Chem., Rapid Commun.*, **11**, 617.
40. Kostromin, S.G., Cuong, N. D., Garina, E.S., and Shibaev, V.P. (1990) *Mol. Cryst. Liq. Cryst.*, **193**, 177.
41. Pugh, C. and Schrock, R.R. (1992) *Macromolecules*, **25**, 6593.
42. Percec, V. and Pugh, C. (1989) In *Side Chain Liquid Crystal Polymers* (ed. C.B. McArdle), Blackie, Glasgow and London.
43. Gray, G.W. (1989) In *Side Chain Liquid Crystal Polymers* (ed. C.B. McArdle), Blackie, Glasgow and London.
44. Le Barny, P. and Dubois, J.C. (1989) In *Side Chain Liquid Crystal Polymers* (ed. C.B. McArdle), Blackie, Glasgow and London.
45. Shibaev, V.P. and Friedzon, Y.S. (1989) In *Side Chain Liquid Crystal Polymers* (ed. C.B. McArdle), Blackie, Glasgow and London.
46. Shibaev, V.P. (1993) In *Liquid Crystal Polymers* (ed. N.A. Platé), Plenum, New York.
47. Shibaev, V.P. (1994) *Mol. Cryst. Liq. Cryst.*, **243**, 201.
48. Kostromin, S.G., Tuan, P. A. and Shibaev, V. P. (1992) *14th International Liquid Crystal Conference, Pisa, Italy, Abstracts*, vol. II.
49. de Gennes, P.G. (1975) *C. R. Acad. Sci. Paris*, **281**, 101.
50. Finkelmann, H., Kock, H.J. and Rehage, G. (1981) *Makromol. Chem. Rapid Commun.*, **2**, 317.
51. Hessel, V. and Finkelmann, H. (1985) *Polym. Bull.*, **14**, 3751.
52. Keller, P., Hardouin, R., Mauzac, M., Achard, M. (1988) *Mol. Cryst. Liq. Cryst.*, **155**, 171.
53. Gray, G.W., Hill, J.S. and Lacey, D. (1989) *Adv. Mater.*, **8**, 292.
54. Kreuzer, F.H. (1981) *Proceedings of the Freiburger Arbeitsagung Flussigekristalle*, paper 5.
55. Kreuzer, F.H., Andrejewski, D., Haas, W., et al. (1991) *Mol. Cryst. Liq. Cryst.*, **199**, 345.
56. Makow, D. (1982) *Leonardo*, **15**, 257.
57. Bunning, T.J. and Kreuzer, F.-H. (1995) *TRIP*, **3**(10), 318.
58. Shibaev, V.P., Kulichikhin, V.G., Kostromin, *et al.* (1982) *Dokl. Akad, Nauk SSSR*, **263**, 152.
59. Casagrande, C. (1984) *Mol. Cryst. Liq. Cryst.*, **113**, 193.
60. Shibaev, V.P., Kozlovsky, M., Beresnev, L., *et al.* (1984) *Polymer Bull.*, **12**, 299.
61. Shibaev, V.P. and Platé, A. (1985) *Pure and Appl. Chem.*, **57**(11), 1589.
62. Decobert, G., Soyer, F. and Dubois, J.C. (1985) *Polymer Bull.*, **14**, 179.
63. Freidzon, Y.S., Boiko, N.I., Shibaev, V., *et al.* (1986) *Polym. Commun.*, **27**, 190.
64. S. Uchida, S., Morita, K., Myoshi, K., *et al.* (1988) *Mol. Cryst. Liq. Cryst.*, **155**, 93.
65. Verrall, M., Beattie, D., Coates, D., *et al.* (1996) *Ferroelectrics*, **181**, 327.
66. Yuasa, K., Uchida, S., Sekiya, T., *et al.* (1992) *Proceedings SPIE*, **154**, 1665.
67. Sekiya, T., Yuasa, K., Hachiya, S., *et al.* (1993) *Liquid Crystals*, **14**(5), 1255.
68. Coles, H.J. and Simon, R (1984) *Mol. Cryst. Liq. Cryst. Letters*, **102**, 75.
69. Coles, H.J. (1985) *Faraday Disc. Chem. Soc.*, **79**, 210.
70. Coles, H.J. and Simon, R. (1986) *Mol. Cryst. Liq. Cryst.*, **3**, 37.
71. Coles, H.J. and Simon, R. (1985) *Polymer*, **26**, 1801.
72. Kelker, H. (1963) *Z. Anal. Chem.*, **198**, 254.

73. Kelker, H. (1963) *Ber. Binsenges. Phys. Chem.*, **67**, 698.
74. Lin, J.-L. and Hsu, C.-S. (1993) *Polymer Journal*, **25**(2), 153.
75. Palffy-Muhoray, P. (1990) In *Liquid Crystals, Applications and Uses*, vol. 1 (ed. B. Bahahur), World Scientific, Singapore.
76. Donald, A.M. and Windle, A.H. (1992) In *Liquid Crystalline Polymers*, Cambridge University Press, Cambridge.
77. Broer, D.J., Finkelmann, H. and Kondo, K. (1988) *Makromol. Chem.*, **189**, 185.
78. Broer, D.J., Lub, J. and Higgings, J.A. (1993) *Macromolecules*, **26**, 1244.
79. Anderson, H., Sahlen, F., Gedde, U.W. and Hulte, A. (1994) *Macromol. Symp.*, **77**, 339.
80. Broer, D.J., Boven, J., Mol, G. N. and Challa, G. (1989) *Makromol. Chem.*, **190**, 2255.
81. Broer, D.J., Hikmet, R.A.M. and Challa, G. (1989) *Makromol. Chem.*, **190**, 3201.
82. Broer, D.J. and Mol. G.N. (1990) *Polymer Eng. Sci.*, **31**(9), 625.
83. Broer, D.J. (1991) European Patent 0428, 213.
84. Broer, D.J., (1995) *Nature*, **378**, 467.
85. Coates, D., Goulding, M.J., Greenfield, S., *et al.* (1996) *SID 96 Applic. Digest*, 67.
86. Hikmet, R.A.M. and Zwerver, B.H. (1992) *Liq. Cryst.*, **12**(3), 319.
87. Yang, D.-K. and Doane, J.W. (1992) *SID Digest Techn. Papers*, **23**, 759.

8 Structural polymers

I. HAMERTON and J.N. HAY

8.1 Introduction

The development of polymers with fine mechanical properties has led to their use in many structural applications which have traditionally been occupied by materials such as metals, wood and glass, etc. Hence, although this chapter details a wide variety of polymers, the common themes running through their selection are that they all find use in what might broadly be termed 'structural applications' and, importantly, that they have stimulated commercial interest. Following an introduction to multicomponent polymer composites, the reader will find reference made to both thermoset and thermoplastic materials, and this forms the basis of the primary distinction. It is our intention to address both the chemistry *and* the properties of the structural polymers covered, for understanding the fundamental chemistry may allow the rational development of new and improved systems. The prediction of future directions for scientific research is always fraught with difficulties as many commercial developments are rapidly overtaken by economic changes, leading to some very short-lived products. We have highlighted a couple of burgeoning areas upon which interest is becoming increasingly focused at the current time. A full discussion of each of the polymers featured here is outside the scope of this work and so for more information the reader is directed in the Further Reading section to more extensive, specific, literature references.

8.2 Introduction to composites

8.2.1 Background

In general terms, a composite material may be defined as a material made up of two or more constituent components, each of which retains a recognizable identity. A structural composite is usually defined as a combination of a matrix (polymer, ceramic, etc.) with a reinforcement which is normally fibrous. Examples of common reinforcements include carbon, glass and aramid fibres which may be either chopped short or long fibres or a continuous fibre reinforcement. Inorganic fillers such as oxides, carbonates, etc. are also often present, sometimes in the absence of fibres, to provide various effects such as improved mechanical properties, reduced cost (e.g. calcium carbonate) or improved fire properties (e.g. alumina trihydrate, ATH).

The properties of composites may be either isotropic or anisotropic depending on factors such as end-use requirements, design parameters and characteristics of the processing method selected. For many structural applications, loads may not be distributed evenly throughout a structural member and so anisotropy can be introduced into the composite deliberately so that the greatest reinforcement is provided where it is most needed. Computer-aided design (CAD) and computer-aided manufacturing (CAM) are often used to assist in this process. In continuous fibre composites the anisotropy can be controlled by varying the lay-up, that is the orientation of fibre layers with respect to each other in the composites. This can be achieved through use of prepregs (see next section). In short and long fibres composites, the processing method and part design can dictate the degree of anisotropy, since regions of elongational flow will tend to orientate the fibres in the flow direction.

Selection of the processing method is thus a key factor in obtaining composites with the desired properties. In addition, the choice of resin may be defined by the processing route chosen, or vice versa.

8.2.2 Processing methods

A wide variety of composite processing methods is available depending on the requirements of the end use, the resin composition, the economics of the process, etc. Some of these are preferred for use with thermosets, some for thermoplastics and some are routinely used for both.

Moulding compounds are a class of complex formulations whose principal members are bulk moulding compounds (BMC) and sheet moulding compounds (SMC). BMCs are also called dough moulding compounds (DMC) in Europe. A number of other types of moulding compounds have been developed in recent years for specific purposes. Moulding compounds contain resin (usually unsaturated polyester), glass fibres, various inorganic fillers, cure initiators and other additives such as low-profile additives (shrinkage-control agents), pigments and chemical thickening agents in the case of SMC. The composition of moulding compunds is discussed in more detail below. The dough-like BMCs are processed into components using either injection moulding or compression moulding. In thermoset injection moulding, a ram or a screw is used to inject the BMC into a heated mould where cure takes place rapidly. Care needs to be taken to avoid fibre attrition as this can affect final properties. The process is well suited to producing large numbers of small components such as electrical connectors and circuit breakers. Compression moulding is used to manufacture parts from both BMC and SMC. Large, relatively flat panels are well suited to this production method which has therefore been used to form body panels for cars, trucks and buses. A hot mould is used and the press close rate is important.

Prepregs consist of continuous fibres which have been impregnated with resin (usually using thermosets). The fibres may be woven or unidirectional (bundles of fibres aligned in the same direction). For woven fibres, the fibres are normally impregnated from a solution of the resin, followed by evaporation of the solvent in a drying tower. Undirectional prepregs are commonly prepared by a hot-melt or pseudo-hot-melt route. In the latter, a little solvent is added to reduce viscosity. Bundles ('tows') of fibre are passed through the molten resin and sandwiched between release papers to form a sheet. Alternatively, a resin film is applied to the fibres.

Prepreg plies are laid up in the desired orientation and thickness and the resultant stack cured by compression moulding or in an autoclave to form a composite laminate. In compression moulding, matched dies are used and the mould placed in a heated press. In autoclave moulding, the prepreg stack is enclosed in a vacuum bag, with bleeder cloths to absorb excess resin. A vacuum is applied to remove volatile by-products, and gas pressure is applied to consolidate the laminate. Other methods are available for forming components from thermoplastic prepregs.

Reaction injection moulding (RIM) is a technique used to process rapid curing thermosets or, occasionally, thermoplastics (e.g. nylon). Two resin precursor streams are mixed very rapidly in an impingement mixer and injected into a mould, which may be heated. Cure is initiated on mixing and continues in the mould. Materials with gel times as short as a few seconds can be handled in this way. The technique is well suited to rapid production and high throughput. Reinforced reaction injection moulding (RRIM) and structural reaction injection moulding (SRIM) are used when additional reinforcement is required in the moulding. In RRIM a reinforcement such as a filler or short fibres is incorporated into one of the reactant streams. In SRIM a preformed fibre mat is placed in the mould prior to injection of the resin. Polymers used in RIM, RRIM and SRIM include isocyanate polymers such as polyurethanes (PU) and polyureas, epoxy resins, polydicyclopentadiene (RIM) and nylon 6. Products are used in transport applications and machinery.

Resin transfer moulding (RTM) has some similarities to SRIM, but in this case a single reactant stream is injected into a hot mould. As a result, a wider range of thermoset resins can be employed in RTM. Composite components have been manufactured from unsaturated polyesters, vinyl esters, urethane acrylates (Modar™), epoxies, bismaleimides and cyanate esters. Cycle times are generally longer than for RIM resins. Applications are similar to SRIM products, but also extend to aerospace.

Thermoset injection moulding has been mentioned above, but injection moulding is also widely used to manufacture thermoplastic composites. Typically, pellets of thermoplastic polymer containing short fibres are fed into a heated screw 'extruder' which injects the molten composition into a cooler mould where it solidifies again. Cycle times are short and throughputs high.

Pultrusion is worthy of mention as a relatively high-volume, continuous process for producing profiles of constant cross-section. In the case of thermoset pultrusion a continuous bundle of, for example, glass fibres is pulled through a resin bath containing monomers, initiators, etc. After the fibres have been wetted by the resin they are pulled through a heated die where resin cure occurs. The resultant pultruded composite rod or other profile is then cut to the required length for supply to end-users. Unsaturated polyesters, vinyl esters and epoxies have all been pultruded successfully. The composite profiles are used in cable trays, conduits, as building stock and in other civil-engineering applications.

8.2.3 Carbon fibre

Carbon fibre is widely used as the reinforcement in composites where high specific modulus and strength are demanded by the final application. Carbon fibre might itself be loosely described as a structural polymer. It is formed by heat treatment of polyacrylonitrile (PAN) or mesophase pitch. The method for conversion of PAN to carbon fibre was developed in the 1960s. It involves oxidizing PAN fibres in air under tension at 220°C followed by carbonization in an inert atmosphere at very high temperatures. Carbonization leads to formation of graphitic structures in the fibres. The carbonization temperature affects the final mechanical properties of the carbon fibre. Very high temperatures, up to 2500°C, favour high modulus, whereas temperatures around 1500°C favour high tensile strength. The high specific properties arise from the high degree of alignment of the graphite planes along the fibre axis. PAN-based carbon fibres are used in high-value composite applications. Pitch-based carbon fibres have lower mechanical strength than PAN-based fibres and find uses in sports goods such as fishing rods, squash rackets, etc.

Other organic fibres used in composites include aramid (polyamide) and polyethylene fibres.

8.3 Thermosets

8.3.1 Unsaturated polyesters and vinyl esters

Both unsaturated polyesters (UPEs) and vinyl esters (VEs) find extensive use in BMC and SMC for structural applications. A UPE prepolymer is made typically by reaction of an acid derivative and maleic anhydride with an aliphatic diol or glycol at 150°C–200°C (Scheme 8.1). A thermoplastic prepolymer of molecular weight around 2000 is formed.

Isomerization of maleate to fumarate occurs, the extent of which determines the properties of the final product. Other acid derivatives and diols can be used to impart a variety of properties to the final cured resin. The fumarate and maleate double bonds in the processable prepolymer provide sites for free radical cross-linking. The prepolymer is usually mixed with a reactive diluent such as styrene or methyl methacrylate and cured thermally with a peroxide initiator. (UPE composite manufacturing plants are smelly!) A cobalt (II) accelerator can be added to reduce the cure temperature. Each cross-link normally contains between one to three styrene units. The main features of the various chemical types of UPEs are summarized in Table 8.1.

UPEs can be processed by RTM, pultrusion, hand lay-up and as BMC or SMC. Cured UPEs have excellent chemical and acid resistance. Mechanical

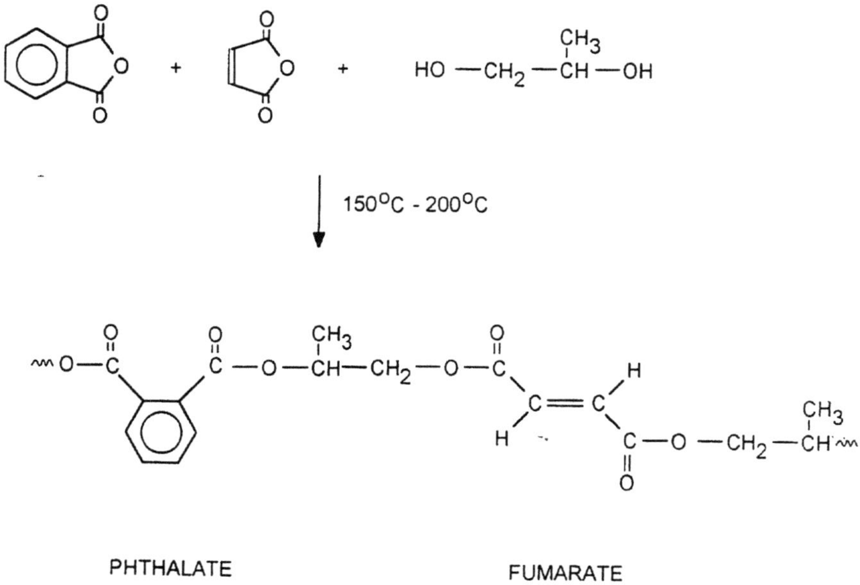

Scheme 8.1 Preparation of an unsaturated polyester prepolymer.

properties are also generally good, but the high cross-link density makes them brittle. For some applications, rubber toughening is required (Section 8.3.8). As mentioned previously, BMCs and SMCs are complex mixtures of resins, fibres and additives. A typical BMC contains 15%–20% w/w glass content with a fibre length of 6–12 mm and *ca.* 200 phr of inorganic filler. Although a high filler content reduces the cost, it can also degrade the mechanical properties. Also, BMC composites can have a specific gravity as high as 2.0. A typical SMC, on the other hand, contains a lower filler content but more than 30% w/w glass of fibre length 25–50 mm. Because SMC is required as a dry non-tacky sheet, a key feature of SMC formulations is the presence of a chemical thickening agent

Table 8.1 Unsaturated polyester chemical types

Chemical type		Cost	Heat distortion	
Acid variation	Diol variation	(relative)	temperature	Other features
Maleates	–	Low	Moderate	–
Fumarates	–	–	High	High reactivity; rigid
*Ortho*phthalic resins	–	Low	Moderate	High strength
*Iso*phthalic resins	–	High	–	High strength; good chemical resistance
	Propylene glycol	Low	–	Flexible; water resistant
	Bisphenol-A	High	High	Corrosion resistant, good mechanical properties
	Neopentyl glycol	–	–	Corrosion resistant; low colour

which is typically a group II metal oxide or hydroxide such as magnesium oxide. The thickening process is believed to proceed via formation of ionic cross-links involving the metal ion and terminal carboxyl groups on the UPE prepolymer. The thickening process is time-dependent and is known as maturation. In addition to controlling the SMC physical properties, the presence of thickening agent also affects moulding behaviour and product consistency. Physical thickening methods based on crystallization have been developed and are claimed to have advantages over chemical thickening. In preparing BMC and SMC, the order of addition of the different components is important in controlling product consistency.

UPE composites find widespread industrial application. They are used as car, bus and truck body panels, in business equipment, as panels, louvres and conduits in construction and also in electrical connectors, switchgear and circuit breakers. Their outstanding acid and chemical resistance has led to their extensive use in chemical plant. They are also used to make artificial marble.

VEs are manufactured by the reaction of an epoxy resin (di-epoxide) with acrylic or methacrylic acid (Scheme 8.2). The resultant VE monomer can be polymerized with styrene, using a peroxide initiator, to give a cross-linked polymer product. While the methacrylates find use in composites, the acrylates are used mainly in coating applications. In general, VE composites have better mechanical properties and improved toughness compared with UPE composites. Their chemical resistance is also greater than that of UPEs, but so is their cost. They find use in applications where their superior properties to UPEs are

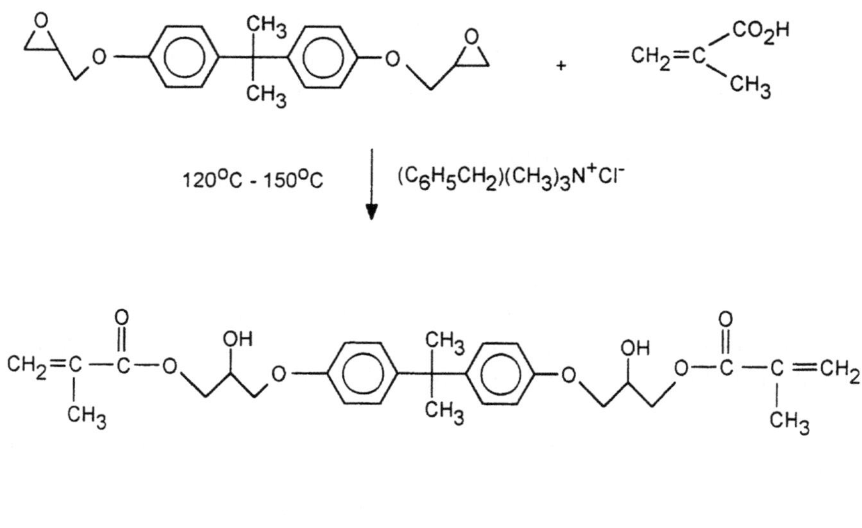

VINYL ESTER MONOMER

Scheme 8.2 Preparation of vinyl ester monomer.

required. VEs are used in both BMC and SMC. VE composites are chosen in preference to UPEs for some automotive and machinery applications and in chemical plant where the environment is too severe for UPEs.

8.3.2 Epoxy resins

Epoxy resins contain at least two epoxy (or oxirane) groups within their structure; a planar, three-membered ring comprising two carbon atoms and one oxygen atom. The substantial ring strain (due to angular distortion from the tetrahedral carbon angle and the electronegativity of the oxygen atom) renders it more reactive than non-cyclic ethers, and those reactions involving either elecrophilic attack on the oxygen atom or nucleophilic attack on one of the ring carbon atoms are of greatest importance for the synthesis and cure of epoxy resins. The most important routes for the synthesis of commercial epoxy resins are via the dehydrohalogenation of halohydrins [Scheme 8.3(a)] and the epoxidation of alkenes with peracids or their esters [Scheme 8.3(b)]. Generally speaking, the glycidyl-ether-type resins derived from epichlorohydrin reactions are used for adhesives or resin matrices, whereas the 'ene-oxide'-type, arising from the second route are more often used in electrical applications.

Generally, epoxy resins will not crystallize, although it should be noted that the pure diglycidyl ether of bisphenol-A (DGEBA) may do so, but is more often a glassy, amorphous solid. The presence of glycidyl groups in an epoxy resin imparts flexibility and reduces the viscosity of the uncured resin and thus enhances its processability, but they also decrease the heat resistance of the cured epoxy resin. DGEBA has been the most important commercial epoxy resin system because of its relatively low viscosity, its physical strength after curing, and its relatively low production cost. Unfortunately, its relatively poor thermo-mechanical properties ($T_g < 120°C$) make it unacceptable for high-performance applications unless speciality aromatic curing agents are formulated into the final compounds. Consequently, many speciality epoxy resins have been developed to meet specific performance requirements (Table 8.2). Both high functionality and low molecular weight (to promote high cross-link density) and the incorporation of rigid structures between two glycidyl groups (such as 9,9-*bis*(4-hydroxyphenyl)fluorene, 3,6-dihydroxyspiro(fluorene-9,9′-xanthane and 10,10′-*bis*(4-hydroxyphenyl)anthrone) are employed to improve the thermal and mechanical properties of the cured epoxy resins. These structures may provide molecular rigidity and offer flame resistance in the cured materials.

Epoxy resins with higher functionality may yield more reactive epoxy monomers and polyepoxy networks with higher cross-link density and hence improved heat and thermal resistance, e.g. triglycidyl isocyanurate (commercially available as Araldite PT-810™, Ciba-Geigy) and low-viscosity liquid triglycidyl resins of *p*-aminophenol (Araldite MY 0500™ and Araldite MY 0510™, Ciba-Geigy). The best-known tetrafunctional epoxy resins are probably those based on *N,N,N′,N′*-tetraglycidyl-4,4′-diaminodiphenylmethane

(TGDDM), known as Araldite MY 720™ and Araldite MY 721™ (Ciba-Geigy). The resulting resins offer good rheology and high functionality and are useful for high-strength composites. More recently, two more commercial epoxies

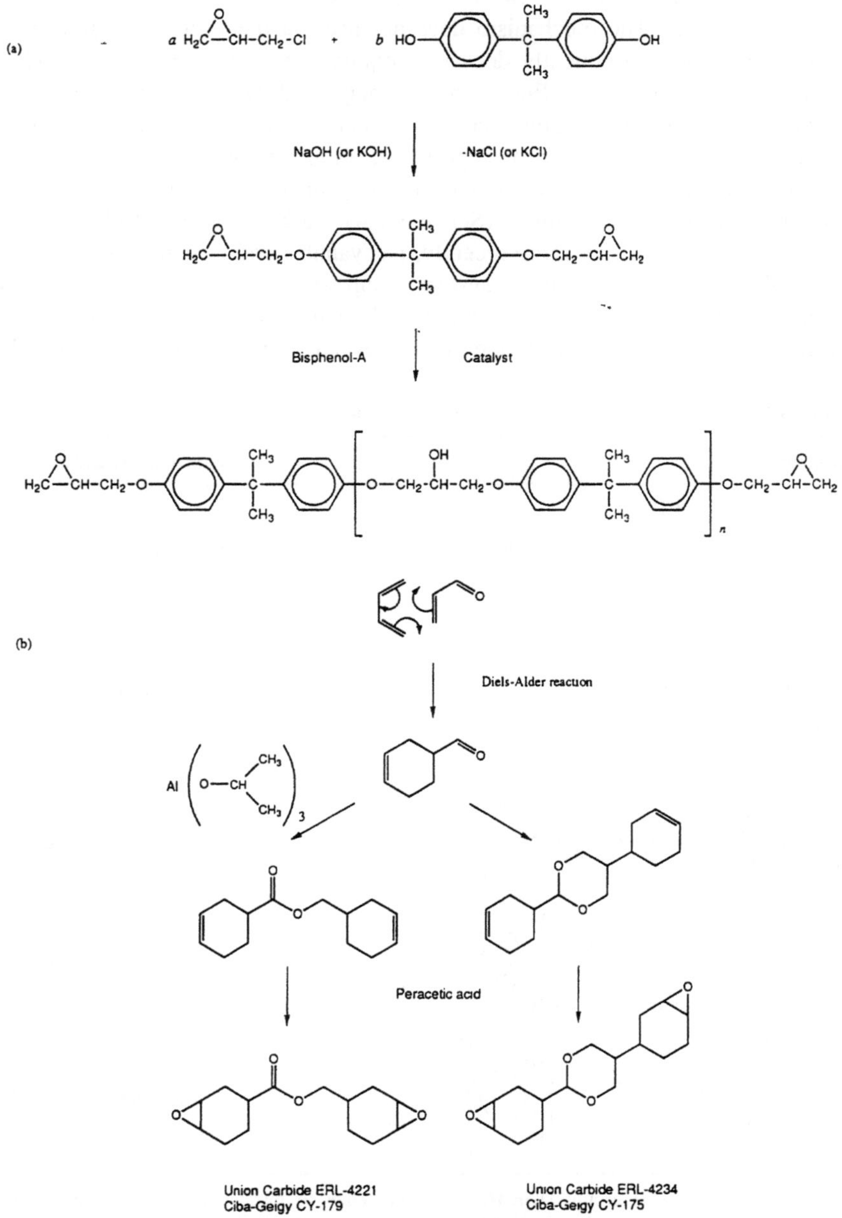

Scheme 8.3 Synthesis of epoxy resins by (a) dehydrohalogenation of halohydrins; and (b) epoxidation of alkenes with peracid or its ester.

Epon HPT 1071™ (Shell) and Epon HPT 1072™ (Shell) have been designed to provide reduced moisture sensitivity.

Epoxy resins cure to form a rigid three-dimensional network from a liquid or thermoplastic epoxy monomer or oligomer to form a hard, insoluble and infusible product. The mechanism does not involve the generation of volatile condensates and consequently does not require pressure to produce void-free mouldings and artefacts. Epoxy resins may undergo homopolymerization (involving the use of catalytic curing agents or initiators) and step-growth polymerization (polyaddition or copolymerization involving the use of curing agents or 'hardeners'). A wide range of curing agents are available, including primary and secondary amines (Scheme 8.4), carboxylic anhydride, amides, phenols, mercaptans and, more recently, isocyanates and cyanate esters.

Usually a commercial epoxy system comprises the resin itself, the curing agent and perhaps a number of additional materials whose presence in the formulation serves to modify the properties and characteristics of both the

Table 8.2 Selected structures of commercial epoxy resins

Structure	Functionality and name[a]
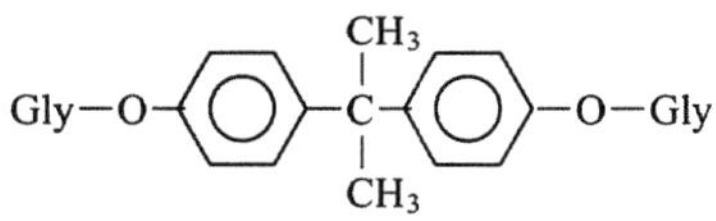	Difunctional, DGEBA
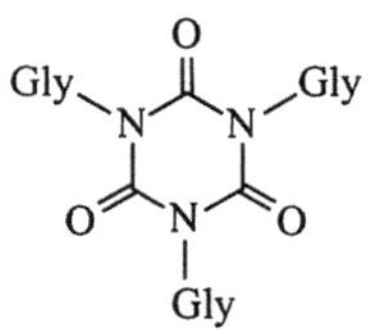	Trifunctional, TGIC
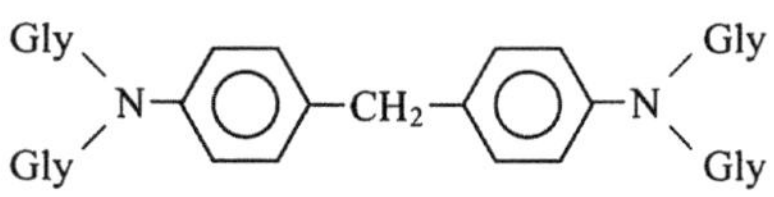	Tetrafunctional, TGGDM
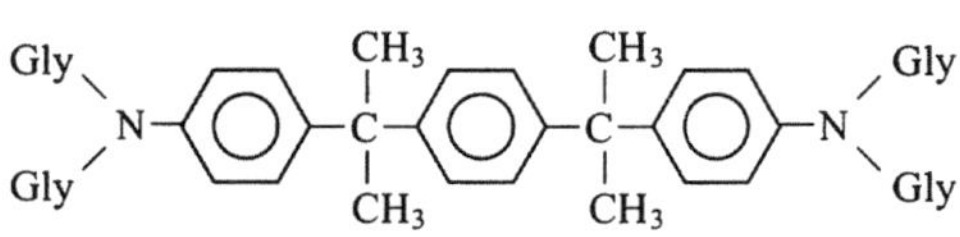	Tetrafunctional, TGBAP
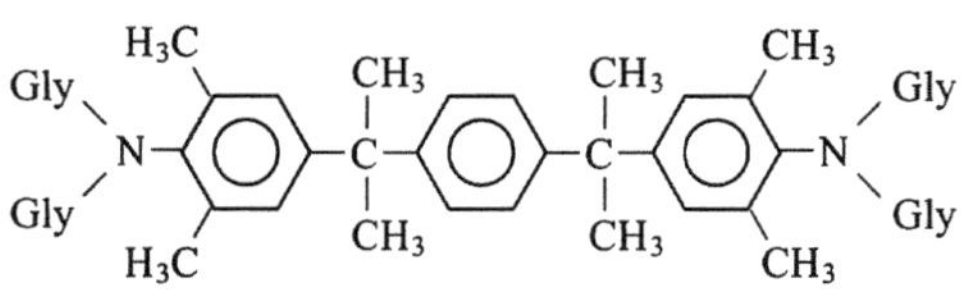	Tetrafunctional, TGMBAP

[a] Gly = glycidyl (epoxy) group; DGEBA = diglycidylether of bisphenol-A; TGIC = triglycidyl isocyanurate; TGDDM = *N,N,N″,N″*-tetraglycidyl-4,4′-diaminodiphenylmethane; TGBAP = *N,N,N″,N″*-tetraglycidyl-α,α′-*bis*(aminophenyl)-*p*-di-isopropylbenzene; TGMBAP = *N,N,N″,N″*-tetraglycidyl-α,α′-*bis*(3,5-dimethyl-4-aminophenyl)-*p*-di-isopropylbenzene.

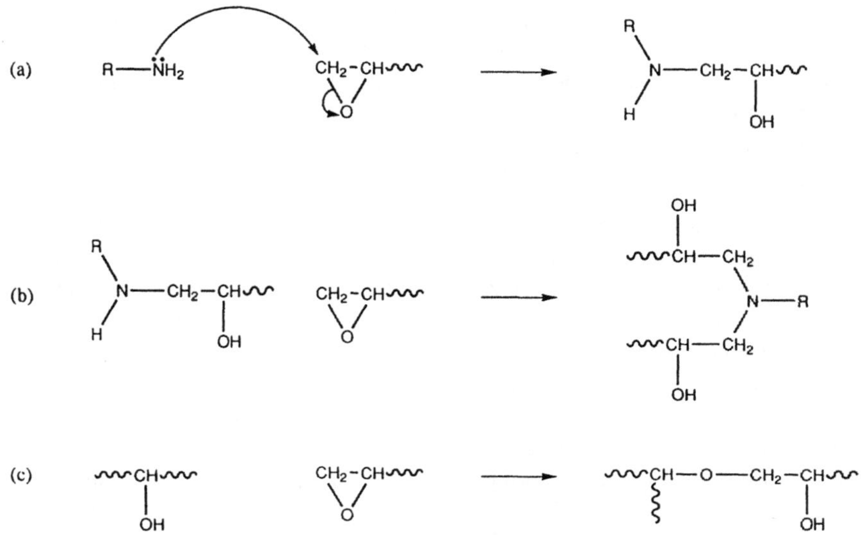

Scheme 8.4 General mechanism for amine cure of epoxy resins involving (a) primary and (b) secondary amines; (c) hydroxyl-catalysed polyetherification reaction.

uncured and the cured forms of the epoxy resin. These additives may include many of the following.

- Diluents: these may reduce viscosity and allow for greater incorporation of other ingredients in the formulation (e.g. fillers) and as a means of improving wetting characteristics. Diluents may be non-reactive (e.g. aromatic hydrocarbons, di-*n*-butyl phthalate, styrene, phenols and castor oil) to provide viscosity reduction or reactive (e.g. mono and polyfunctional epoxies and non-epoxy-based compounds). The selection of a diluent should optimize the diluent effect while minimizing the deleterious effects on the cured resin properties.
- Fillers: these may modify the properties and characteristics of epoxy resins, but also importantly are a means of reducing cost. Some disadvantageous characteristics associated with their use include an increased density (and hence weight) together with an increase in viscosity – with possible processing implications.
- Plasticizers and flexibilizers: these are used to improve properties such as strain tolerance, impact strength, low temperatures crack resistance and adhesive properties such as lap shear and particularly peel strength, but they may reduce chemical and solvent resistance. The use of rubbers and elastomers is widely used to toughen epoxy resins and depends on the nature of the rubber, the volume fraction of the rubber, the particle size and particle size distribution of the dispersed phase and the interfacial bonding. The most widely used rubbers for toughening thermosets are carboxy terminated butadiene-acrylonitrile random copolymers (CTBN, such as BF Goodrich's Hycar series). The more expensive polysiloxanes (which offer excellent low temperature flexibility,

outstanding moisture resistance and excellent resistance to weathering), fluoroelastomers or polyacrylates are reserved for more demanding environments. Preformed core-shell rubbers are also highly effective toughening agents. The use of tough, thermoplastic polymers may offer greater epoxy toughness, in comparison with rubber-modification, but the level of improvement is often generally poor, particularly with relatively low T_g systems.

The consumption of epoxy resins has increased steadily over the past 50 years and is largely because of their increasing use in adhesives, tooling, composites and electronics because of the versatility of their processing procedures and their fine properties (e.g. electrical properties – dielectric permittivity, dissipation factor, dielectric strength and volume resistivity, with relatively low sensitivity to temperature, good adhesion to metals and silicon, low permeability or absorption of water, good stability against chemical and solvent attack, thermal resistance and good mechanical strength). It has been estimated that in 1991 the total world consumption of epoxy resins was in excess of 516 000 tonnes. An average annual increase of 5% was predicted to yield a figure of 660 000 tonnes for epoxy consumption in 1996. The United States is the largest user of epoxy resins (*ca.* 175 000 tonnes in 1991) with the combined European states following closely behind (*ca.* 169 000 tonnes in 1991), although it is worthy of note that Japan accounted for *ca.* 125 000 tonnes during the same period. A dramatic expansion has been observed in the range of applications in which epoxy resins have been employed. The major outlet for epoxy resins is still for surface coatings, which consumes over 50% of all epoxy resins produced, but epoxies are increasingly being used in more advanced applications (Table 8.3).

In addition to the above, multifunctional epoxy matrix systems have also been used from the beginning of the application of composite technology to space vehicles, but more recently attention has shifted to the consideration of engineering thermoplastics and cyanate esters in this application.

Table 8.3 Applications of epoxy resins in the United States and Western Europe. Sources: Ellis, B. (ed.), *Chemistry and Technology of Epoxy Resins*; published by Blackie Academic and Professional, Glasgow, 1993; Hamerton, I., *Recent Developments in Epoxy Resins*, Rapra Review Report 91, 1996, **8**(7)

	United States, 1990		United States, 1991		Western Europe, 1991	
Application	1000 tonnes	%	1000 tonnes	%	1000 tonnes	%
Surface (protective coatings)	89	49	84	51	59	35
Electrical and electronics	25	14	22	13	42	25
Composites (reinforced resins)	14	7.5	13	8	19	11
Bonding and adhesives	13	7.5	12	7.25	17	10
Flooring (construction)	12	6.5	11	6.25	25	15
Tooling and casting	13	7.5	12	7.25	–	–
Other	15	8.3	12	7.25	7	4
Total	181	100	166	100	169	100

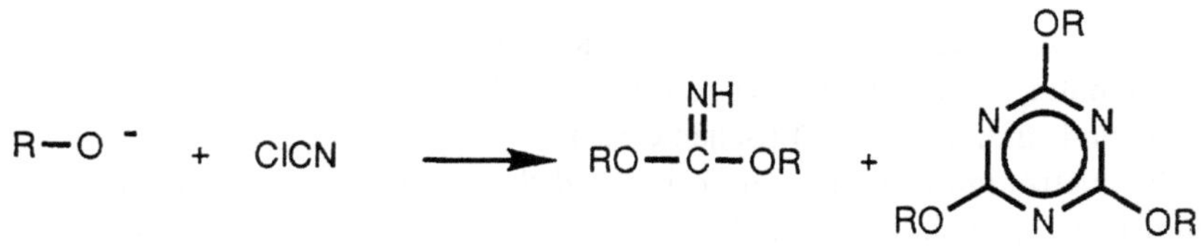

Scheme 8.5 Cloez's preparation of organic cyanate ester.

8.3.3 Cyanate esters

During the past decade, aromatic cyanate esters (CEs) have emerged as a new class of thermosetting resins for use as prepreg matrices, principally in the aerospace and electronics industries. Attempts to prepare organic CEs date back to 1857 when Cloez carried out the reaction of an alkoxide species with cyanogen chloride (Scheme 8.5). This procedure and later attempts with aryloxides were not successful because the use of excess oxide led to the production of mixtures containing both the desired cyanurates and imidocarbonates.

The use of *ortho*-substituted phenols proved successful because of the increased steric hindrance afforded by the substituent, thus preventing the reaction of the excess aryloxide from consuming the product, leading to the isolation of the first *o*-alkylated aryl cyanate. The only synthetic method to achieve any commercial importance for the preparation of high-temperature thermosetting resins has been the reaction between a cyanogen halide and a phenol in the presence of a base (Scheme 8.6) as not only is it successful with monophenols and polyphenols as well as with a number of aliphatic hydroxyl compounds, but crucially it may also be carried out on an industrial scale. This method may be applied to a large number of aryl and haloalkyl cyanates but it is not applicable to alkyl cyanates – other reactions, involving the thermolysis of a thiatriazole intermediate, amy be employed to produce alkyl cyanates (although the resulting products are still apt to undergo exothermic isomerization to the corresponding isocyanate). Cyanogen chloride is the reagent of choice for industrial preparations, when cyanogen bromide is employed diethylcyanamide is produced as a by-product (arising via the competing von Braun reaction between the aryloxide and the organic cyanate). Hence, the stoichiometry of the reaction is important and the use of excess cyanogen halide or triethylamine to extinguish residual phenol will increase the production of diethylcyanamide.

Chemically this family of thermosetting monomers and their prepolymers contain the reactive ring-forming cyanate (—O—C≡N) functional groups. Conversion of polyfunctional CE monomers to thermoset polymers occurs via cyclotrimerization (an addition process) to form three-dimensional networks of

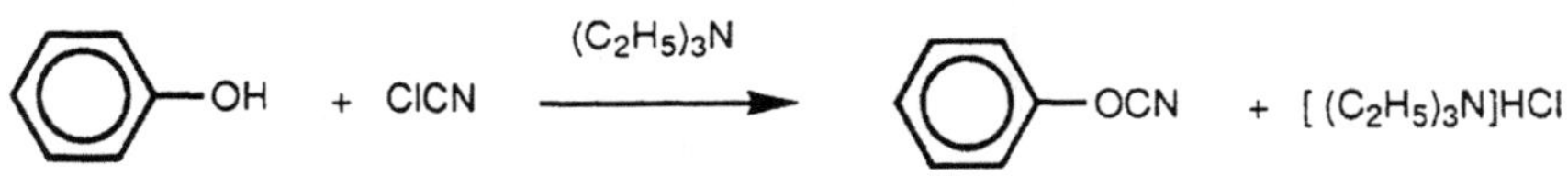

Scheme 8.6 Synthesis of a high-temperature thermosetting resin.

oxygen-linked triazine (or cyanurate) and bisphenyl units, correctly termed polycyanurates (although the resins are variously called 'polycyanates', 'cyanate esters' and 'triazine resins') (Scheme 8.7).

The cyclization reaction can be induced thermally, but this reaction is slow and usually difficult to drive to high levels of conversion; cure catalysts (such as carboxylate salts and chelates of transition metal ions) are commonly used. It is believed that the role of transition metal ions in the polymerization reaction is to facilitate the cyclization reaction of cyanate monomers by formation of coordination complexes. The choice of catalyst may also impart particular properties to the CE (e.g. the use of copper at 100–500 ppm Cu^{2+} to maximize the degree of cure, or zinc to give latent cure properties). The use of a co-catalyst (typically an involatile alkylphenol such as nonylphenol at 2–4 phr) serves both to improve

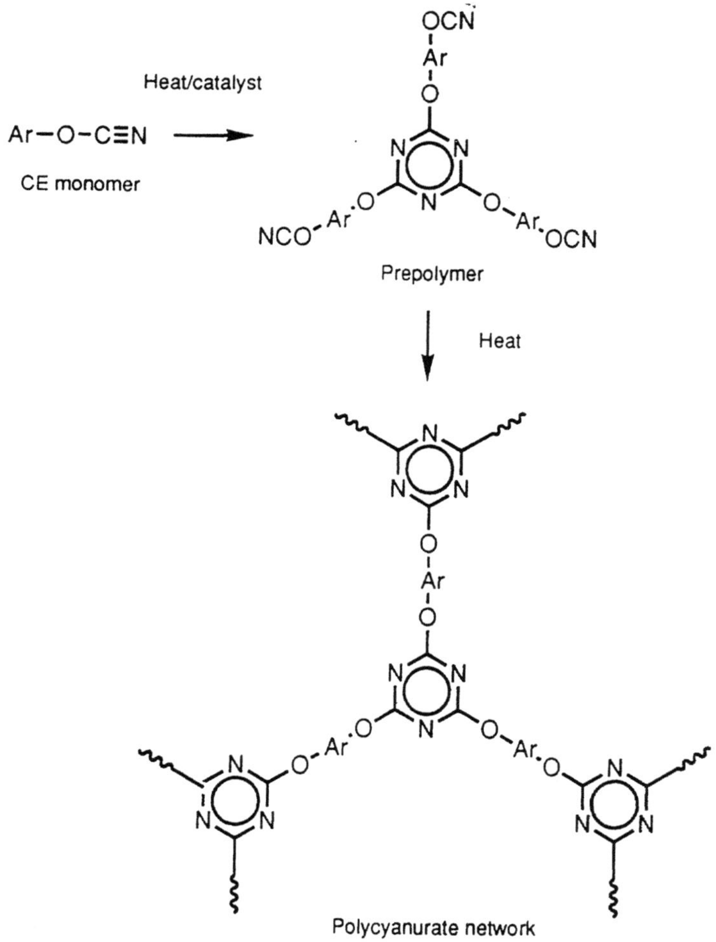

Scheme 8.7 Conversion of polyfunctional cyanate ester (CE) monomer to a thermoset polymer.

the solubility of the transition metal salts in the CE monomer and to complete ring closure of the triazine ring via hydrogen transfer in the highly viscous polymer at higher degrees of polymerization. However, an undesirable side effect of the transition metal catalysts such as the carboxylates is their tendency to increase CE hydrolysis by any water present, leading to the formation of unstable carbamates (zinc salts are particularly problematical in this respect). The storage stability of uncatalysed CE monomers and prepolymers at room temperature is generally excellent, and room temperature storage lives of six months are not uncommon and can be extended by lower-temperature storage. In blends of CEs with epoxy resins, the storage life has been reported to vary from a few days to several weeks at 50°C. Finally, one of the important considerations for materials selection is toxicity and, in contrast to many commercial high-performance monomers, CEs display low toxicity (acute oral LD_{50} (rat) $> 2.5\ \mathrm{g\ kg^{-1}}$ for the bisphenol-A dicyanate monomer) and negative Ames test for all commercial CE monomers.

Industrial CE producers have adopted a variety of synthetic approaches to explore the resulting structure–property relationships in the polycyanurate (Table 8.4), but despite possessing excellent physical properties (making them suitable for high-performance material applications) it is still necessary to broaden the application of CEs by further improvements in performance and costs. Fortunately, CEs lend themselves to a number of processing strategies as they may be prepolymerized to form a range of oligomers with differing viscosities and/or blended with reactive diluents such as epoxy resins or toughening agents such as engineering thermoplastics or elastomers. A variety of systems based on an understanding of CE chemistry has been developed in recent years, including polyphenols, polyamines, cyanamides, epoxy resins, acid anhydrides, benzocyclobutenes and unsaturated olefins – demonstrating the scope of the modification of properties which is available.

CEs offer an attractive selection of physical and mechanical properties. The high aromatic content in monomers and the cured network is responsible for the relatively high T_g (which dominates service temperature in composite applications), inherently low smoke generation and good flame retardancy; the relatively low cross-link density, thought to arise from the inclusion of triazinebicyclophane rings arising from intramolecular reaction, and high free volume are responsible for the good toughness (e.g. strain energy release rate and tensile elongation); the balanced dipoles in the cured state and absence of hydrogen-bonding functionality are responsible for the low dielectric properties, low melt temperature and low moisture absorption. As a result, cured CEs are considerably more hydrophobic then are epoxies and bismaleimides (BMIs), and neat resin specimens exhibit a moisture absorption of 0.5%–2.5%. Partly for this reason, by the late 1980s cyanate esters had carved a niche for themselves between epoxies and BMIs, whilst CE resins have been used as the matrix in continuous fibre-reinforced laminates with carbon (in aerospace composites), glass and aramid (in high-speed multilayer circuit boards, their primary application). Recent

Table 8.4 Chemical structure of commercial or developmental polycyanurates available in monomer and/or prepolymer form

Polycyanate monomer structure or precursor	Trade name and supplier	Physical state	Homopolymer property[a]			
			T_g (°C)	H_2O (%)	D_k at 1 MHz	G_{IC} (J m^{-2})
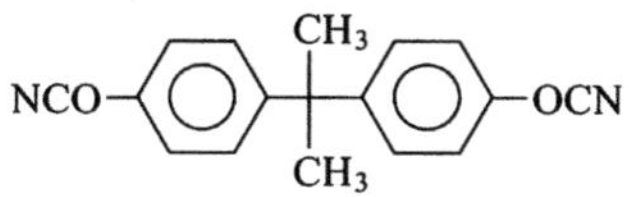 Bisphenol-A	AroCy B, Ciba-Geigy BT-2000, Mitsubishi GC	Crystal	289	2.5	2.91	140
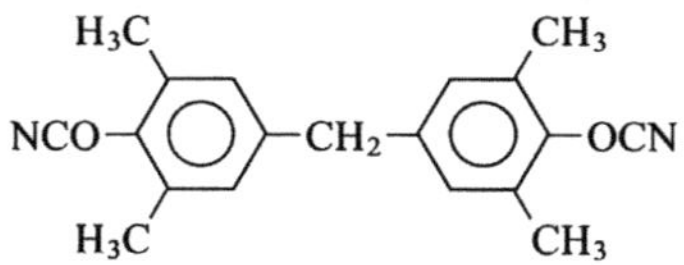 Tetramethylbisphenol-F	AroCy M, Ciba-Geigy	Crystal	252	1.4	2.75	175
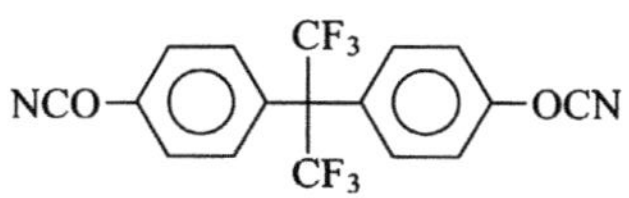 Hexafluorobisphenol-A	AroCy F, Ciba-Geigy	Crystal	270	1.8	2.66	140
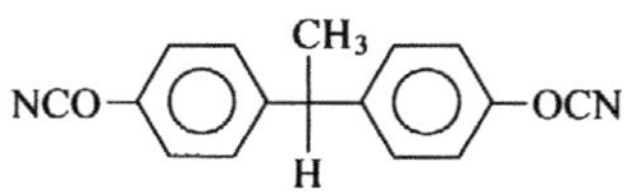 Bisphenol-E	AroCy L, Ciba-Geigy	Liquid	258	2.4	2.98	190

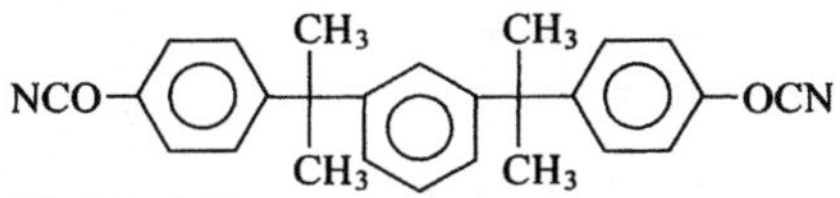 Bisphenol-M	RTX-366, Ciba-Geigy	Semi-solid	192	0.7	2.64	210
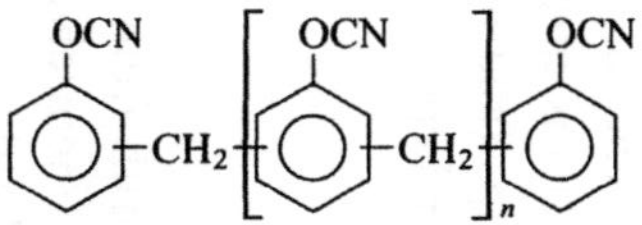 Novalac resin	Primaset PT, Allied Signal REX-371, Ciba Geigy	Semi-solid	270–>350	3.8	3.08	60
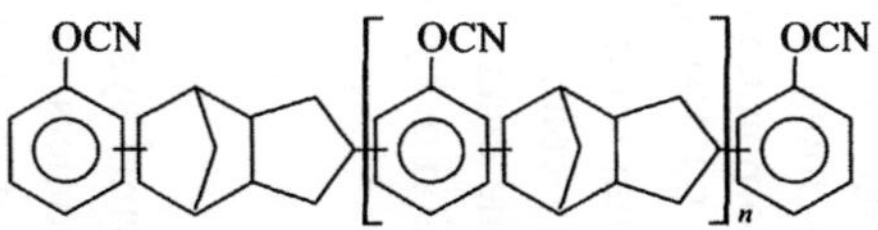 Dicyclopentadienyl bisphenol	XU-7187, Dow Chemical	Semi-solid	244	1.4	2.80	125

[a] Cured state properties:
T_g, glass transition temperature (DMA);
H_2O (%), absorption of water at saturation (100°C);
D_k, dielectric constant;
G_{IC}, fracture energy by the double torsion method.

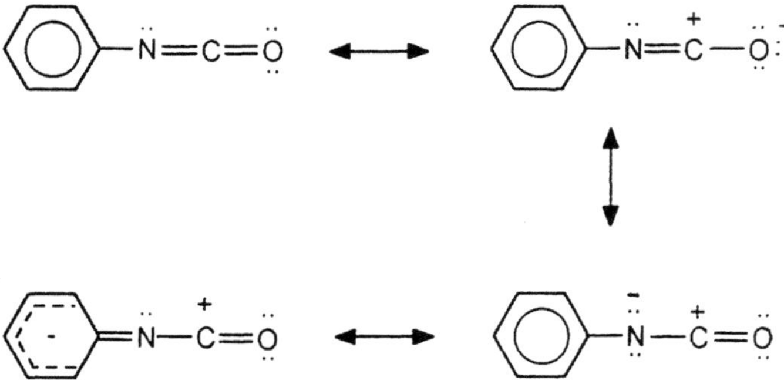

Scheme 8.8 Resonance structures of the isocyanate group.

developments in molecular architecture have produced second-generation cyanate ester resins with performance in the T_g range 190°C–290°C, with inherently more toughness than either epoxies or BMIs, lower moisture absorption and, the common denominator for circuitry design, lower dielectric loss properties (Table 8.4).

Now, in the 1990s, increasingly hi-tech applications make everincreasing demands upon candidate materials for use in both terrestrial and extraterrestrial environments, and CEs have already flown in prototype radomes and high-gain antennae, with possible applications in primary and secondary structures of future high-speed civil transport systems. Cyanates do suffer from the drawbacks of relatively high cost (the bisphenol-A dicyanate prepolymer currently costs *ca.* £31 kg^{-1}, in contrast to *ca.* £15 kg^{-1} for a basic epoxy) and a high sensitivity to the effects of moisture absorption (the action of water serves to initiate a degradation route for unreacted cyanate groups at temperatures in excess of 200°C, leading to the evolution of carbon dioxide and the blistering of laminates). This phenomenon makes correct choice of catalyst and curing schedule an important consideration in order to maximize the degree of polymerization. The relatively high cost of these resins makes their use in blends with cheaper resin systems more desirable, and common examples of two approaches involve their incorporation with epoxy resins and thermoplastics (in a product such as Fibredux 927™) and BMIs (BT2160™). In these systems, particularly those incorporating several components, the cure chemistry is often complex. In the case of epoxy-containing systems cure may involve not only the homocyclotrimerization of the cyanate groups (and polyetherification of epoxy groups) but also the co-reaction of cyanate and epoxy group via the formation of oxazolidinone moieties.

8.3.4 Isocyanate polymers

A number of important polymer systems result from the reaction of isocyanate monomers with other monomers or with themselves. The reactivity of the isocyanate group is a result of the bond polarities represented by the resonance structures shown in Scheme 8.8. The carbon atom in the isocyanate group carries a partial positive charge which makes it susceptible to nucleophilic attack. It is thus not surprising that isocyanates react readily with alcohols and amines. The most common example of this is the reaction of di-isocyanates with diols and polyols to form polyurethanes, which are widely used in foams, adhesives, coatings, etc. It is arguable whether polyurethanes are truly structural polymers and so they will not be considered in detail here. Reinforced polyurethanes produced by RRIM find applications in vehicle parts, including bumpers and fascias. Hybrid systems such as vinyl ester–urethanes have potential uses as structual foams. Poly(urethane isocyanurates) have greatly improved thermal stability compared with the polyurethanes, as a result of the high stability of the isocyanurate ring

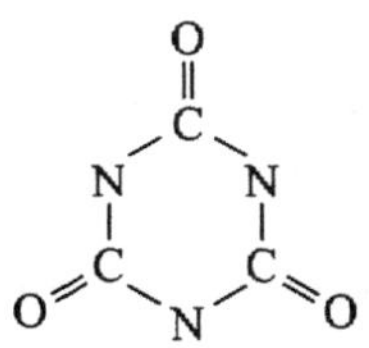

which is formed by catalysed cyclotrimerization of the isocyanate group. The foams have been used to insulate liquid-natural-gas tankers, as pipe insulation and as insulation for the NASA Space Shuttle fuel tanks. It is worth noting at this point that concern over the damage caused to the Earth's ozone layer by chlorofluorocarbons (CFCs) has resulted in the phasing out of CFCs as the preferred blowing agent in these types of foam. Isocyanate polymers have also been processed by SRIM to form components which are used in automotive applications.

Polyureas result from the reaction of di-isocyanates with diamines and polyether diamines. The reaction is a step-growth polymerization, with no evolution of volatile by-products. Analogous to polyurethanes, the use of short-chain diamines leads to 'hard' polyurea blocks, whereas 'soft' blocks result from use of more flexible polyether amines. Phase separation of soft and hard blocks leads to a system with a fracture toughness higher than otherwise would be expected. The soft blocks act as rubber toughening agents for the thermoset polyurea.

The di-isocyanate–diamine reaction to form polyureas is very fast, with gel times as low as 1 s. Processing of polyureas can therefore be problematic, even when using the RIM process. Despite this, polyurea components have found use in automotive parts, where a higher temperature and mechanical performance than that of polyurethane is required. Fairly recently, a way of modifying the polyurea system reactivity and viscosity build-up has been reported, allowing scope for greater control in the manufacture of structual components from the resins. The process is illustrated in Scheme 8.9 and results in formation of a poly(urea-isocyanurate.)

In RIM processing, one reactant stream consists of the di-isocyanate monomer whereas the other is a mixture of the diamine, polyamine and catalyst (DABCO). The presence of an excess of di-isocyanate results in formation of a controlled viscosity isocyanate-terminated urea oligomer during mixing. When the oligomer fills the hot mould (90°C), the catalyst initiates cyclotrimerization of the isocyanates to form isocyanurate cross-links. Mould filling takes about 1.4 s and the gel time is approximately 4 s. The processing and mechanical and thermal properties of the products can be tailored by varying reactant ratios. Manufacture of components by RRIM and SRIM is also feasible. In the above systems, the thermal stability of the cured resin generally decreases in the series

isocyanurate > urea > urethane.

The similarity of the isocyanurates to the cyanate esters (triazines) described in the previous section should be noted.

Isocyanates will co-react with epoxy resins at elevated temperatures in the presence of a catalyst such as tetramethylammonium iodide to give cross-linked resins via a combination of cycloaddition and trimerization reactions. The cycloaddition reaction leads formally to formation of poly(2-oxazolidone)s. Commercial systems have sometimes been known as EPIC resins. The resin systems may be processed readily at 160°C, followed by a post-cure, leading to

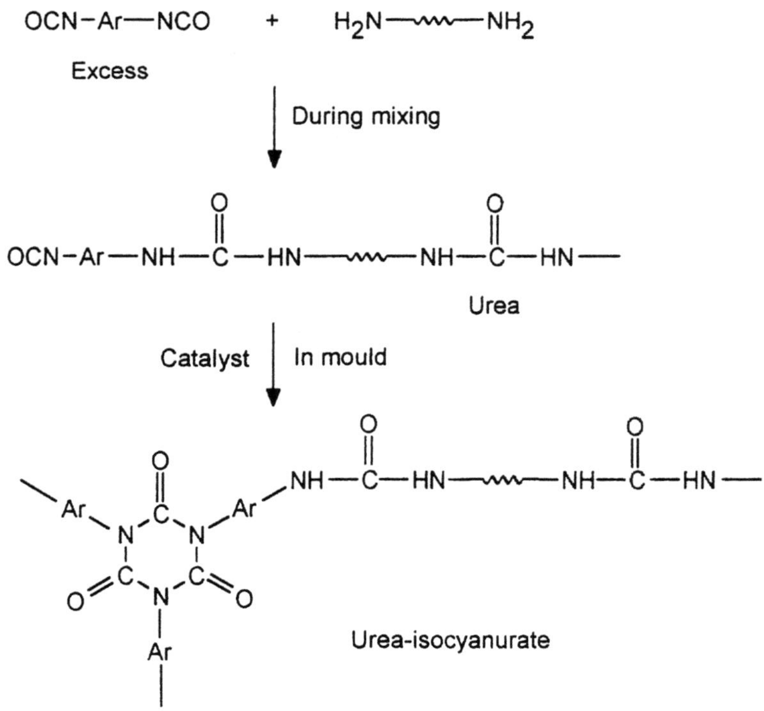

Scheme 8.9 The formation of poly(urea-isocyanurate).

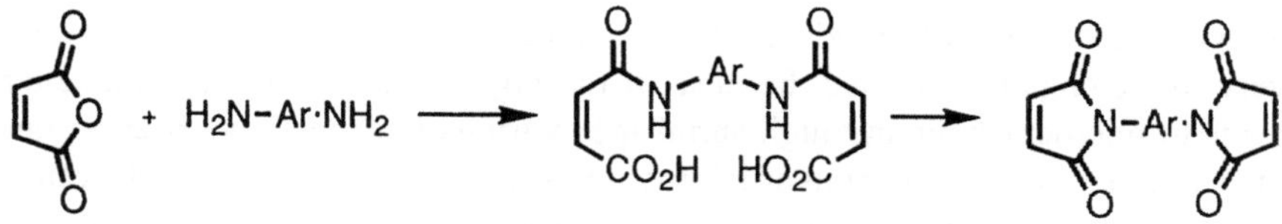

I II

Scheme 8.10 Synthesis of *N,N'*-arylene bismaleimides.

thermosets with T_g values in excess of 200°C. The precursors can have low viscosities, and latent catalysis can also be achieved. The resins have been recommended for use in electrical and electronic applications.

Finally, the reaction of di-isocyanates with dicarboxylic acids and dianhydrides to give polyamides and polyimides, respectively, can be useful synthetically. These products are discussed in Sections 8.4.1 and 8.4.2.

8.3.5 Bismaleimides

BMIs are a leading class of thermosetting polyimides which are finding application as matrix resins for fibre reinforced composites and are rapidly gaining acceptance as engineering materials for high-performance structural applications. The synthesis of *N,N'*-arylene BMIs follows a two-step route (Scheme 8.10) which involves the opening of an aromatic diamine to yield an amic acid intermediate I. This intermediate is then imidized via a cyclodehydration reaction in the presence of acetic anhydride and sodium acetate to yield the corresponding BMI, II. Typically, the yield of recrystallized BMI is 65%–75% when sodium acetate is used as a catalyst (side reactions include the formation of isoimides, acetanilides and maleimide-acetic acid adducts, etc. Most industrail BMI formulations are based on the BMIs of methylene dianiline (MDA), a relatively cheap and common diamine (although possible toxicity hazards have already been highlighted), 2,4-diaminobenzene, 1,3-diaminobenzene and some aliphatic BMIs based on *n*-alkylamines and *iso*-alkylamines. BMI monomers tend to be high melting crystalline solids, which often undergo polymerization directly after melting. The double bond of the maleimide group is relatively electron deficient, due to the proximity of the two carbonyl groups, which may be regarded as both inductively and mesomerically electron-withdrawing. For this reason the maleimide double bond is susceptible to attack by nucleophiles (although the reactivity of the BMI is also influenced by the chemical nature of the residue between the maleimide rings and the purity of the BMI monomer), radical homopolymerization, ene reactions, diene additions and cycloadditions.

Low molecular weight, bisimide prepolymers, end-capped with reactive maleimide rings are cured by thermally or anionically induced polymerization to give highly cross-linked, void free network polymers having good physical properties, thermal stability, better fire resistance and lower water absorption than currently used epoxy resins. One of the serious shortcomings of the BMI system is that when cured poly BMIs are high-modulus, low-strength materials with a very low elongation at break because of the high cross-link density, the latter leads to microcracking and high shrinkage. Various methods have been examined as a means of toughening or improving the impact resistance of BMI resins, and diamine terminated oligomers can be easily transformed into oligomeric BMIs in order to reduce the cross-link density of the resulting network. The reaction of BMIs, via a Michael addition, with a variety of nucleophiles (e.g. diamines, diphenols, dicarboxylic acids, bisthiols, bisthiophenols, biscyanamines and barbituric acid) has produced processable resins, although most (with the exception of polyaspartimide adducts involving diamines) are not commercially available. Kerimid 601™ is formed from the co-reaction of BMI–MDA and MDA (in ratio of 2.5 : 1). Two commercial BMI resins based on a similar concept (but involving BMI–MDA and aminobenzoic hydrazide) are Compimide 183™ (for use in printed circuit boards) and Compimide 796™ (for low-pressure autoclave moulding).

The introduction of amine- and carboxyl-terminated elastomers into BMI systems have been shown to increase short beam shear strength or fracture energy – BMI resins containing carboxyl-terminated acrylonitrile-butadiene (CTBN) elastomers exhibit improved toughness without significant reduction in heat resistance. For example, Compimide 353™ (a low-melting eutectic BMI from Shell-Technochemie) may be mixed with Hycar 1300X8™ (from B.F. Goodrich) to form low-viscosity resins, although Hycar 1300X8™ is not compatible with the BMI at room temperature nor at elevated temperatures. However, during cure the rubber chemically reacts with the BMI through alkenyl bonds in the main chain and pendant groups of the rubber to yield a morphology similar to CTBN-rubber-modified epoxy. The properties of CTBN-modified BMI depend on the chemical composition of the rubber. The acrylonitrile content influences BMI–CTBN rubber compatibility and therefore the morphology (and fracture toughness) of the modified resin. For example, high acrylonitrile rubbers are more soluble and compatible with the BMI resin and therefore provide a lower degree of phase separation. The fracture toughness of BMIs was increased between 300 J m^{-2} and 1000 J m^{-2} when copolymerized with CTBN rubber. The principal drawback of CTBN-rubber-toughened BMI is the significant loss of high-temperature properties, such as T_g and elastic modulus, and this is associated with poor long-term ageing stability at 200°C. The reaction of BMIs with styrene, hydroxyethyl methacrylate has also yielded tougher materials, with fracture toughness values of 170 J m^{-2} as compared with 30–50 J m^{-2} for unmodified BMI.

High T_g thermoplastics (such as polyhydantoins, polyethersulphones, polyimides, polyetherimides and polyaryleneethers) are increasingly used for the toughening of BMI resins to improve their damage tolerance without compromising other high-temperature properties. All (with the exception of polyhydantoins) provide polyphase morphologies and at *ca.* 20% (by weight of thermoplastic), phase inverted systems with the thermoplastic as the continuous

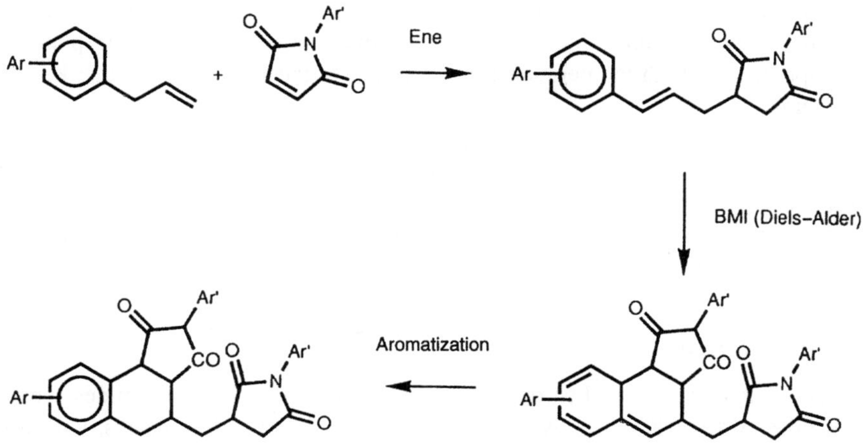

Scheme 8.11 Copolymerization of bismaleimides with the unsaturated monomers *ortho*-allylphenols.

phase are obtained; higher thermoplastic levels provide increased toughness. Thermoplastic modified BMIs are used mainly as matrix resins for advanced composites, and fracture energy, G_{IC}, values of 400–640 J m^{-2} have been obtained with commercial BMI systems toughened in this manner.

Ene and diene reactions have also been employed to overcome the brittleness of the BMI networks, and commercial BMIs are formulated with comonomers, chain extenders, reactive diluents and sometimes viscosity modifiers. BMIs have been copolymerized with unsaturated monomers such as *ortho*-allylphenols (Scheme 8.11), *ortho*-allylcyanates or *ortho*-propenylphenoxy oligomers. The first steps of the reaction have been identified as an ene reaction between the allylic group and the maleimide unsaturated bond. Significant increases in fracture toughness have been reported for BMIs co-reacted with oligomeric allyl phenyl compounds (Table 8.5) or alkenyl-terminated polyaryl ether ketones or sulphones. It has been reported that a co-continuous network of phase-separated

Table 8.5 Physical and mechanical properties of DDM-BMI/*o*,*o'*-diallylbisphenol-A (DABA) copolymers

	BMI : DABA molar ratio			
Property	1.2 : 1	1.5 : 1	2 : 1	3 : 1
Flexural strength (MPa)	186	188	174	131
Flexural modulus (GPa)	4.02	3.94	4.05	4.14
Deflection at breaking (%)	7.78	7.30	5.53	3.50
Stress-intensity factor at onset of crack growth, K_{IC} (MN m$^{-3/2}$)[a]	0.97	0.86	0.80	0.64
Fracture energy, G_{IC} (J m^{-2})[a]	197	158	133	83
Glass transition temperature (°C)	279	282	288	288

[a] By the double torsion method.

components (0.05–0.5 μm domain size) of high-performance thermoplastic resins is essentail for achieving significant toughness without a corresponding compromise in high-temperature characteristics of the BMI alkenyl ether ketone systems.

BMIs offer a balance of versatile processability and fine thermal and mechanical properties which have made them extremely popular in electronic and electrical materials such as printed circuit boards and insulators, as matrix resins for structural composites in aircraft (e.g. motor cases, nose cones for jet engines) and as thermal insulation materials.

8.3.6 PMR resins

PMR resins are a member of the family of thermosetting polyimides which includes BMIs. They occupy the high temperature end of thermosets, with use temperatures of up to in excess of 300°C. The acronym PMR stands for polymerization of monomeric reactants, an uninspiring description which refers to the fact that the monomers are polymerized *in situ* during processing, usually as a fibre reinforced composite. The materials were developed originally by NASA for aerospace use, with the most common and still most widely used member of the PMR family being PMR-15. The chemistry of PMR-15 is illustrated in Scheme 8.12. The dianhydride-derived monomer, BTDE, reacts with the diamine, MDA, with the norbornene acid ester (NE) present to act, in part, as the end-cap in the condensation reaction. Elimination of water and methanol leads to formation of a norbornene-terminated imide oligomer with a molecular weight of 1500 (hence PMR-15).

At elevated temperatures (*ca.* 300°C) and pressures, the norbornene end-group can decompose via a retro Diels–Alder reaction with subsequent cross-linking to form a thermoset product. Although the cross-links are normally written as cyclopentene cross-links formed by reaction of cyclopentadiene with a maleimide double bond, several studies have demonstrated that the real situation is much more complex.

PMR-15 is normally supplied as a carbon or glass fibre prepreg where the fibre is impregnated with a mixture of the three monomers and perhaps a little solvent (methanol). The prepreg has to be stored at −20°C to limit resin advancement, that is, premature reaction of the monomers to produce various adducts which are inevitably present to some extent. Components may be laid up as described previously and cured either in two stages or one. In the former, the prepreg stack is prepolymerized (staged) in a vacuum oven and then compression moulded. More commonly, both stages of the reaction are carried out sequentially in an autoclave. Careful choice of temperature ramps and holds and application of vacuum (to remove condensation volatiles and unreacted cyclopentadiene) and pressure are required to produce high-quality, low-voidage laminates. For PMR-15, a cure temperature up to 316°C is used, often followed by a higher post-curing temperature, and an applied gas pressure of *ca.* 200 psi.

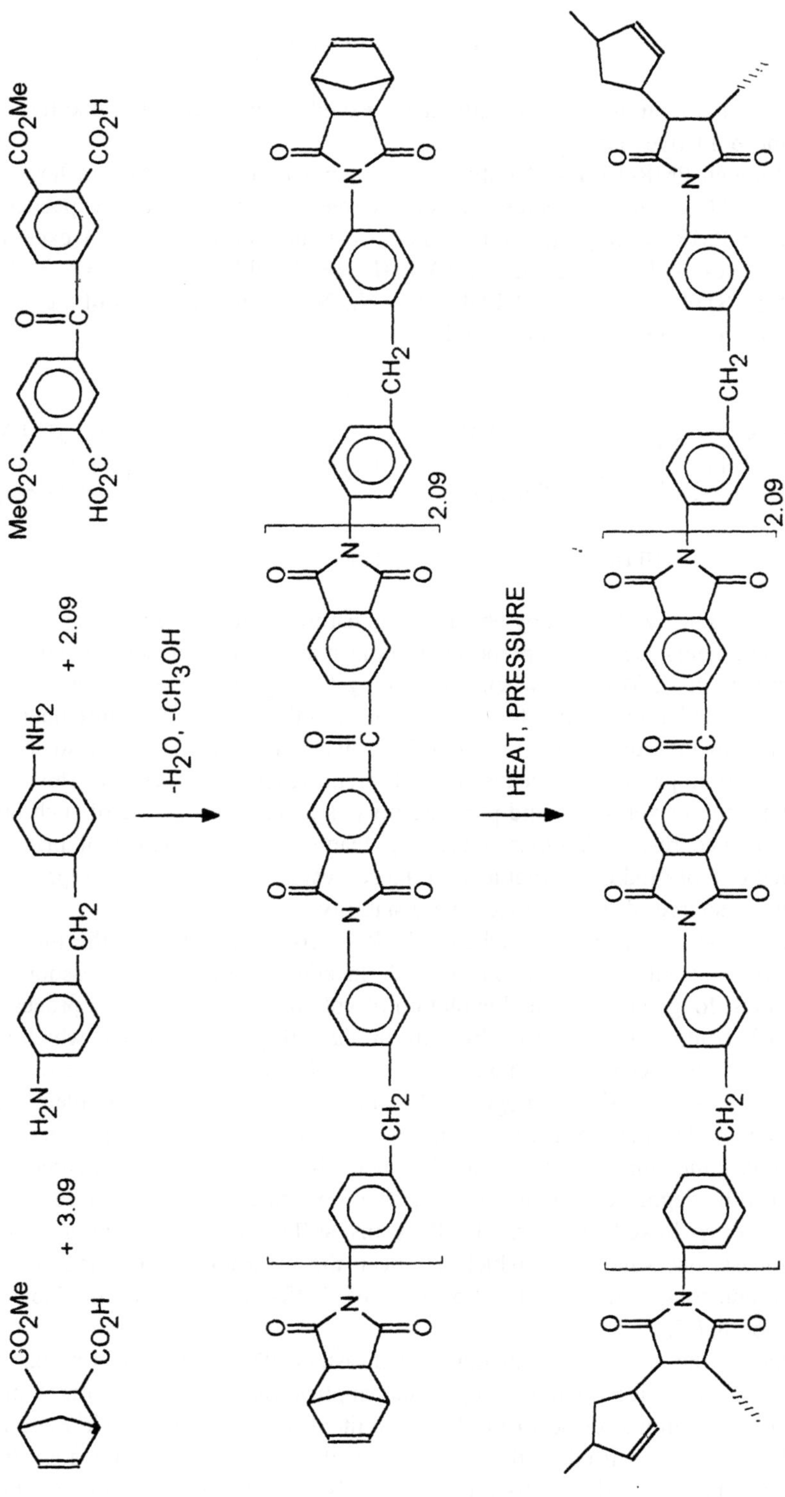

Scheme 8.12 Synthesis of PMR-15.

The resultant laminates generally have excellent mechanical properties and thermo-oxidative stability.

Although PMR-15 has a continuous-use temperature (thousands of hours) of around 300°C much research and development effort has been expended to produce PMR resins with even higher use temperatures, up to in excess of 350°C. Resins developed include PMR-II (HFDE–PPD–NE), V-CAP (HFDE–PPD–PAS) and AFR 700B (HFDE–PPD–NE/NH_2). The structures of the monomers used in these resins are as follows:

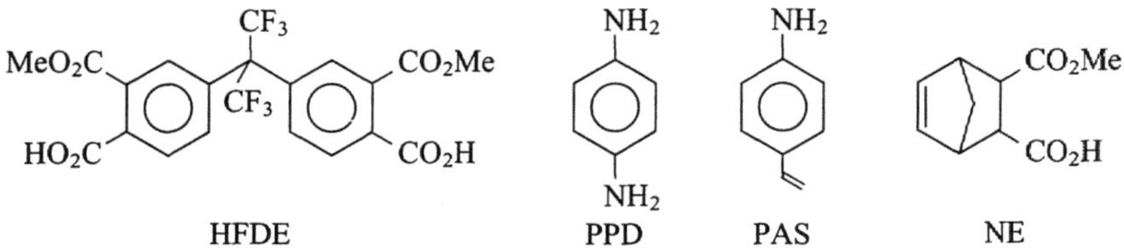

The versatility afforded by changing the nature of the monomers and end-cap and parameters such as the prepolymer molecular weight represents a significant advantage of PRM resins when formulating for high-temperature applications; however, and importantly, the very changes which result in high-temperature stability in the cured resin (e.g. highly aromatic monomer units) also lead to difficulties in processing because of high prepolymer melt viscosity. This trade-off between performance and processability is a common feature of such high-performance resins. Another interesting feature is the presence of primarily aliphatic cross-linking structures in these resins, which clearly represent a potential source of thermo-oxidative instability.

A number of other drawbacks of PMR resins have exercised the minds of aerospace scientists over a long period. In addition to high processing temperatures, long cure times and evolution of volatiles during cure, the processing of PMR composites can suffer from irreproducibility, the origins of which may not be obvious. Diamine monomers such as MDA are often toxic and in many cases are suspected carcinogens. Possible solutions to this problem have included replacement of the monomer by a less toxic diamine and production of prepregs containing the preformed oligomer. The properties of PMR composites can also be degraded by microcracking due to thermal cycling. Because of their highly cross-linked structure, PMR resins suffer generally from a lack of toughness, one solution to which has been the incorporation of thermoplastic polyimide toughening agents. Properties of PMR-15 are shown in Table 8.9 (Section 8.4.5).

Despite the above apparent litany of disasters, PMR composites remain the materials of choice for high-temperature applications. Use temperatures up to 370°C are claimed, close to the likely limit of stability for organic materials. PMR-15 has been used in the engine duct of the US F-18 fighter aircraft, in nacelles, fairings and in parts for the Space Shuttle. PMR resins are also high-temperature adhesives and find use in high-temperature bearings because of their excellent friction properties.

8.3.7 Benzocyclobutene resins

Fundamentally, benzocyclobutene (BCB) technology involves a family of polymerizable monomers which contain one or more BCB groups per molecule and, such is the growing interest in this particular family of resins that they have recently appeared in several reviews; that of Kirchoff and Bruza (1994) is particularly extensive providing a thorough examination of the polymers produced from BCB.

The first report of the synthesis of a BCB derivative was that of Finkelstein (1910), and the parent BCB hydrocarbon (structure III, Scheme 8.13) is the starting material for almost all of the monomers and polymers. The polymerization mechanism depends largely on the nature of the monomer molecule and in turn has a significant effect on the resulting structure of the polybenzocyclobutene. If reactive unsaturation is present then the monomer may react via Diels–Alder 4+2 cycloaddition reactions; lack of such a reactive site leads to the thermolysis of BCB followed by self-oligomerization of *o*-quinodimethane (structure VI). Following studies involving model compounds, a series of pericyclic reactions have been proposed for the BCB homopolymerization (Scheme 8.13). It is generally agreed that the first step in the polymerization of a BCB monomer occurs via a thermally induced electrocyclic ring opening reaction to yield a highly reactive *o*-quinodimethane intermediate (structure IV) which forms a thermally labile spiro dimer (structure V). The geometry of structure V ensures that it may undergo rearrangement to yield both dihydroxybibenzocyclooctadiene (structure VI) and dihydroxydimethylstilbene (structure VII). The formation of tetrahydronaphthalene (structure VII), occurs as a result of the Diels–Alder cycloaddition of *o*-quinodimethane and dihydroxydimethylstilbene (structure VIII) whilst a second (simultaneous) process occurs by further cycloaddition of structures III and IV to yield a labile spirodiene (structure IX). This intermediate undergoes a retroene rearrangement, followed by sigmatropic shifts to give the substituted stilbene (structure X). A subsequent cycloaddition of structures III and IV yields the oligomer (structure XI).

Generally, however, the polymerization mechanism of BCB monomers is poorly understood, as is the structure of the resulting polymers. However, the reactive *bis*-(*o*-quinodimethane) intermediate is key to all polymerization routes, and highly cross-linked networks result from its polymerization. BCB is relatively stable to cycloreversion at temperatures below *ca.* 200°C and as a result should have an extremely good shelf-life at room temperature. However, it should be noted that the presence of either electron donating or withdrawing groups in the four-membered ring significantly lowers the temperature at which ring opening occurs (e.g. 25°C when the 1-position bears an amino group, 150°C when a carbonyl group is present and 200°C for the unsubstituted BCB).

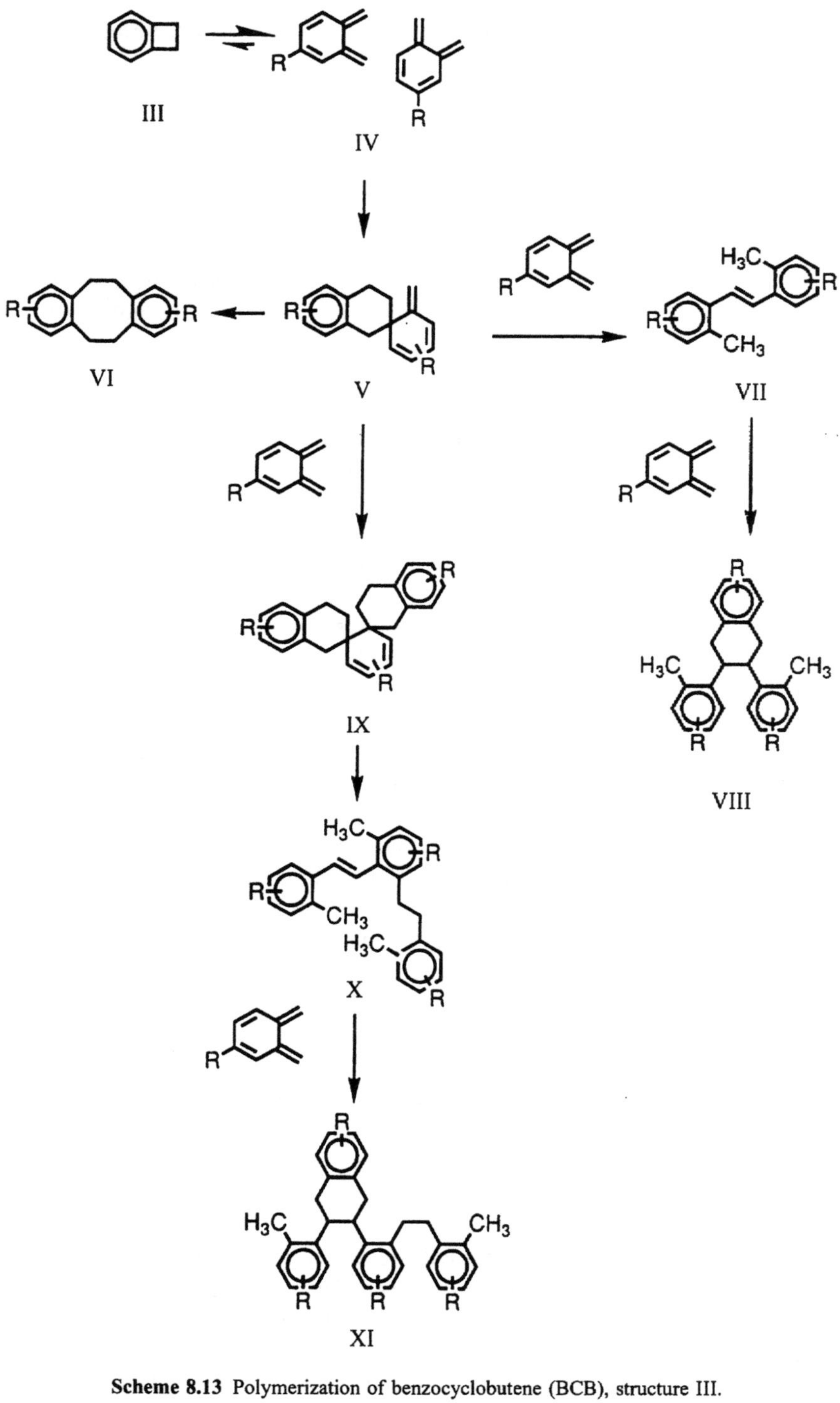

Scheme 8.13 Polymerization of benzocyclobutene (BCB), structure III.

Polymers derived from BCB monomers are generally of high short-term thermal stability (zero weight losses have been reported up to 475°C depending on the nature of the bridging group; Table 8.6), including those which contain relatively large proportions of aliphatic linkages. It has been noted that the bridging linkages in BCB polymers may take an active role in the polymerization process to modify the final polymer architecture (and hence the final properties such as thermal stability). The polymers also exhibit an excellent retention of their room-temperature mechanical properties (at least to 200°–250°C), and long-term ageing studies of the polyester from BCB (structure XIII, Table 8.6) and the polydiketone from BCB (structure XVI) at 250°C suggest that this is the maximum continuous-use temperature for these polymers. BCB polymers generally offer high T_g values owing in no small part to the highly cross-linked nature of the networks and they also display good resistance to a variety of common solvents (with the notable exceptions of methyl ethyl ketone, *N,N'*-dimethyl formamide and trichloroethylene). Furthermore, long-term ageing studies of the polyester from BCB (structure XIII) and the polydiketone from BCB (structure XVI) at 250°C suggest that this is the maximum continuous-use temperature for these polymers. In many cases, mechanical properties of BCB polymers are quite similar to other high-temperature thermosets, making them attractive propositions for further study.

BCBs have been used in a variety of Diels–Alder polymerizations, although few have led to high molecular weight polymers with commercial potential. One of the main problems which preclude the widespread use of BCBs in this way has been the lack of suitably reactive and readily accessible bisdienes which are not prone to homopolymerization (a side reaction which leads to termination of the growing polymer chain and this limits the molecular weight). The most widely used approach to prepare polymers derived from BCB monomers has been to use BCB either with a second BCB molecule (as discussed above) or with a molecule containing additional reactive sites of unsaturation (i.e. AB copolymers). A number of studies have focused on the reaction of BCB with activated diacetylenes, alkenes, bismaleimides and cyanate esters. AB benzocyclobutene monomers can be polymerized to form polymers with a broad range of desirable thermal and mechanical properties. These properties depend not only upon the nature of the reactive functionalities, but also on that of the linking group between functionalities (Table 8.6).

AB benzocyclobutene monomers which combine both maleimide and BCB groups have been actively studied and developed by Dow Chemical Company. In particular, the homopolymer of structure XIV appears to have a good combination of thermal and mechanical properties (Table 8.6) and offers ease of fabrication. When used as a matrix resin for the preparation of composite panels by resin transfer moulding (RTM) with Celion™ G30–500 8HS fabric it appears that the inherent fracture toughness of the matrix resin ($G_{IC} = 455$ J m^{-2}, $K_{IC} = 1.26$ MPa m$^{1/2}$) carries over to the composite panels. The open-hole compression (OHC) values under hot–wet test conditions would seem to

Table 8.6 Properties of selected poly(benzocyclobutenes). Source: Kirchoff, R.A. and Bruza, K.J., Polymers from benzocyclobutenes, in *Advances in Polymer Science, Volume 117: High-Performance Polymers* (ed. P.M. Hergenrother), pp.3–66; published by Springer, Berlin, 1994

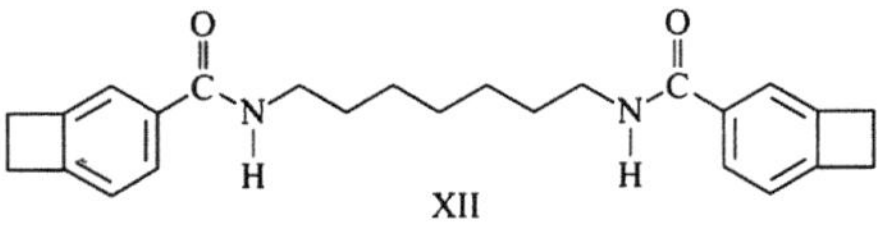

XII

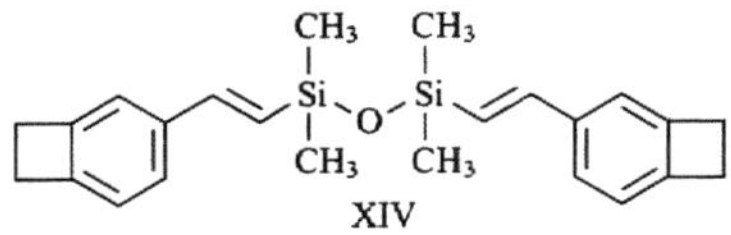

XIV

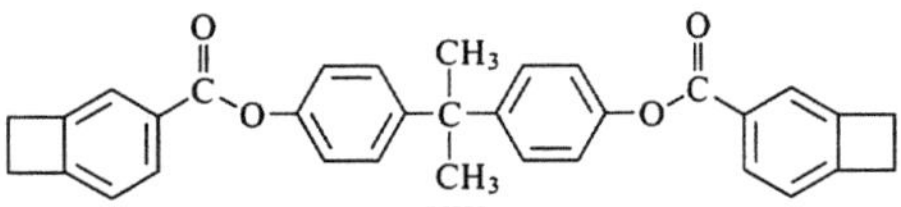

XIII

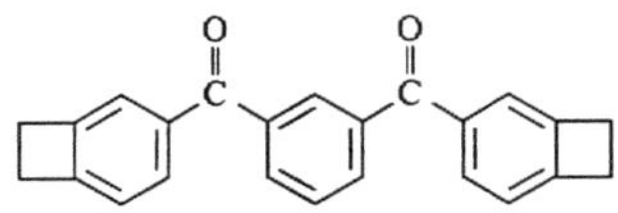

XV

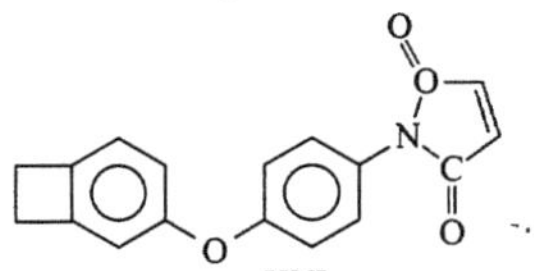

XVI

	Poly(benzocyclobutene)				
Property	XII	XIII	XIV	XV	XVI
Tensile strength (MPa):					
25°C	82.8	27.6	–	85.0	–
150°C	41.4	–	–	–	–
Tensile modulus (GPa):					
25°C	2.48	2.66	–	–	–
250°C	1.32	1.86	–	–	–
Shear modulus (GPa):					
25°C	0.95	1.00	–	–	–
250°C	0.50	0.70	–	–	–
Flexural strength (MPa)	–	–	207	–	151.7
Flexural modulus (GPa):	–	–	3.24	2.0[a b]	4.25
300°C, DMA	–	–	–	–	2.94
Elongation at break (%)	–	–	6.0	6.0	–
Adhesion (tensile lap shear) (MPa) on steel:					
25°C	13.5	–	–	–	–
200°C	11.4	–	–	–	–
Coefficient of thermal expansion (μm m°C^{-1}):					
25°C–250°C	4.1×10^{-5}	8.0×10^{-5}	43[c]	–	35[c]
250°C–235°C	9.8×10^{-5}	–	193[d]	–	150[d]
Moisture uptake after 48 h at 100°C (%)	1.4	0.9	4.2	0.2	–
Dielectric constant (1 MHz)	–	–	3.15	2.7	–
Dissipation factor (1 MHz)	–	–	0.0026	0.0008	–
Fracture energy, G_{IC} (J m^{-2})	–	–	780	–	80
T_g of cured polymer (°C)	>270	>270	317	>350	310
0% weight-loss temperature (°C) (10 K min^{-1} under nitrogen)	385	400	470[e]	–	445[e]

[a] 25°C.
[b] 30°C.
[c] Above glass transition temperature.
[d] Below glass transition temperature.
[e] 5% weight-loss temperature (°C).

indicate that the composite has very good retention of its mechanical properties at both 177°C and 203°C (compressive strength after impact = 332 MPa, OHC strength = 293 MPa dry, and 202 MPa at 177°C, wet).

8.3.8 Toughened thermosets

The lack of toughness of highly cross-linked thermoset resins is a common feature which has limited the application of these in many areas. Much effort has therefore been expended in developing routes to improve the impact resistance of thermosets so that they may compete more effectively with thermoplastics in applications where impact performance is critical. Most work on toughened thermosets has focused on epoxy resins, but similar approaches have been used to toughen unsaturated polyesters, cyanate esters, BMIs, PMR resins and others. Since good resistance to fracture under impact is a feature of rubbers and thermoplastics (TPs) it is perhaps not surprising that most efforts to improve thermoset fracture toughness have involved use of rubbers and TPs as toughening additives.

Early work concentrated on the use of organic rubbers, in particular reactive liquid rubbers such as CTBN. The liquid rubbers tend to be soluble in the epoxy monomer formulations but phase separate during cure to produce a cross-linked network containing approximately micrometre size spherical rubber particles. The rubber particles are highly effective in absorbing and dissipating impact energy by a number of mechanisms. Large increases in thermoset fracture toughness may therefore be observed. Rubber toughening of epoxy resins is used extensively in epoxy compositions, particularly in adhesives. Other end groups may be used in the rubbers to improve compatibility with epoxy or other resins. End groups include epoxy, amine and vinyl. Rubbers have also been shown to toughen higher-temperature thermosets such as cyanate esters, but for elevated temperature use, addition of an antioxidant may be necessary. As well as reducing end-use temperature, rubbers reduce the overall resin modulus and tend to be ineffective in toughening highly cross-linked thermosets such as aerospace epoxies. To overcome these problems, much recent work has been focused on the use of TPs as toughening agents.

Poly(ether sulphone)s were the TPs used in early studies, but a wide range of TPs have since been used as toughening agents, including polyimides, polyarylene ethers, polyesters, polyhydantoins, etc. As with liquid rubbers, the TP is normally dissolved in the thermoset monomers, with phase separation occurring during cure. The improvement in fracture toughness observed depends on a number of factors including the nature of the TP (chemistry, molecular weight, end groups), TP concentration, cured resin morphology, the nature of the thermoset system and the cure conditions. The ultimate cured resin morphology depends to a greater or lesser extent on all the other parameters mentioned above. The thermoset/TP blends commonly display lower critical solution temperature behaviour during cure, with the shift in the phase boundary as the thermoset molecular weight increases resulting in the system moving from the single-phase to the two-phase region of the phase diagram. Phase separation results. As the concentration of TP increases, the resin morphology changes from phase-separated (continuous thermoset phase; TP $<$ *ca.* 20%) to co-continuous (TP *ca.* 20%–30%) to phase-inverted (continuous TP phase; TP $>$ *ca.* 30%). In some cases, the resin fracture toughness is critically dependent on the resin morphology, with maximum toughness observed for co-continuous structures, whereas in other cases fracture toughness has been found to be fairly independent of morphology. Several studies have suggested that the presence of reactive end groups on the TP can lead to greater improvements in fracture toughness than use of unreactive TPs. Reactive groups should modify the interface between the phases. In general, TPs can toughen thermosets effectively while retaining resin modulus and end-use temperature. Toughening mechanisms include plastic deformation of the TP and crack bridging with ductile tearing of the TP phase.

These types of TP-toughened resins are used extensively to improve the fracture toughness of structural composites. Full transfer of resin toughness to the composite is not always observed, probably because of the presence of the fibres. The exact cause is still a matter of debate. One route developed to attempt to overcome this problem is interleaving or interlaminar toughening (ILT). TP film or powder is applied between the fibre layers, producing TP-rich layers in the interlaminar regions of the cured laminate. The laminates have greatly improved mode II fracture toughness and compression after impact, as well as reduced delamination area due to impact.

Other thermosets

Other thermoset resins with potential applications in structural applications have been introduced or are under development, and the following represent some of the less heavily commercialized systems.

Acetylene-terminated (AT) resins have been examined in view of their processability and ease of cure upon heating without yielding void-causing by-products. Several curing mechanisms have been proposed, including a combination of simple cyclotrimerization of three terminal acetylene groups into an aromatic cross-link, Glaser coupling of two acetylene groups to generate diacetylene segments and Strauss coupling of two acetylene groups to produce acetylene-vinyl structures. The resulting conjugated functional linkages arising from the latter two mechanisms may participate in Diels–Alder reactions with terminal acetylene and/or the aromatic polymer backbone. In addition, acetylene groups may also undergo conventional free radical polymerization. AT resins with various functionalized backbones have been developed for a variety of applications where properties such as excellent thermal and heat resistance, high service temperature, reduced defects, low moisture uptake, and high strength are important. Commercial AT resins have included wholly hydrocarbon

AT-oligomers (such as the short-lived H resins by Hercules) based on *p*-diethynylbenzene and prepared to a low degree of polymerization, but discontinued in the 1970s, as well as AT oligoimides such as HR600 (Hughes Aircraft Corporation, subsequently Gulf and later National Starch and Chemical) based on *m*-ethynylaniline and commercialized as Thermid MC-600™. Thermid MC-600.

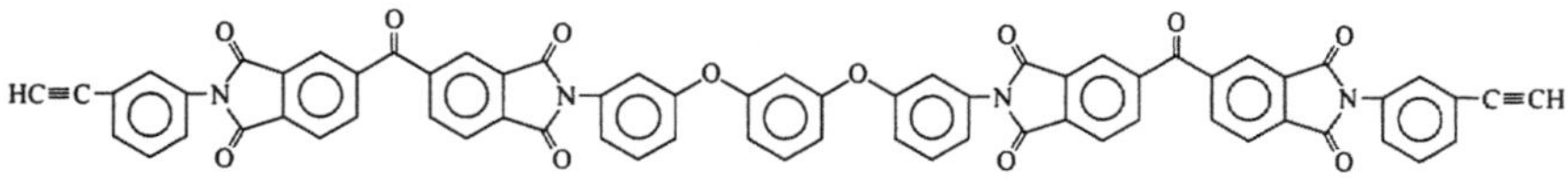

is soluble only in aprotic solvents (e.g., NMP, DMF and DMSO) and thermally curable above the softening temperature of 157°C–210°C. Thermid™ undergoes a slow cure at temperatures in excess of 315°C, because the low segmental mobility of the molecule, and requires a post-cure of 371°C to develop optimal properties. These cure conditions yield a densely cross-linked resin which has fine thermal properties (the cured resin undergoes decomposition above 500°C), high stiffness, low moisture absorption and good electrical properties at high temperatures. Whilst these basic characteristics and excellent property retention at high temperatures commend AT resins for high-performance composite, adhesive and electronic applications, the most significant weakness of this resin system has been the relatively expensive and complex preparation route, and this remains a stumbling block to their more widespread use.

Bisphtalonitrile resins are another group of thermosetting polymers which have seen extensive development, by the US Naval Research Laboratory, Washington, DC. The basis of the system comprises solid monomers bearing four nitrile groups which may undergo polymerization to yield a highly cross-linked product whose exact nature is not known, although it is believed to consist of a complex mixtures of phthalocyanine, triazine and isoindoline moieties, depending on the reaction conditions and co-reactants used (e.g. the use of a redox co-reactant such as hydroquinone promotes the formation of a phthalocyanine structure, whilst in the absence of co-reactants, isoindoline is believed to predominate in the network structure; triazine formation occurs only at high temperatures (Scheme 8.14).

The monomers have an indefinite shelf life and may be easily melt processed. The cross-linking reaction, which typically takes place at 315°C (for 16 h), may be catalysed by acids (organic, inorganic or Lewis), amines, metals or metals salts. Cured phthalonitrile resins develop fine mechanical properties (T_g values of at least 300°C; high tensile strength, 94 MPa; and fracture toughness, $G_{IC} = 120\ \mathrm{J\ m^{-2}}$ and $K_{IC} = 0.45\ \mathrm{MN^{-3/2}}$) and extremely good fire properties, making them of potential use in advanced composite, adhesives and electronics applications. The imminent commercialization of phthalonitrile resins is expected in the United States where they will be competing directly against high-temperature polyimides such as PMR-15.

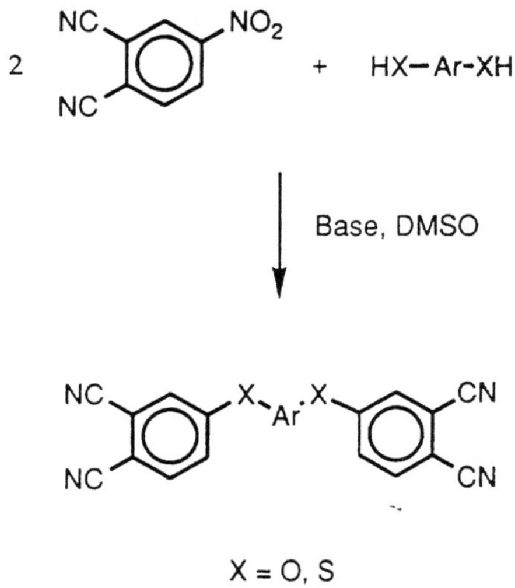

Scheme 8.14 Synthesis of bisphthalonitrile resin.

The so-called EPIC resins (based on the catalysed reaction of epoxy resins with isocyanates) were introduced by Bayer under the name Blendur™. The complex cure mechanism involves a combination of cyclotrimerization and cycloaddition reactions and yields a highly cross-linked thermoset resin with T_g values greater than 200°C. The uncured resins may be of low viscosity and may be cured (using latent catalysts) at temperatures of 160°C, with a post-cure step at 250°C to yield ultimate properties. Again, the proposed application areas for EPIC resins are in the electrical and electronic arena.

Bisoxazoline-phenolic resins were developed by Culbertson and Tufts in the early 1980s and commercialized by Ashland under the name Arotech 6000™. The basis of the resin system is the cross-linking reaction between the hydroxyl groups of the phenolic resin and the oxazoline groups of an aromatic bisoxazoline [such as 1,3-phenylene bisoxazoline (1,3-PBOX)] to form poly(ether amide) bridges (Scheme 8.15).

The catalysed cure may be effected at temperatures of 150°C–180°C, and post-cure at 225°C–240°C yields optimum physical and mechanical properties. As the cure involves ring-opening reactions, cure shrinkage (and void formation) is low, thus enchancing their physical properties and long-term durability. The cured resin T_g values increase quickly with increasing bisoxazoline (e.g. 1,3-PBOX) content. For example, a value of T_g *ca.* 125°C is obtained with 40 wt% PBOX, whereas T_g is *ca.* 275°C with 70 wt% PBOX. However, this is accompanied by a slight decrease in decomposition temperature (values of 275°C and 125°C, respectively, have been reported for the same compositions). In general, cured neat resins exhibit excellent mechanical properties which may be tailored to meet a variety of high performance applications. For example, the

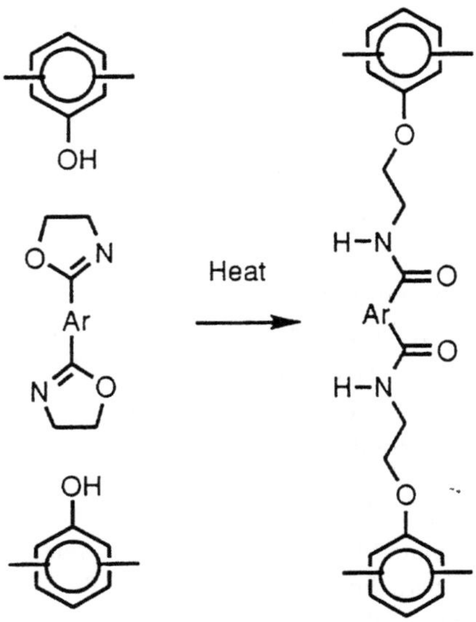

Scheme 8.15 The cross-linking reaction between the hydroxy group of a phenolic resin with the oxazoline groups of an aromatic bisoxazoline.

low flammability of the cured resins has been exploited in their use in composites for aircraft interiors.

Other thermoset resins with potential applications in structural applications have been examined but are outside the scope of this chapter and the reader is directed to the Further Reading section for more information.

8.4 Thermoplastics

8.4.1 Polyamides

The research that led to the large-scale production of synthetic fibres from polyamides commenced with the work of Wallace Carothers in the late 1920s and led to the commercialization of nylon fibres in the early 1930s. The continuing interest in polyamides over a period of decades led to the development of high-strength aromatic polyamides known generically as aramid fibres. Nomex™ is a rigid chain aramid based on poly-*m*-phenylene isophthalamide (MPD-I):

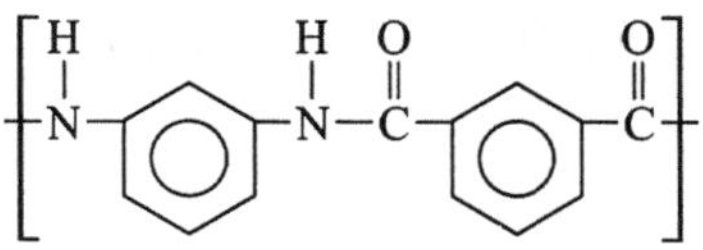

It has high thermal and electrical properties and finds application in both apparel and electrical insulation. It is also transformed into honeycomb for strong, lightweight panels. The *para*-substituted aryl polyamide, poly(*p*-phenylene terephthalamide (PPD-T),

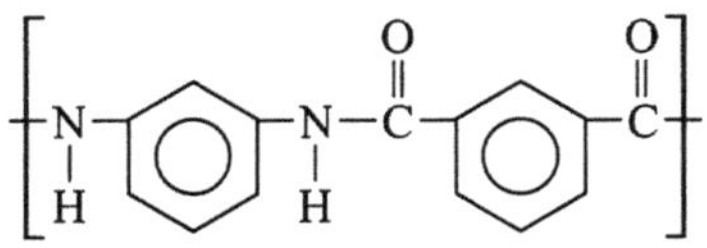

is a rigid molecule with limited flexibility and bond rotation and as a result may form highly-orientated fibres which are significantly stronger than MPD-I.

Kevlar™ aramid fibre is based on PPD-T. It may be prepared by a variety of synthetic routes, but the classical synthesis (Scheme 8.16) involves the low-temperature polycondensation of *p*-phenylene diamine (PPD) and terethaloyl chloride (TCl) in a dialkyl amide solvent (e.g., hexamethylphosphoramide – a suspected carcinogen, *N*-methylpyrrolidinone and dimethylacetamide, etc.). The resulting polymer is isolated by precipitation with water neutralized and subsequently washed and dried. For fibre preparation, PPD-T polymer is generally synthesized at high molecular weight to ensure high fibre properties and is insoluble in conventional solvents. In the commercial process employed by Du Pont to produce Kevlar™, PPD-T is dissolved in concentrated sulphuric acid and dry jet-wet spun into fibres (Akzo spin Twaron™ from a methyl pyrrolidinone/$CaCl_2$ solution).

Kevlar™ aramid fibre (Table 8.7) has a much broader range of applications than other high-performance fibres (e.g. carbon and glass fibres, largely limited

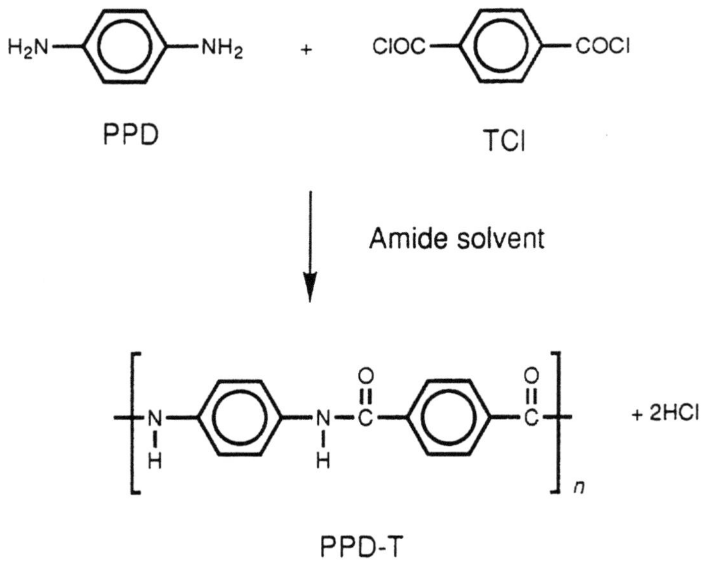

Scheme 8.16 Polycondensation of *p*-phenylene diamine (PPD) and terethaloyl chloride (TCl) in a dialkyl amide solvent to produce poly-*p*-phenylene terephthalamide (PPD-T).

Table 8.7 Typical properties of Kevlar™ aramid yarns. Note: data in parentheses are from epoxy impregnated strand. Source Yang, H.H., *Kevlar Aramid Fibre*, published by John Wiley, Chichester, Sussex, 1993

Yarn properties	Kevlar 29™	Kevlar 49™	Kevlar 68™	Kevlar 119™	Kevlar 129™	Kevlar 149™
Tensile strength (Kpsi)	420 525	420 (525)	420 (525)	440 (610)	485 (–)	340 (500)
Initial modulus (Mpsi)	10.3 (12)	17.4 (18–19)	14.4 (16)	8.0 (16)	14.0 (–)	21.0 (25–26)
Elongation (%)	3.6 (4.4)	2.8 (2.9)	3.0 (–)	4.4 (–)	3.3 (–)	1.5 (1.9)
Density (g cm^{-3})	1.44	1.45	1.44	1.44	1.45	1.47
Moisture regain at 25°C, 65% relative humidity (%)	6.0	4.3	4.3	–	–	1.5

to the reinforcement of rigid composites, and high-strength polyethylene in rigid and flexible structures) and is finding increasing use in a number of applications, but principally in five major areas:

- fibre-reinforced composites;
- fibrous armour systems;
- fibre-reinforced rubber goods;
- ropes and cables;
- Kevlar aramid pulp.

Thermoset resins (such as epoxy, polyurethanes, vinyl esters and unsaturated resins, etc.) and thermoplastic polymers [e.g. PEEK, polyether ketoneketone (PEKK) and polyaimides, etc.] are used as matrices for Kevlar™-reinforced composites: thermoset composites are generally stiffer, more thermally stable and brittle than their thermoplastic counterparts: thermoplastic composites offer greater impact resistance and processibility. High-modulus Kevlar 49™ and Kevlar 149™ aramid fibres are suitable reinforcement for composites in the form of unidirectional, fabric-reinforced and laminated composites. Kevlar™ has been employed as chopped fibre moulding compound, bulk moulding compound and short fibre reinforced thermoplastics.

Kevlar 29™ and Kevlar 49™ aramid filament yarns can be woven into fabrics which provide the unique properties of high-tensile properties, light weight, excellent thermal and dimensional stability, flame resistance, cut resistance, high impact resistance and damage tolerance and rapid vibration damping in reinforced composite structures. These properties make Kevlar™ fabrics suitable for use in many high-performance applications, such as protective apparel, soft armour (e.g. blankets, curtains and liners) and reinforced rubber goods (e.g. pneumatic rubber tyres). Flame resistant textile fabrics can be prepared from the blend of Kevlar™ and Nomex™ staples for use in protective clothing.

Kevlar™ aramid pulp (an engineered short fibre with a high degree of fibrillation) was originally developed for industrial paper and elastomeric reinforcement in which it is mixed with resins, elastomers and fillers to form a uniformly reinforced material. Recently, a number of new applications have emerged, including brake linings and clutch facings, gaskets, thixotropic sealants, adhesives and coatings.

8.4.2 Polyimides

Thermoplastic polyimides have been known since 1908 and available commercially for nearly 40 years. They are often known as condensation polyimides and in some cases do not demonstrate true thermoplastic behaviour as they may decompose below the softening point. By varying the chemical structure of the polyimides (PIs) it is possible to vary the glass transition temperature, T_g, from below 200°C to above 400°C. Both amorphous and semi-crystalline PIs may be obtained, depending on the chemical structure. Approximately one quarter of the world market for high-temperature polymers is accounted for by PIs (including thermosets), partly as a consequence of the versatility noted above.

The most common route used to synthesize PIs is the reaction of a diamine with a dianhydride (Scheme 8.17). Mechanistically, the reaction proceeds in two steps. The first step is the polymer-forming reaction leading to a polyamic acid (PAA) with no elimination of any by-products. Solutions of PAAs in the polar aprotic solvents in which they are made are reasonably stable and can be stored, although the molecular weight distribution may change with time. The second step is a cyclodehydration of the PAA, with elimination of water, to give the PI. This may be carried out thermally (azeotropically) or chemically by using a dehydrating agent such as acetic anhydride and triethylamine. The polar aprotic solvents used in the polymerization are normally amides such as *N*-methylpyrrolidinone (NMP) and *N,N*-dimethylacetamide (DMAc). The two steps can be carried out in a single process by heating a solution of the monomers in the presence of an azeotroping co-solvent such as toluene. Because of their inherent rigidity, many PIs have poor solubility in even the powerful polar aprotic solvents. Alternative approaches are then used to form the final

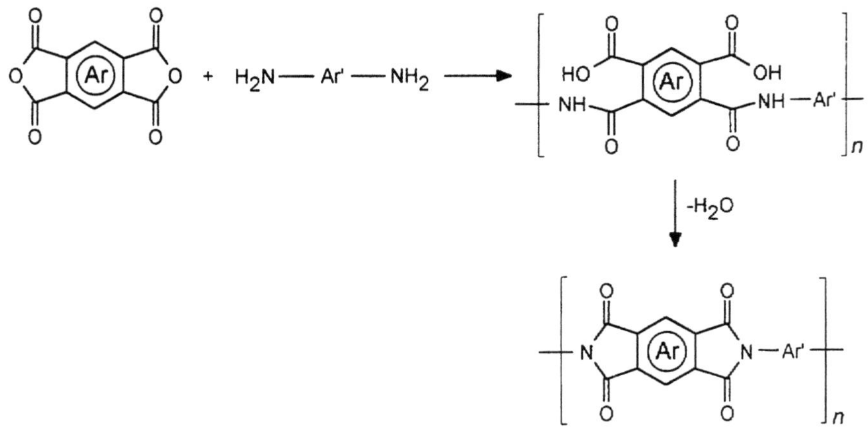

Scheme 8.17 Synthesis of a polyimide.

product. For example, films such as Kapton™ (pyromellitic dianhydride/ 4,4′-diaminodiphenyl ether) are made by casting a film from the soluble PAA followed by thermal imidization to the intractable PI.

By choice of the appropriate stoichiometry and/or use of end-capping (monofunctional) monomers, the diamine/dianhydride synthesis can provide a route to controlled molecular weight PIs with amine, anhydride, or non-reactive end-caps. Variants of the method have also been developed to overcome drawbacks of the conventional routes in particular cases. These includes use of acid esters (*cf.* the PMR route), thioanhydrides, *N*-silylated diamines and di-isocyanates. The reaction of thioanhydrides with diamines is relatively facile, but hydrogen sulphide is the by-product. The di-isocyanate/dianhydride reaction has been used to form soluble, low-colour PIs in cases where the presence of a ketonic group in the anhydride led to gelation on reaction with diamines.

An important subset of the polyimides is the polyether imides (PEIs). The most important member of the PEIs is the commercial polymer Ultem™. A key reaction in the synthesis of PEIs is a nucleophilic aromatic substitution reaction (S_NAr). PEIs can be synthesized by one of two main routes (Scheme 8.18). In the first of these, a nitrophthalimide is reacted with a bisphenoxide to give a bisimide which is converted to a dianhydride by hydrolysis and cyclodehydration. The dianhydride can then react with diamines to form PEIs. The complementary route involves direct polymerization of a bis(nitrophthalimide) with a bisphenoxide to form the PEI. The key to success in these reactions is the activation of the nitro group towards nucleophilic displacement by the presence of the powerfully electron-withdrawing imide functionality. In the synthesis of the dianhydride in Scheme 8.18, the nitrophthalic anhydride starting material is first protected as the phthalimide. Recently, it has been shown that the dianhydride can be formed directly by reaction of 4-fluorophthalic anhydride with a bisphenol in the presence of a base (Scheme 8.19).

The synthetic versatility of PIs allows their properties to be tailored to a range of applications. It is possible to vary T_g, thermo-oxidative stability, dielectric constant, thermal expansion coefficient, mechanical properties and processability (e.g. melt viscosity), bearing in mind that some trade-off between desired properties may be necessary. Typical PEI properties are tabulated in Table 8.9 (Section 8.4.5). PIs (including PEIs) can be used to form films, fibres, composites, coatings, adhesives, etc. The use of PIs as toughening agents for thermosets has already been described. Prepregs and fibre-reinforced moulding compounds containing PIs are available commercially. These are processed readily by compression moulding or injection moulding. Composites find uses in aerospace and electrical and electronics applications. Three-dimensional printed circuit boards have been made from PEI composites. The good fire, smoke and toxicity properties have led to their use in aircraft interiors. The application of PIs may be extended even further by use of blends, such as those with polycarbonate or polyphenylene sulphide, or by use of copolymers. Recent PI developments include the semi-crystalline NASA product, LARC-CPI, which is

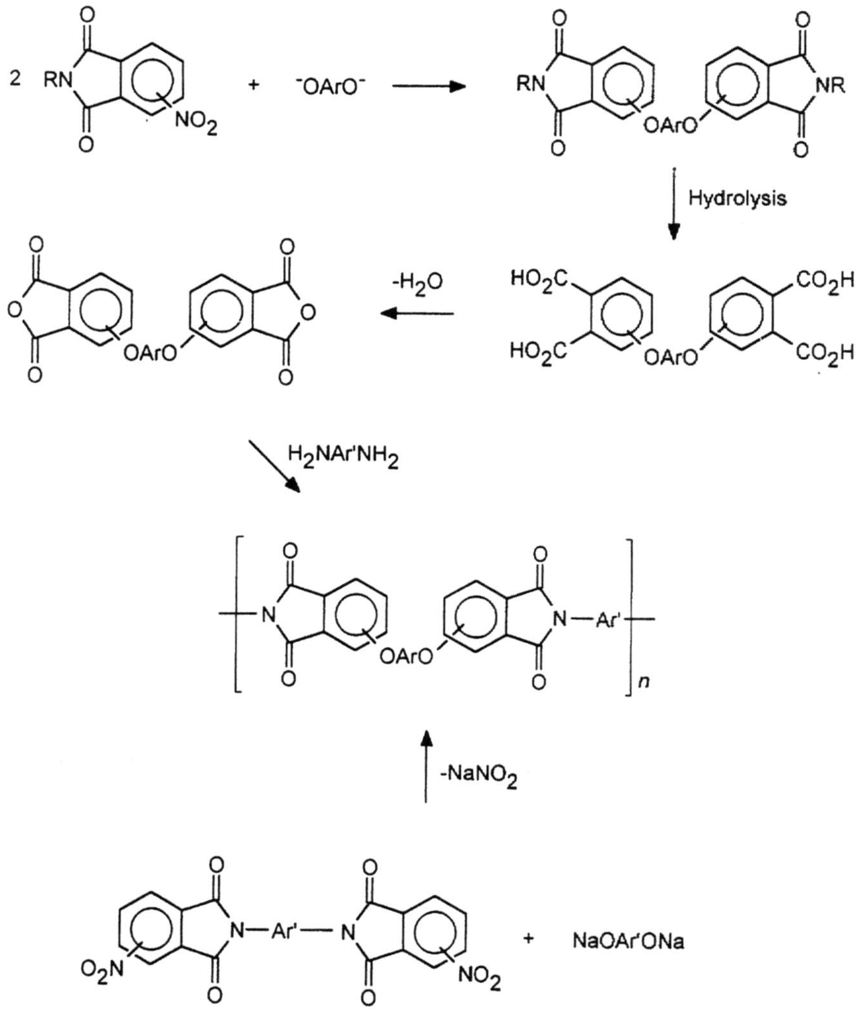

Scheme 8.18 The two main synthetic routes of polyether imides.

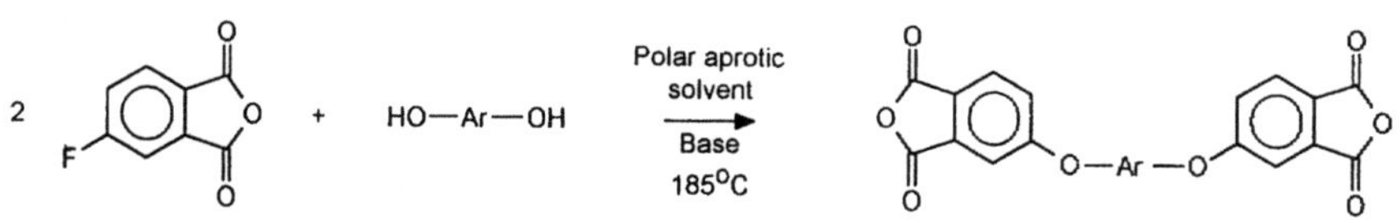

Scheme 8.19 Formation of dianhydride for use in the synthesis of polyether imides.

synthesized from 3,3′,4,4′-benzophenone tetracarboxylic dianhydride (BTDA) and the following diamine:

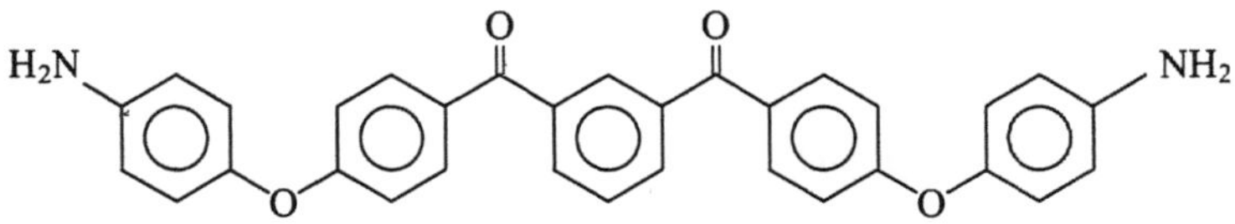

8.4.3 Polysulphones and polyketones

These polymers are members of a family of polyaromatics which also includes polyphenylene sulphide (Section 8.4.4). They are commonly found in polymers in conjunction with ether linking groups, in which case they are known generically as polyarylene ethers. A typical polyaromatic structure is $-\!\!\left(Ar_1-X-Ar_2-Y\right)\!\!-_n$ where X and Y are sulphone, ketone, ether, etc., and where the aryl groups may be joined by either *meta* or *para* linkages to the connecting groups. Polymers of this type have been in commercial production for over 40 years. The properties of the polymers can be tailored by changing the nature of the groups present. *Para* linkages tend to lead to higher modulus and greater crystallinity than do *meta* linkages. Conversely, processing tends to be more facile for the *m*-linked polymers. Polyether sulphones (PES) are commonly amorphous and soluble in organic solvents, whereas the polyether ketones (PEK) are semi-crystalline and have a lower T_g than do the PES. The ether links lend improved processability to the polymers. The absence of easily oxidized groups in the polymers leads to fairly high thermal stability.

The syntheses of the PES and PEK polymers are similar in that they both involve a nucleophilic aromatic substitution reaction (*cf.* PEIs, Section 8.4.2) as shown in Scheme 8.20. The mechanism is an addition–elimination reaction involving an intermediate Meisenheimer complex which is resonance stabilized. Elimination of the fluoride ion is more facile than the elimination of a chloride ion, which in turn is easier than either a bromide ion or an iodide ion. The reaction is activated by the presence of a powerfully electron-withdrawing group *para* to the leaving group. Sulphone and carbonyl groups both act as good activating groups. The reaction is normally carried out in an anhydrous polar aprotic solvent such as NMP, DMAc or supholane. A base such as potassium carbonate is required to remove the hydrogen halide produced when the starting material is a bisphenol. Alternatively, a preformed bisphenoxide is used. The synthesis of a polyetheretherketone (Victrex™ PEEK) is illustrated in Scheme 8.21. The same product can be obtained starting from a precursor where the fluorine atoms have been replaced by nitro groups, as in the PEI synthesis described earlier. PEEK was developed by ICI and is a semi-crystalline polymer with a T_g of 143°C (416 K) and a melting temperature, T_m, of 340°C (613 K). Note that the oft-quoted relationship T_m (K)/T_g (K) = 1.5 holds goods in this case. The high T_m means that high processing temperatures, approaching 400°C

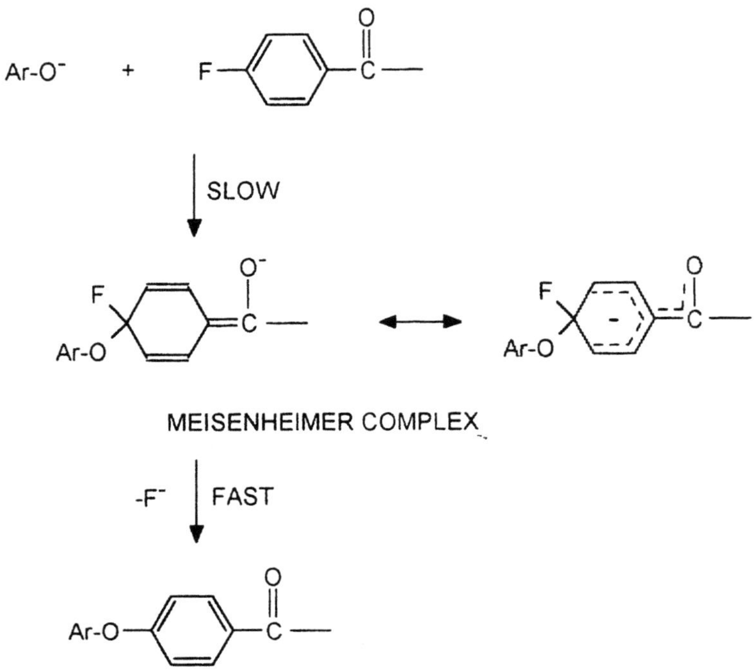

Scheme 8.20 Synthesis of polyether sulphones and polyether ketones.

in some cases, must be used to fabricate components from PEEK. This has cost implications in terms of both energy consumption and ancillary materials required. PEEK components may be made by the usual thermoplastic processing routes such as thermoforming and injection moulding. Composite processing routes include the use of prepregs which can be made by melt impregnation of the fibres, by powder coating or by interweaving of glass/carbon fibres and polymer fibres. Compounds for injection moulding typically contain 30% glass

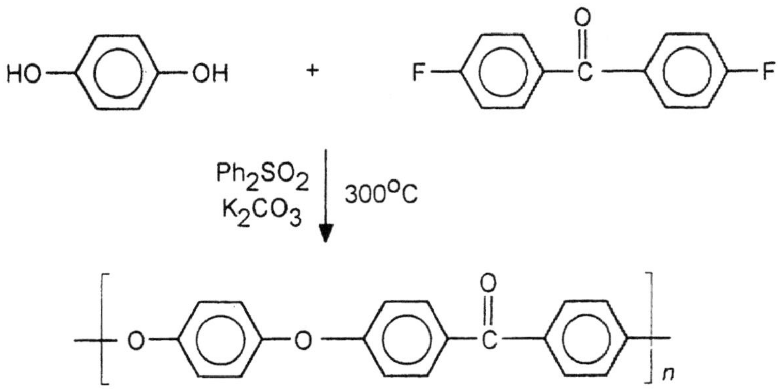

Scheme 8.21 Synthesis of a polyetherether ketone.

or carbon fibres. During processing, the effect of the conditions, particularly cooling rate, on polymer crystallinity needs to be considered.

PEEK has a range of outstanding properties which allows it to find widespread applications, although the relatively high cost may be a consideration. Many of the properties are a result of the crystallinity of the polymer. The properties of PEEK are shown, along with those of PES, in Table 8.9 (Section 8.4.5). PEEK and its composites have a service temperature up to 250°C and excellent fire resistance. The chemical and solvent resistance of the polymer is outstanding and it is also resistant to gamma radiation. The polymer and its composite are tough (polymer $G_{IC} \approx 4.8$ kJ m^{-2}) and abrasion resistant and have good fatigue properties. The applications of PEEK are diverse, ranging from small connectors for high-pressure liquid chromatography systems to aerospace components. It also finds use in valves for hot-water systems, in bearings and in electrical connectors.

PES is analogous to PEEK, but the polymer backbone contains sulphone groups instead of carbonyl groups. This change is sufficient to ensure that PES polymers are amorphous in contrast to the semi-crystalline PEEK. A typical synthesis of PES is illustrated in Scheme 8.22. Note that, depending on the stoichiometry, the polymer may be end-capped by chlorophenyl, by phenolic groups or by a combination of the two. This is an important point as in some cases, for example the toughening of thermoset epoxy resins by PES (Section 8.3.8), the nature of the end groups has been found to have a major influence on the properties. The T_g of amorphous PES is *ca.* 230°C, higher than that of PEEK but well below the T_m of PEEK. The maximum service temperature is therefore reduced to around 200°C. The processing temperature is lower than that of PEEK, with temperatures of 315°C–360°C being typical. The polymers are less solvent resistant than are semi-crystalline polymers, but this allows the option of prepregging via solution impregnation. PES and its composites can be processed by the usual routes for thermoplastics. The products have good fire properties and dimensional stability and outstanding creep resistance. Fracture toughness is also good, with a neat polymer G_{IC} of *ca.* 3.0 kJ m^{-2}. PES and its composites find use in electrical connectors, aerospace components and as toughening agents for thermoset resins. It is interesting to note than some manufacturers of PES have ceased manufacture in recent years, perhaps indicating that the market has not developed as originally forecast.

The general route to synthesizing polyarylene ethers can be applied to a wide range of different structures. One example is the use of preformed heterocyclic monomers to produce heterocyclic polyarylene ethers (Scheme 8.23). A quinoxaline-containing bisphenol is reacted with an activated difluoride under basic conditions to give the polymer.

8.4.4 Polyphenylene sulphide

The group of polymers known as polyphenylene sulphide (PPS) constitute a few per cent of the world market for high-temperature polymers. The synthesis of

PPS is relatively straightforward and involves the reaction of 1,4-dichlorobenzene with sodium sulphide (Scheme 8.24). The same product can, in principle, be made from 4,4′-dichlorophenyl sulphide and sodium sulphide. Chain branching can occur during the manufacturing process, so improved processes were developed to produce highly linear PPS. As sodium chloride is the by-product of the synthesis, it is essential to reduce the level of ionic impurities in the PPS to a minimum, particularly for electrical and electronic applications. PPS properties are given in Table 8.9 (Section 8.4.5). PPS is a semi-crystalline

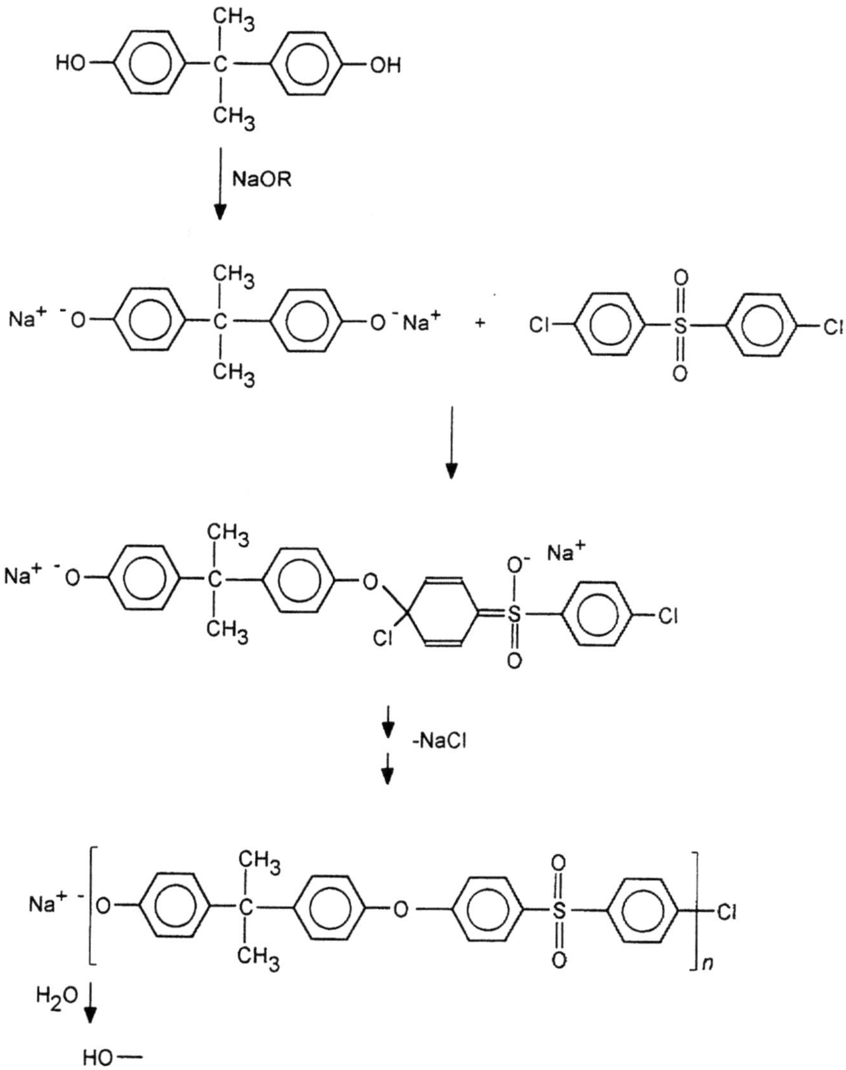

Scheme 8.22 Synthesis of a polyether sulphone.

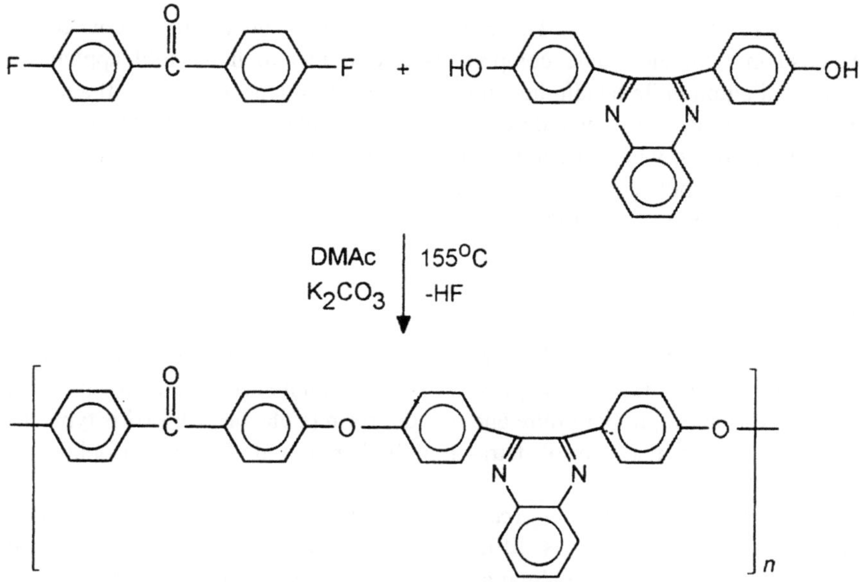

Scheme 8.23 Synthesis of heterocyclic polyarylene ethers.

polymer with a T_g of 85°C and a T_m of 285°C and therefore requires high processing temperatures. PPS is intrinsically flame retardant and passes the Underwriters Laboratories UL-94 V-0 test. It also has outstanding chemical and solvent resistance. PPS is normally used in the form of mineral- or fibre-reinforced composites. The mechanical properties of the composites are very good and include good dimensional stability and high tensile strength (particularly glass-fibre-reinforced composites). The toughness and strength of glass-fibre-reinforced PPS are greater for linear than for branched PPS. The continuous-use temperature of PPS composites can be as high as 240°C. Worldwide consumption of PPS has been forecast to rise to 40 000 tons per annum by 1997.

PPS composites can be processed by a number of routes including injection moulding, laminating, thermoforming, stamp moulding and pultrusion. Films and fibres are also produced from PPS. Injection moulding is by far the most important processing route for PPS composites. Glass fibre contents up to 40% are common. During processing of PPS it is important to take account of the crystallization rate, which tends to be fairly fast. PPS composites are found in a

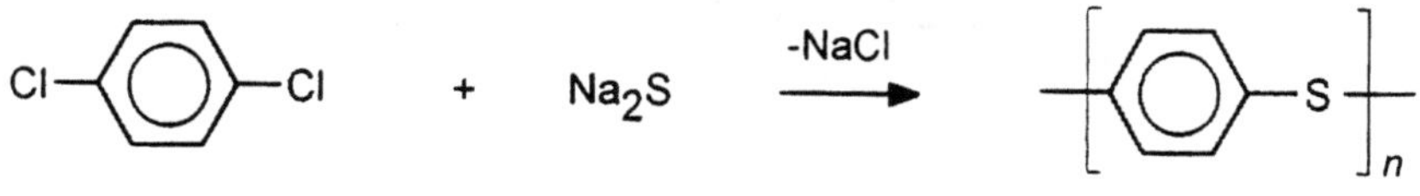

Scheme 8.24 Synthesis of polyphenylene sulphide.

wide range of applications. These include switches and electrical connectors, encapsulants for electronic components, mouldings in household appliances, metalized automobile headlights and a number of under-bonnet (hood) automotive applications. These include carburettor parts, starter motor components and fuel system parts, where the high-temperature capability and fuel resistance are crucial. Newer polyarylene sulphides (PAS) have been developed to raise the service temperature still further.

8.4.5 Polybenzimidazoles

Poly[2,2′-(*m*-phenylene-5,5′-benzimidazole)], polybenzimidazole (PBI), is a rigid-rod heterocyclic aromatic polymer system which is commercially available (Hoechst-Celanese) in both fibre and powder form (Celazole™). PBI is typically prepared by the reaction of tetra-aminobiphenyl and diphenyl isophthalate (Scheme 8.25).

PBI displays a very high T_g (*ca.* 430°C) and possesses excellent mechanical properties (Table 8.9), but is difficult to process into large parts owing to its rigid molecular structure and high degree of aromaticity. Celazole™ PBI may be processed via a proprietary sintering process to yield moulded components which exhibit desirable properties (Table 8.8). The polymer is non-combustible in air and exhibits exceptionally high temperature resistance (moulded materials may survive short excursions to temperatures as high as 740°C) although PBI

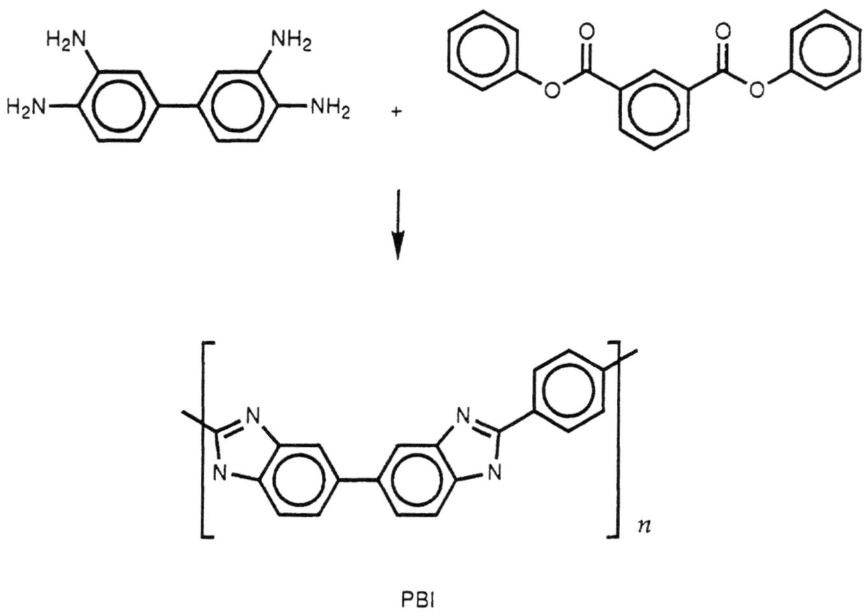

Scheme 8.25 Synthesis of polybenzimidazole.

Table 8.8 Structures and properties of candidate high-temperature matrix polymers. Sources: Arnold, F.E. Jr and Arnold, F.E., *Adv. Polym. Sci.*, **117**, 1994; Pilato, L.A. and Michno, M.J., *Advanced Composite Materials*, published by Springer, Berlin, 1994

Properties	PBI[a]	PEI[b]	6FCoPI[c]	PBI/PEI[a,b] (85 : 15)	PBI/6FCoPI[a,c] (10 : 90)
Density (kg m^{-3})	1.3×10^3	–	–	–	–
T_g (°C)	420–425	220	340	–	–
Tensile properties:					
strength (MPa)	100–160	108	97.2	158	103
elongation (%)	1.8–3	33	4.4	3.4	4.8
modulus (GPa)	5.68–5.9	3.18	3.36	5.34	3.36
Flexural properties:					
strength (MPa)	100–220	143	153	248	156
modulus (GPa)	6.32–6.5	3.39	3.83	5.73	3.90
Compressive properties:					
strength (MPa)	340–400	–	193	300	187
modulus (GPa)	6.46	–	3.71	5.18	3.81
Notched izod (J m^{-1})	30	–	–	–	–

[a] PBI, Poly[2,2′-(*m*-phenylene-5,5′-benzimidazole)].
[b] ULTEM 1000™ (General Electric), a polyether imide.
[c] 6FCoPI, copolyimide containing 3.5 mol% 4,4′-hexafluoroisopropylidene diphthalic anhydride (6FDA).

generally has poor thermo-oxidative stability in continuous exposure to temperatures above *ca.* 260°C and displays high moisture regain. PBI does exhibit good solvent resistance and an unusually high compressive strength (Table 8.8) and these properties have stimulated interest in PBIs for aerospace, military and electronic applications. It is widely accepted that miscible polymer blends and copolymerization offer complementary routes to polymer systems with tailored properties and may have important implications for cost-saving. PBI forms miscible blends with a broad range of thermoplastic polyimides (PI 2080™, Ultem 1000™, or LARC-TPI™), polyarylates (PAr), polyamideimides (PAI) and poly(4-vinyl pyridine) (PVPy) making them potentially of great interest in a wide variety of high-performance applications.

Solution blends of PBI and thermoplastic polyimides (of 20%–25% by weight) in DMAc may be processed using conventional techniques (e.g. dry spinning and film casting and precipitation) to produce blend fibre, film and powder. PBI/PI blends yield excellent compressive properties (Table 8.8) and offer dramatic improvements in solvent resistance over the PI homopolymers, although the rate of degradation of the blend is faster than that of either homopolymer, indicating that the strong intimate interaction present within the blend exacerbates the degradation route. However, techniques which have been used to improve the thermo-oxidative stability of PBI (e.g. protonation of the imidazole ring by treatment with phosphoric acid) have also been successful with PBI/PEI blends, and evaluation of PBI/PEI blends suggests them to be viable candidates for matrix use up to *ca.* 260°C.

PAr are wholly aromatic polyesters derived from aromatic dicarboxylic acids and diphenols or their derivatives. As PAr tend to be soluble in similar solvents (e.g. NMP, DMAc, DMSO, etc.) to PBI, stable miscible solutions may be formed allowing films to be cast. The solvent resistance (e.g. to dichloromethane) of PAr is improved by blending with PBI, and spectroscopic studies have suggested that intermolecular hydrogen bonding occurs between PBI and PAr in the film blend, serving to enhance the physical integrity of the film. Dry-spun PBI/PAr has very impressive mechanical properties which are at least comparable with PBI/PEI fibres. PBI/PAr blend fibres have higher tensile moduli and strengths than those of standard PBI fibres, although their elongations at break are inferior. They have been suggested as candidates for engineering and aerospace applications.

PAI are generally characterized by excellent high-temperature properties with T_g values typically above 270°C and continuous-service temperatures of about 230°C. Solution PBI/PAI blends (using DMAc at 15%–20% solids concentrations) are apparently homogeneous with no visible insoluble material (and no precipitation for at least 24 h). Transparent blend films may be cast from solution blends of PBI and 10% 6F-PAI (based on ODA) which display

Table 8.9 Typical properties of structural polymers. T_g = glass transition temperature; T_m = melting temperature; G_{IC} = fractive energy; abbreviations and details of trade name materials are given at the beginning of this book

Polymer	T_g (°C)	T_m (°C)	Modulus (GPa)[a]	Strength (MPa)[a]	Elongation (%)	G_{IC} (J m^{-2})
Thermosets						
DGEBA (Epon 825™)	262	–	3.6	90	1.8	53
TGDDM–DDS	250	–	3.6	90	1.8	53
TGDDM–PEI (70 : 30)	215, 265	–	3.5	–	3.7	488
AroCy B™	230–270	–	3.1	155	7.7	140
Primaset PT™	<400	–	4.7	97	2.5	–
XU-71787™	250–265	–	3.4	124	2.7	180
Compimide 796™	>300	–	4.6	76	1.7	63
Compimide 796™–TM123™ (60 : 40)	261	–	3.7	132	3.7	439
Matrimid 5292™	273	–	4.0	186	–	197
PMR-15	340	–	–	176	1.5	87
Thermid MC-600™	320	–	4.5	145	1.5	–
Cyclotene 3022™ (BCB)	>350	–	2.0	–	6.0	–
Phenol/Formaldehyde	127	–	5–7	80–100	1–1.5	–
Phenol/Formaldehyde 60/40	159	–	5.1	190	1–1.8	190
1,3 PBOX						
Phtalonitrile resins	>300	–	–	94	–	120
Thermoplastics						
PEI (Ultem™)	220	–	3.3	110	–	3700
Kevlar 49™	–	538	–	2.7	2.8	–
Kevlar 149™	–	–	–	3.5	1.5	–
PES	230	–	2.4	83	–	3000
PEEK	143	340	3.8	103	–	4800
PEKK	156	338	4.5	102	–	1000
PPS	85	285	3.8	65	–	210[b]
Celazole™ PBI	420	–	6.4	143	1.8	–

[a] Values of modulus and strength are quoted in flexure for thermosets and in tension for thermoplastics.
[b] Highly crystalline PPS.

acceptable mechanical properties. There is evidence of synergistic miscibility between the blend components since the tensile strength and modulus (tensile strength, 15 ksi; modulus, 0.46 msi) of each corresponding component is equivalent to or exceeds the simple rule of mixtures.

PBI/PVPy blends show single T_gs values intermediate between the two pure polymers, although some blends containing low molecular weight PBI may phase separate at temperatures above their T_gs value blends with high molecular weight PBI degrade at temperatures below their phase separation temperature. The thermal stability of these blends is similar to the stability of pure PVPy and hence cannot be considered truly high-performance blends. Table 8.9 details the physical and mechanical properties of PBI and selected examples of a wide variety of structural polymers (both thermosets and thermoplastics).

8.5 Recent and future directions

8.5.1 Molecular composites

In conventional fibre-reinforced composites, the maximum achievable mechanical properties are limited by a number of factors, including the nature of the interface between the fibre and the resin and the presence of defects in even the best macroscopic spun fibre. The concept of a molecular composite involves reinforcement of a matrix by *in situ* formation of aligned rigid molecular fibres (i.e. rigid-rod polymer chains). In this way, it should be possible to translate more of the inherent strength and stiffness of the rigid-rod polymer molecule into the properties of the composite. Such molecular composites hold the promise of a new generation of materials with properties markedly better than current state of the art materials. The main problem hindering progress in this area is the difficulty in processing the composites while avoiding aggregation of the rigid polymer molecules. Molecular-level dispersion of the rigid rods is required to maximize the reinforcing effect.

The principal rigid-rod polymers which have been investigated as reinforcements include heterocyclic polymers such as poly(*p*-phenylenebenzobisthiazole):

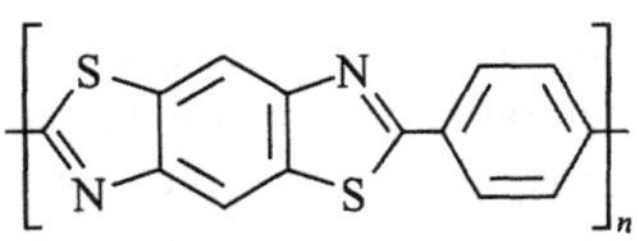

poly(*p*-phenylenebenzobisoxazole):

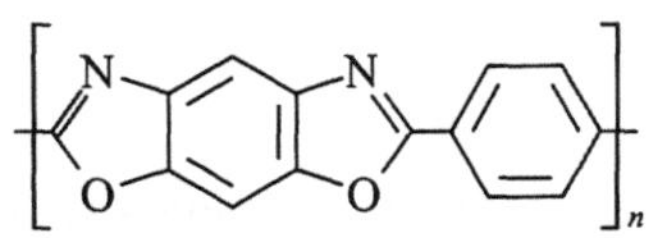

and polyimides, as well as aromatic polyamides.

Approaches used to improve the processability of molecular composites have included:

- attachment of flexible, solubilizing units to the rigid polymer chain (e.g. PBT-g-polyetherketone); these grafts can improve compatibility with a separate matrix, or may themselves act as the matrix phase;
- *in situ* polycondensation to form the rigid-rod polymer in the presence of the matrix;
- use of more processible precursors such as polyamic acids.

The last two both suffer the disadvantage that a small molecule is eliminated.

None of these methods has resolved the difficulties in producing monolithic molecular composites with effective reinforcement. Success has been realized only in the production of fibres and films with very good mechanical properties. An alternative approach to molecular composites which has attracted attention is the use of star-shaped rigid rods.

8.5.2 Stereocontrolled polyolefins

Since the development of Ziegler–Natta olefin polymerization catalysts more than 40 years ago, there has been great interest in controlling the stereochemistry (tacticity) of the resultant polyolefins, the main reason being the wide range of properties (T_g, mechanical properties, etc.) exhibited by chemically identical polymers of different tacticity. The most exciting approach is undoubtedly the use of metallocene and other so-called single-site catalysts. The most commonly used metallocenes are those based on zirconocene (Cp_2ZrX_2; Cp = cyclopentadienyl, X = Cl, Me, etc.) The cyclopentadienyl ligand may be fused to another ring, and the rings may be linked together via covalently bonded bridging groups. A co-catalyst is needed to increase catalyst activity, and the most common of these is methylaluminoxane (MAO; $[MeAlO]_n$). MAO-free catalyst systems have also been developed. The metallocene–MAO catalyst systems have permitted the production of the entire range of stereocontrolled polypropylenes, including stereoblock polymers.

Although these polypropylenes have a range of interesting and important properties they are not structural polymers in the sense represented by the other polymers described in this chapter. One polyolefin which has great potential for high-performance structural applications is syndiotactic polystyrene:

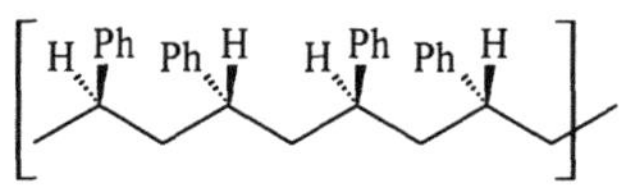

Syndiotactic polystyrene can be made by using a single-site catalyst such as

Note that this is not a metallocene catalyst. A co-catalyst is also required. Syndiotactic polystyrene is highly crystalline and this in turn has a marked influence on its properties. The crystalline melting point is around 265°C. The polymer has good dimensional stability and excellent chemical and solvent resistance in contrast to conventional atactic polystyrene. Syndiotactic polystyrene is a true engineering thermoplastic and can be processed by the usual methods, including injection moulding. The polymer can also be made into fibres, which show exceptional tensile strength, comparable to aramid fibres.

Further Reading

Background reading

Allen, G. and Bevington J.C. (eds) (1989) *Comprehensive Polymer Science*, vol. 5, Pergamon, Oxford.
Billmeyer, F.W. Jr, (1984) *Textbook of Polymer Science*, 3rd edn, John Wiley, New York.
Cowie, J.M.G. (1991) *Polymers: Chemistry and Physics of Modern Materials*, 2nd edn., Blackie, Glasgow.
Dyson, R.W. (ed.) (1990) *Engineering Polymers*, Blackie, Glasgow.
Ebdon J.R. and Eastmond G.C. (eds) (1995) *New Methods of Polymer Synthesis*, vol. 2, Blackie Academic and Professional, Glasgow.
Fawcett, A.H. (ed.) (1991) *High Value Polymers*, The Royal Society of Chemistry, Cambridge.
Hergenrother, P.M. (ed.) (1994) High-performance polymers. In *Advances in Polymer Science*, vol. 117, Springer, Berlin.
Jones, F.R. (ed.) (1994) *Handbook of Polymer-Fibre Composites*, Longman, Harlow, Essex.
Lin, S,-C, and Pearce, E.M. (1993) *High-performance Thermosets: Chemistry, Properties, Applications*, Hanser, Munich.
Macosko, C. (1990) *Fundamentals of Reaction Injection Moulding*, Hanser, Munich.
Mark, H.F., Bikales, N.M., Overberger, C.G. and Menges, G. (eds) (1985) *Encyclopedia of Polymer Science and Engineering*, John Wiley, New York.
Nicholson J.W. (1991) *The Chemistry of Polymers*, The Royal Society of Chemistry, Cambridge.
Pilato, L.A. and Michno, M.J. (1994) *Advanced Composite Materials*, Springer, Berlin.
Saunders, K.J. (1988) *Organic Polymer Chemistry*, 2nd edn, Chapman and Hall, London.
Simpson, W.G. (ed.) (1995) *Plastics and Resin Compositions*, The Royal Society of Chemistry, Cambridge.
Wright, R.E. (1991) *Molded Thermosets: A Handbook for Plastics Engineers, Molders and Designers* Hanser, Munich.

Polyesters

Meyer, R.W. (1987) *Handbook of Polyester Moulding Compounds and Moulding Technology*, Chapman and Hall, New York.

Epoxy resins

Ellis, B. (ed.) (1993) *Chemistry and Technology of Epoxy Resins*, Blackie Academic and Professional, Glasgow.

Hamerton, I. (1996) Recent Developments in Epoxy Resins. *Rapra Review Report.*
Lee, H. and Neville, K. (eds) (1967) *Handbook of Epoxy Resins*, McGraw-Hill, New York.
May, C.A. (ed.) (1988) *Epoxy Resins – Chemistry and Technology*, 2nd edn., Marcel Dekker, New York.

Cyanate esters

Hamerton, I. (ed.) (1994) *Chemistry and Technology of Cyanate Ester Resins*, Blackie Academic and Professional, Glasgow.

Benzocyclobutenes

Finkelstein, H. (1910) *Chem. Ber.*, **43**, 1528.
Kirchoff, R.A. and Bruza, K.J. (1994) *Adv. Polym. Sci.*, **117**, 3–66.

Toughened thermosets

Cantwell, W.J. and Morton, J. (1991) *Composites*, **22**, 347.
Hay, J.N., Woodfine, B. and Davies, M. (1996) *High Perform. Polym.*, **8**, 35 (and references cited therein).
Riew, C.K. and Kinloch, A.J. (eds) (1993) *Toughened Plastics I* (*Adv. Chem. Ser.* 233), American Chemical Society, Washington, DC.

Polyamides

Yang, H.H. (1993) *Kevlar Aramid Fibre*, John Wiley, Chichester, Sussex.

Polyimides

Davies, M., Hay, J.N. and Woodfine, B. (1993) *High Perform. Polym.*, **5**, 37.
Schwartz, W.T. (1990) *High Perform. Polym.*, **2**, 189.
Wilson, D., Stenzenberger, H. and Hergenrother, P.M. (eds) (1990) *Polyimides*, Blackie, Glasgow.

Polysulphones and polyketones

Cotter, R.J. (1995) *Engineering Plastics, A Handbook of Polyarylethers*, Gordon and Breech, Basel.

Polyphenylene sulphide

Kohlhepp, K. (1993) *Kunststoffe German Plastics*, **83**, 44.
Reinking, K. (1990) *Kunststoffe German Plastics*, **80**, 10.

Stereocontrolled polyolefins

Horton, A.D. (1994) *Trends Polym. Sci.*, **2**, 158.
Küber, F. (1993) *New Scientist*, 14 August, p. 28.

Index

acetylene 85
acetylene terminated resins 230–233
acetyl sulphate 142
acroloyl chloride 171
acrylamide(s) 79, 82
2-acrylamido-2-methyl-propane sulphonic acid 82
acrylic acid and derivatives 78, 82, 85, 86, 124, 125, 205
acrylonitrile 125
addition polymerization 19–23
agar 72
alginates 72
alkyl amines 219
aluminium hydroxide 45
aluminosilicate polyacrylic acid 153
amino-acids 75
ampholyte 122
amylose 72
amyl pectin 72
anionic polyacrylamide 83
anionic polymers 75
anisotropic networks 192, 193, 194, 195
antimony oxide 45
antioxidants 41
aramid fibres 233
aziridine 155

benzocyclobutadiene resins 225–229
  properties of 228
β-mercaptopropionic acid 139
bi-layer carboxylate membranes 147
biopolymers 158
birefringence 167, 186, 195
bis [catena-poly(methyl arsenic)] 103
4,4′-bis (3,4-dicyanophenoxy) biphenyl 92
bismaleimide polymers 202, 219–222
bisphthalonitrile resins 231
bis-oxazoline-phenolic resins 232
bond dissociation energies 14, 37, 38, 47
bulk moulding compounds 201, 203
burning of polymers 41–43

carbon fibre 203
carboxylated polysulphones 143
carboxy methyl cellulose 75, 76
carboxy polymers 125
carrageenan 73
casein 75
cationic polymers 125
cationic polyacrylamide 83

cellulose 75, 76
cellulose acetate 148
cellulose derivatives 75, 76
cellulosic ion exchange polymers 152
charge mosaic membrane 131
chiral nematic phase 162, 163, 165, 166, 188, 195
chlor-alkali cell 145, 146, 147, 151
chlorendic acid 44
chlorosulphonated polyethylene 141–142
cholesteric phase 162
cholesteryl polymers 182
chromatography 189
Cloez reaction 211
collagen 75, 81
compensation films 196
conductive coatings 100, 101
conductive polymer applications 100, 101
copolymer types 11, 12
corona poling 107, 108
counter ions 123
creep curves 55
cross-linking 24,25, 75, 126–129, 193
crystal reorientation 183
crystal transition temperatures 166
cyanate esters 202, 211–216,
cyanogen halides 211
4′-cyano-4-hydroxy biphenyl 171
cyclic polysiloxanes 181

degradation 15, 37
de-ionisation 130, 133
desalination 133
diacrylate 197
diallyl dimethylammonium chloride 82
dicarboxy polybutadiene 153
1,4 dichloro benzene 242
4,4′ dichloro benzene 242
1,2 dichloroethane 142
dichroic dyes 185
dielectric properties of LCP 167
dimethylaminoethyl acrylate 82
dimethylaminoethyl methacrylate 82
de-mineralisation 133
diamino benzene 219
discotic phase 162
dispersants 134
divinyl benzene 149
Donnan dialysis 131, 132
Donnan exclusion 130, 133
doping of polymers 95, 96, 97, 99, 101

double stand polymer 65
dough moulding compounds 210, 203

elastomeric ionomers 138–142
elastomeric LCP 179
electrical conduction mechanisms 92–93
electrical conductivities of materials 94
electrochemical polymeristion 98
electroclinic effect 165
electrochromicity 51
electro dialysis 130
electromagnetic radiation 15
electro optic cell 184
electron beam lithography 118
electroreprographic processes 13
EPIC resins 232
epoxide LCP 191
epoxides (epoxies) 202, 203, 206–211
epoxide synthesis 206, 207, 208
epoxide applications 210
epoxide curing agents 208
ethylene glycol diacrylate 85
ethylene glycol dimethacrylate 80
ethylene propylene rubber 125
ethyl hydroxy ethyl cellulose 75
ethylidene norbornene 140

ferroelectric properties 165
fillers in plastics 40
fire retardants 44, 45
flammability and chain structure 44
flocculating agents 135
fluoropolymers 47–48
fluorosilicone elastomers 47
formaldehyde 80
free radical polymerization 20–22, 87
fuel cell membranes 145

gallium arsenide 111
gelatin 75
gel times 216, 217
germanium 111
glass transition temperature 8, 13, 128, 178, 182
glow discharge polymerization 103
glucose 71
glutaraldehyde 80
glycidylmethacrylate/ethyl acrylate copolymers 118
glycidylmethacrylate/3-chlorostyrene copolymers 118
glycollic acid 124
glyoxal 80
graft copolymeriazation 87
gums 71–74

halotelechelic polymers 141, 153, 154, 155, 156

heat distortion temperature
  test 46
  data 40, 47, 53, 54, 55, 56, 57, 58, 61, 63
hexafluoroethylenepropylene polymers 48
hydrogels 78–82
hydrogel casting 81
hydrogen bonded cross-link 129
hydrophilicity 132–135
hydroxy ethyl cellulose 75
hydroxy ethyl methacrylate 78, 79, 80, 81
hydroxy propyl cellulose 75
hydroxy propyl methyl cellulose 75

isocyanate polymers 202, 216–219
index materials 104
indium tin oxide coatings 185
initiators (polymerization) 20, 21
interlaminar toughening 230
inverse emulsion polymerization 87
ionenes 123, 154
ionic cross-linking 126–129
ion exchange 129–132, 149–152
ion exchange membranes 130
ion exchange resins 150, 151
ionic polymerization 22–23
ionomers 123, 125
Izod impact values 40, 55, 56, 57, 58, 61

ladder polymer 65
liquid crystal displays 183–85, 196
Limiting Oxygen Index values 43, 53, 55, 56, 57, 58, 61, 63
liquid crystal polymers 50, 56, 57
lithium carboxylate 142
lyotropic crystals 168

magnesium hydroxide 45
magnetic properties of LCP 168
maleic anhydride 125
maleimide rings 219
m-ethynyl benzene 231
melting point 8, 9
Menschutkin reaction 155
metallocene catalysts 248
methacrylic acid and derivatives 78, 82, 124, 124, 135, 205
methacroloyl chloride 171
methyl acetylene 95
methyl cellulose 75
methylene dianiline 219
methylene-bis-acrylamide 80, 85
2-methyl-4-nitroaniline 105
microbial gums 73
microlithography 111, 112
modulus values 40, 53, 55, 56, 57, 58, 61, 63
molecular susceptibilities 189
molecular weight distributions 10, 11
MOSFET 110

naphthoquinone diazides 114, 115
natural rubber 113
negative photoresists 113–114
nematic phase 162, 163, 167, 183, 188, 195
non-ionic polymers 75
non-linear optical properties 189
norbornene LCPs 180
novolac resins 114
nylon 4 6 58
nylon 6 7, 202
nylon 6 6 7, 58
nylon 11 108
N-methyl pyrrole 98
N-vinyl-2-pyrrolidone 85
N-vinyl-2-pyrrolidinone 85
N,N′ arylene 219
N,N,N′,N′ tetramethylethylenediamine 142

optical data storage 186, 187
optical properties of LCPs 167
optic fibres 103 ff
order parameter 166

p-bromostyrene 142
p-diethynylaniline 231
permselective membranes 130
phase transitions – LCP 179, 180
phenol formaldehyde condensates 151
photoconduction 101
photoconductive polymers 101
photolithographic process (schematic) 113
photopolymerization 190, 191
photo resists 110–119
photosensitivity 187
piezodialysis 131, 132
piezoelectric applications 109–110
plasticizers 143, 20
PMP resins 222–225
polarisers 196
poling 106, 107
poly-2-chloroethyl methacrylate 118
poly-2-fluoroethyl methacrylate 118
poly (2-oxyazolidone) 218
poly (2 2′-m-phenylene 5 5′
    bibenzimidazole) 224–247
poly (4-vinyl pyridine) 156, 245
poly (acene-quinone radical) 91
polyacetals *see* polyoxymethylene
polyacetylene 93, 96
polyaniline 98, 99
polyacrylamides 82–84
polyacrylic acid and derivatives 81, 88, 152,
    171, 181, 186
polyacrylonitrile 80, 108
polyamides 49 (*see also* nylons)
polyaramids 59, 60
polyaromatic ionomers 148,149
polyarylates 56, 245, 246
polyarylene ethers 220, 229
polyarylimides 245, 246
polybenzimidazoles 244–247
polybenzimidazole blends 245
polybenzoxazole 65
polybenzothiazole 65
poly [bis(2-naphthyloxy) phosphazene] 103
poly [bis(p-tolyloamino) phophazene] 103
poly (bisbenzimidazo
    benzophenanthroline) 66
polybutadiene 124, 138
polybutadiene ionomers 138
polybutyleneterephthalate 55
polycarbonate 6, 40, 50, 55, 104
poly (chloromethyl styrene) 118
polychlorotrifluoroethylene 48
poly (*cis*-isoprene) 113
polycyanates 212
polycyanurates 212, 213, 214
poly diacetylene 105
polydicyclopentadiene 202
poly dimethyl tetrafluoro propyl
    methacrylate 118
polyelectrolytes 122, 123, 124, 127, 130,
    149, 150, 151, 152
polyelectrolyte complexes 156, 157
polyesters 49
polyester LCP 174
polyetherimides 50, 220, 237, 238, 245
polyetherketones 50, 239, 240
polyetheretherketones 61, 235, 239, 240, 241
polyetherketoneketonel 235
polyetherketones 61
polyethers 52
polyethersulphone 6, 50, 54, 220, 229
polyethylacrylate 157
poly (ethyl acrylate co-4 vinyl pyridine)
polyethylene 6, 7, 37, 41
poly (ethylene-co-sodium methacrylate) 127
polyethylene ionomers 135–138
polyethylene naphthenate 50
polyethylene terephthalate 7, 49, 55
polyhydantoins 220, 229
polyhydrogen methylsiloxane 172
polyhydroxy benzoic acid 56
polyimides 62–64, 219, 220, 229, 236, 237
polyitaconic acid 152
polyketones 49, 59
poly-m-phenylene isophthalamide 233
polymethacrylic acid and derivatives 81, 152,
    171, 172, 186, 189
polymethyl methacrylate 6, 104, 117, 118
poly-N-epoxy-3-bromo carbazole 102
poly-N-epoxypropyl carbazole 102
poly-N-epoxypropyl 1,3,6
    tribromo-carbazole 102
poly-N-methyl pyrrole 98
poly-N-vinyl carbazole 101, 102, 103

poly-N-vinyl pyrrolidone 79, 84, 85
polyoxy methylene 7, 40, 55
polyolefine sulphones 118
polyoxyethylenes 174
poly (p-phenylene benzobisoxazole) 247
poly (p-phenylene benzobisthiazole) 247
poly-p-phenylene terephthalamide 234, 235
poly (p-vinyl) phenol 116
poly (para hydroxystyrene) 114
polyparaphenylene polymers 96, 97
polyphenylene 48
polyphenylene ethers (*see* PPO)
polyphenylene oxide 6, 40, 52
polyphenylene selenide 99
polyphenylene sulphide 49, 53, 99, 241, 242, 243
polyphenylene telluride 99
polyphenylene vinylene 51
polypropylene 6, 39
polypropylene oxide blends 157
polypyrrole 97, 98
polypyrrole toluene sulphonate 97
polysaccharides 71–74
polysiloxanes 172, 173, 176, 181, 189
polystyrene 6, 37, 39, 104, 118, 124, 142, 143, 157
polystyrene LCP 173, 174
polysulphonates 125
polysulphone 49, 54, 55, 125, 239, 240, 241
polysulphur nitride 99
polystyrene-cross-linked 149
polystyrene ionomers 142–144
polytetrafluoroethylene 9, 47, 53
polytetrafluoroethylene ionomers 144–147
polyureas 217
polyurethanes 24, 202
polyurethane isocyanate 217
polyvinyl acetate 32, 80
polyvinyl alcohol 32, 80, 84, 87
polyvinyl carbazole 103
polyvinyl chloride 14, 37, 39, 41, 44, 87, 108, 143
polyvinylidene cyanide 108
polyvinyl ethers 174–176
polyvinylidene fluoride 14, 106–110
polyvinylidene fluoride-trifluoroethylene copolymers 108
polyvinyl pyrene 102
polyvinyl pyridine 79, 81
polyviologens 155
polyxylenes 52
POSFET 110
positive photoresists 114–115
post-functionalisation 124
prepregs 210, 237
proteins 74–75
pultrusion 202
pyrrole 95, 97, 98
pyropolymers 91

reaction injection moulding 202, 217, 218
refractive index 167
relative temperature index values 46
resin transfer moulding 202, 203
rubber toughening 229, 230

Schottky barrier 96
seaweed 72
semiconductor fabrication 110 ff
sheet moulding compounds 210, 203
Shrakawa polymers 95
silane cross-linking 24, 25
silicone devices 115–119
silicone polymers 46
single site catalysts 248, 249
smectic phase 162, 164, 165, 166, 185, 188
smoke emission data 43
smoke supressants 45
sodium acrylate 85, 86
sodium amino-phosphonate 151
sodium carboxy methyl cellulose 75, 76, 77, 78
sodium methoxide 135
sodium polysilicate 127
sodium p-styrene carboxylate 127
softening behaviour 8–9
solubility 16, 17
sorbitan esters 88
starches 70–71, 87
step growth polymerization 23–25
styrene/4-chloro styrene copolymers
sucrose fatty esters 88
sucrose methacrylate 81
sulphonated EPDM 141
sulphonated polystyrene 156
sulphonic acid ionomer 125
sulphonyl chloride 125
superabsorbant polymers 85–89
switching 185
syndiotactic polystyrene 248, 249

tacticity 6, 7
telecommunications – polymers in 103, 104
tensile strength values 40, 53, 55, 56, 57, 58, 61, 63
tetraallyloxethane 88
tetrabromo-bisphenol-A 45
tetrafluoro ethylene 125
tetra methyl-6,3-ionene bromide 154
thermal oxidation of polymers 37, 38
thickening agents 134, 205
thioglycollic acid 139
tourmaline 105
transfer agents 21
transfer constants 21
triazine resins 212

1.1.1-trimethylol propane triacrylate 85, 86
2,4,7-trinitro fluorenone 102

unsaturated polyesters 202, 203, 204
urethane acrylates 202

vinyl esters 202, 203, 205
vinyl ester-urethane 217
vinyl sulphonic acid 82
von Pechman condensation 147

water soluble polymer applications 68, 69

x-ray lithography 117, 118

zincocene 248

MIX
Papier aus verantwortungsvollen Quellen
Paper from responsible sources
FSC® C105338

If you have any concerns about our products,
you can contact us on
ProductSafety@springernature.com

In case Publisher is established outside the EU,
the EU authorized representative is:
Springer Nature Customer Service Center GmbH
Europaplatz 3, 69115 Heidelberg, Germany

Printed by Libri Plureos GmbH
in Hamburg, Germany